Biodiesel - Feedstocks, Production and Applications

Biodiesel - Feedstocks, Production and Applications

Editor

Sabbas Radley

scitus academics

Biodiesel - Feedstocks, Production and Applications
Edited by **Sabbas Radley**

ISBN: 978-1-68117-219-4
Library of Congress Control Number: 2016934759

SCITUS Academics LLC,
www.scitusacademics.com
Box No. 4766, 616 Corporate Way,
Suite 2, Valley Cottage,
NY 10989

Notice

Printed in United States of America on Acid Free Paper

Preface

Biodiesel is a replacement for clean, renewable diesel fuel that creates jobs, improves fuel safety and provides clean breathing air. Biodiesel is usually made from vegetable oils such as canola oil, animal fats or recycled greases such as used cooking oil. Other feedstocks being developed include algae, pomgania, jatropha, camelina and dry land juncea. The fatty acid feedstock is combined with alcohol that condenses water molecules and leaves behind them the rich, pure fatty acids. Transesterification is the process in which these oils are converted into biodiesel. The most appropriate source of oil is from oil crops such as rape, palm or soybean. Most currently produced biodiesel comes from waste vegetable oil sourced from restaurants, chip shops, industrial food producers. Vegetable oil waste can often be obtained for free or at low prices. Biodiesel has lots of environmentally friendly properties. The main advantage of biodiesel is that it can be described as carbon neutral. This means that fuel does not generate a net carbon dioxide output (CO2). Biodiesel is used in standard diesel engines and is therefore separate from vegetable and waste oils used in diesel engines converted to fuel. Biodiesel can be used alone or mixed with petrodiesel in any proportion. Mixtures of biodiesel can also be used as fuel oil.

This book highlights the advances of biodiesel systems from the production and processing of feedstocks to the full application of biodiesel and by-products. It will be an invaluable tool for researchers, academics, scientists and technologists.

Table of Contents

1

CHAPTER

SYSTEMATIC SUSTAINABLE PROCESS DESIGN AND ANALYSIS OF BIODIESEL PROCESSES

Seyed Soheil Mansouri, Muhammad Imran Ismail, Deenesh K. Babi, Lida Simasatitkul, Jakob K. Huusom and Rafiqul Gani *

CAPEC, Department of Chemical and Biochemical Engineering, Technical University of Denmark, Building 229, Søltofts Plads, DK-2800 Kgs. Lyngby, Denmark

ABSTRACT

Biodiesel is a promising fuel alternative compared to traditional diesel obtained from conventional sources such as fossil fuel. Many flowsheet alternatives exist for the production of biodiesel and therefore it is necessary to evaluate these alternatives using defined criteria and also from process intensification opportunities. This work focuses on three main aspects that have been incorporated into a systematic computer-aided framework for sustainable process design. First, the creation of a generic superstructure, which consists of all possible process alternatives based on available technology. Second, the evaluation of this superstructure for systematic screening to obtain an appropriate base case design. This is done by first reducing the search space using a sustainability analysis, which provides key indicators for process bottlenecks of different flowsheet configurations and then by further reducing the search space by using economic evaluation and life cycle assessment. Third, the determination of sustainable design with/without process intensification using a phenomena-based synthesis/design method. A detailed step by step application of the framework is highlighted through a biodiesel production case study.

Keywords: biodiesel; superstructure; process intensification; phenomena based; simulation; design

1. INTRODUCTION

Today fossil fuels make up around 80% of the global primary energy consumption, of which nearly 58% is solely used by the transportation sector [1]. Fuel consumption is growing dramatically, resulting in greenhouse gas emissions that affect global warming. Progressive depletion of conventional fossil fuels due to increasing energy consumption have led to a move towards alternative, renewable, sustainable, efficient and cost-effective energy sources and production processes with lesser emissions [2].

One of most promising fuel alternatives is biodiesel, which can be obtained from renewable resources such as palm oil. Biodiesel is a green fuel and has desirable characteristics compared to conventional diesel fuel. For example, it is non-toxic, biodegradable and has a very low concentration of sulfur. In addition it has a high cetane number (above 60 as compared to 40 for regular diesel), a high flash point (<130 °C) and it emits ~70% fewer hydrocarbons, ~80% less carbon dioxide and ~50% less particulates [3]. The interest in biodiesel can also be understood from its global production in recent years. Europe is currently the largest producer, while Brazil has the highest increase in production rate in the last years when compared to Europe and the United States—an increase from 736 m^3 in 2005 to 2,400,000 m^3 in 2010 [4].

Biodiesel can be produced from different feedstocks, which is dependent on the geographical location and climate. Common feedstocks are vegetable oils, for example, sunflower oil, rape seed oil and palm oil [5]. Vegetable oil is the oil that comes from all-vegetable sources and it can be a blend of oils from several different plant sources. Note that not all vegetable oils are edible oils. Examples of non-edible vegetable oils are oils from Jatropha curcas and Pongamia glabra (karanja) while edible vegetable oils are oils from rape seeds, palm seeds and sunflower. The most important obstacle to biodiesel industrialization and commercialization is that these feedstocks are regarded as belonging to the food-chain. Therefore, the usage of waste edible oils that can also be used for biodiesel production is recommended [6,7]. Waste edible oils are normally characterized by relatively high free fatty acid (FFA) and water contents and these have a significant effect on the transesterification reaction, whereas vegetable oils have a very low FFA content.

Biodiesel can be produced using different reaction routes, depending on the feedstocks and catalysts used. Amongst them, the homogenous/heterogeneous acid-/alkali-catalyzed transesterification, enzymatic transesterification, supercritical methanol transesterification, microwave assisted and ultrasound assisted transesterification have been reported. The most commonly used approach, however, appears to be catalytic (alkaline-based) transesterification [8,9]. Different flowsheet alternatives for biodiesel production exist and therefore it is important to study their sustainability issues in order to find the most promising and sustainable alternative. One way to achieve this objective is to employ a systematic computer-aided framework for sustainable process design [10,11,12], which provides the flexibility of evaluating/screening different flowsheet alternatives for selecting the optimal (more sustainable) design and combine it with options to evaluate opportunities for process intensification [13]. A recent review [14] on the role of process synthesis related to bio-process synthesis and design have also highlighted the need for a systematic synthesis/design framework.

West et al. [15] have assessed four biodiesel production processes that employed homogeneous acid/alkali catalysts, a heterogeneous acid catalyst and a supercritical method to convert waste vegetable oil (waste Canola oil was represented by triolein and oleic acid) into biodiesel. They showed that process employing heterogeneous acid catalyst and the supercritical method have the smallest number of unit operations and heterogeneous acid catalyzed process has the smallest capital investment.

Martin and Grossmann [16] have proposed a mathematical approach to determine the optimal production of second-generation biodiesel using waste cooking oil (they assumed that the oil can be characterized as triolein and oleic acid) and algae oil (they assumed that oleic acid with no impurities represents the oil) by simultaneous optimization and heat integration. They considered five different technologies for the transesterification of the oil (homogeneous acid- or alkali-catalyzed, heterogeneous basic-catalyzed, enzymatic, and supercritical uncatalyzed). They formulated the optimization problem as an MINLP problem. Their aim has been to perform simultaneous optimization and heat integration for the production of biodiesel from each of the different oil sources in terms of the technology to use and the operating conditions to apply. Process intensification options were, however, not included.

In this work, a systematic computer-aided hierarchical framework for combined process synthesis and process intensification is proposed for determining more sustainable process designs. This framework has been applied to identify a more sustainable process for the production of biodiesel using pure and waste palm oil as the feedstock. The choice of this feedstock is based on the information that in countries like Thailand and Malaysia, excess palm oil is available together with waste (cooking) palm oil. The objective (or target) for the more sustainable process design with/without intensification is to overcome the bottlenecks of the base case design through more sustainable process alternatives, which may include a reduction in the number of unit operations through the inclusion of intensified operations.

The framework consists of three parts where each part (sub-method) consists of its own set of calculation-analysis steps.

- Part-1 defines a superstructure. Here, information on processing routes for the product-feedstock is collected from available data and literature, which is then used to create a generic superstructure of processing routes. The superstructure describes a network of process configurations from which the optimal flowsheet can be found.
- Part-2 identifies a base case design. The superstructure from part-1 is evaluated in terms of a sustainability analysis that identifies and orders structural (process) bottlenecks within it in terms of those having the highest impact for improvement with respect to the measures of sustainability. Based on this analysis the number of process alternatives within the superstructure is screened and from the reduced set, a base case design is selected. This step is necessary only if a "reference" base case design is not available.
- Part-3 determines a more sustainable process design with/without process intensification. Here, the identified process bottlenecks of the base case design from part-2 are addressed through the generation of new more sustainable alternatives that target resolving the identified bottlenecks. In order to apply the phenomena based process intensification, the base case design is represented in terms of tasks (operations) that need to be performed and the phenomena associated with each task are then identified. Applying the phenomena based method [13], new process alternatives with/without PI-options are generated. These are then further reduced through economic and life cycle assessment (LCA) analysis to determine the most promising alternative that best matches the specified design targets, thereby improving the performance criteria.

2. SUSTAINABLE DESIGN AND INTENSIFICATION FRAMEWORK

The process synthesis problem is represented mathematically by Equations (1–8). Equation (1) represents the objective function, which can either be maximized or minimized. Equations (2,3) define a system of linear and non-linear equations, for example, mass and energy balance (algebraic) equations representing a steady state process model. Equations (4,5) represent the physical constraints and design specifications, respectively; and Equation (6), because process intensification is also included in the process synthesis problem, represents a set of constraints that the intensified process must satisfy [13]. Equations (7,8) represent and define the bounds on the design variables, x (real) and decision variables y (binary-integer), respectively.

$$\min C = C(x, y, d, z, \Theta) \quad s.t. \tag{1}$$

$$g(x, z, \Theta) = 0 \tag{2}$$

$$f(x, y, d, z, \Theta) = 0 \tag{3}$$

$$b^l \leq B_1 x + B_2 y \leq b^u \tag{4}$$

$$h^l \leq h(x, y) \leq h^u \tag{5}$$

$$v^l \leq v(x, y) \leq v^u \qquad (6)\backslash$$

$$w^l \leq u(x, y) \leq v^u \tag{7}$$

$$y_j = 0/1, j = 1, 2, \ldots n_y, x \geq 0 \tag{8}$$

The optimization problem given by Equations (1–8) represents a MINLP problem, which can be difficult to solve if the process model together with its constitutive equations are highly non-linear. The size of the MINLP problem may also be an issue. In order to manage this complexity, an efficient and systematic solution approach is used where the problem is decomposed into a set of sub-problems that are solved according to pre-defined calculation order. This method is referred to as the decomposition based solution method [17]. Most of the sub-problems require bounded solution of a sub-set of equations. The final sub-problem is solved as a set of NLP or MINLP. Equation (4) is used to find the feasible combinations of phenomena and/or flowsheets, the generated combinations are reduced using physical and logical constraints (Equation (4)). The set of feasible phenomena combinations are then used to generate process flowsheet alternatives from which the feasible ones are identified by simultaneously solving the process model equations (Equations (2–5)) and operational constraints (Equation (7)). The feasible alternatives are then evaluated using a set of performance related constraints (Equation (6)). For the surviving process alternatives, the objective function (Equation (1)) is calculated and ordered. The highest or the lowest can be easily located. This is the "generate and test" option. Alternatively, if the number of feasible alternatives is not too high, a well-defined MINLP problem for a reduced size of the vector y is solved. In another variation, a set of NLPs for a fixed set of y can also be solved. The obtained sustainable process flowsheet is the best according to the problem definition, the selected performance criteria, constraints, availability of data, parameters and models [13]. A global optimal solution cannot be guaranteed with this method. Note that in this work only the "generate and test" option has been tested.

Figure 1 illustrates work-flow of the three-part systematic and computer aided framework for process design, with/without process intensification.

2.1. Part 1–Superstructure Generation

2.1.1. Step 1.1: Collect Data/Information on Available Technologies

The objective of this step is to collect data/information on feedstocks, processing steps, products, catalysts, reaction conversions, etc., so that a superstructure of process alternatives can be generated.

2.1.2. Step 1.2: Create Superstructure Configuration

Using the data from step 1.1 and relating tasks to processing routes and unit operations to tasks, a superstructure is created. Each task may involve more than one unit operation and each processing route may involve more than one task. The resulting superstructure is generic, and represents all possible flowsheet alternatives. At this stage the supersctructure has M1 flowsheet alternatives.

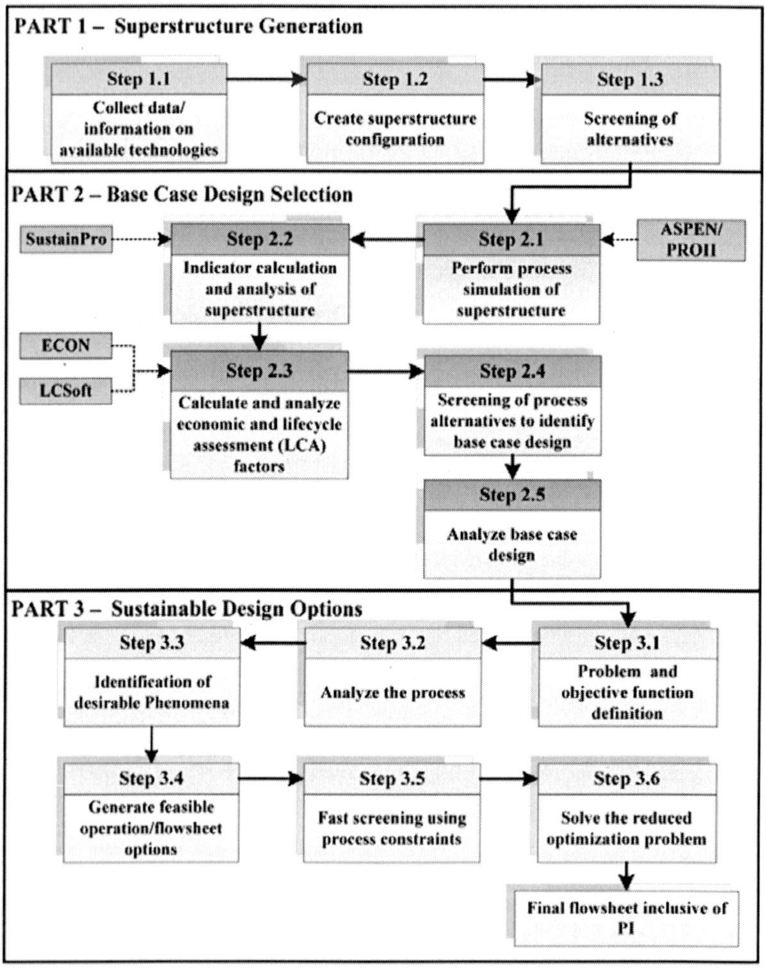

Figure 1. Sustainable design and intensification framework.

2.1.3. Step 1.3: Screening of Alternatives

The objective here is to simplify the superstructure from step 1.2 by employing the screening method also used by Simasatitkul et al. [9]. The screening method is divided into four sub-steps. First, the number of alternatives is reduced based on the type of feedstock where the criteria for selection are their availability and the cost of the raw materials. Second, the number of the processing routes is screened based on productivity of the catalyst. Since the productivity depends on the type of catalysts and operating conditions, the process that gives the highest productivity of main product for each catalyst type is only retained. Third, the number of alternatives is further reduced with respect to alternative downstream separation routes employing thermodynamic insights [18,19]. Here the number of separation tasks are determined first from the number of compounds present in the reactor effluent and the number of compounds to be recovered. Next, for each separation task, the ratios of pure compound properties of adjacent binary pair of compounds within the system are calculated. If the ratios are far from unity, the separation task can be linked to a corresponding separation process. For example, if for pair A-B, the ratio of vapor pressures is high, then separation of A-B (provided they do not form azeotrope) by distillation is feasible. Similarly, for the same pair, if the ratio of molar volumes is high, separation by pervaporation may also be feasible. This data would generate a very large number of alternatives. However, by adding the condition that for each separation task only the process that employs the largest available driving force would be employed, the number of alternatives is reduced significantly. In the final step, the alternatives are screened in terms of limitations of the separation technique (process). For example, separation by adsorption is removed if the temperature of operation is too high for the known adsorbents that are usually quite sensitive to the temperature, or, a suitable adsorbent could not be found. Membrane-based separation is limited by flux and fouling problems. At the end of this step, the number of feasible flowsheet alternatives without PI-options is reduced to M2 < M1.

2.2. Part 2–Base Case Design Selection

2.2.1. Step 2.1: Perform Process Simulation of Superstructure

In this step the superstructure representing the M2 flowsheets is simulated through a process simulator with appropriate choices of process and thermodynamic models. For phase equilibrium calculations involving the separation processes, an appropriate equilibrium constant model (a model for the liquid phase activity and a model for vapor phase fugacity) needs to be selected (for example, the NRTL or UNIQUAC models for liquid phase activity). The reactors are modeled as simple stoichiometric conversion reactors while for solid separations a simple compound splitter is employed. For separations involving vapor-liquid or liquid-liquid equilibrium, the appropriate unit operation model together with the selected equilibrium constant model is needed.

The total superstructure is simulated by setting all binary variables to 1 and allowing equal split for all split streams. This ensures no bias for any specific route and all alternatives are evaluated on the same basis. There could be other ways to perform this simultaneous simulation of the total superstructure. For different types of catalysts, different reactors with conversions obtained from published reaction kinetics data are used. The result from this step is a stream summary, at steady state, consisting of temperature, pressure, compound-flows, enthalpy, etc., for all streams in the superstructure as well as the duties for all equipments requiring energy addition/removal.

Tools: For process simulation PRO/II, Aspen, ICAS or any appropriate process simulation software.

2.2.2. Step 2.2: Indicator Calculation and Analysis of Superstructure

In this step steady state mass and energy balance data from Step 2.1 are used as input for a sustainability analysis, which is based on the method of Carvalho et al. [10]. The sustainability analysis uses an indicator-based methodology where a set of calculated closed- and open-path indicators are used to identify the structural bottlenecks within any process flowsheet (in this case, the superstructure). Through an open-path, mass (of a compound) and energy (carried by a compound) come in and go out of the system (process), while in a closed-path, they go around in cycles without going out (they represent recycle streams). The indicators therefore indicate whether a material or its energy equivalent loose value as they leave a process or if a large amount go around in cycles without going out. Through the values obtained for the indicators, it is possible to identify the location within the process where the mass/energy "path" faces "barriers" with respect to costs, benefits and accumulation [10]. Through a sensitivity analysis, the process (design) variables that have the largest effect on the indicators are identified and the needed change in the indicators is set as a design target. Since the indicators are directly related to the sustainability metrics, achieving the design target also means a more sustainable process has been found. This also means that the carbon footprint would be reduced and the profit would be increased.

The details of the work-flow and the sustainability analysis method can be found in a series of articles published by Carvalho et al. [10,11,12]. Using the sustainability analysis, the number of feasible flowsheet alternatives within the superstructure is further reduced (to M3 < M2) by selecting those having the lowest sustainable impact based on the following sustainability indicators:

1. Material value added (MVA): This indicator gives the value added between the entrance and the exit of a given compound, that is, the

value generated or lost between the start and the end point of an open-path. Negative values of this indicator show that the compound has lost its value in this open-path and therefore point to a potential for improvement.

2. Energy and waste cost (EWC): This indicator is applied to both open- and closed-paths. It takes into account the energy costs (EC) and the costs related with the compound treatment (WC). The value of EWC represents the maximum theoretical amount of energy that can be saved in each path within the process. High values of this indicator show high consumption of energy and waste costs and therefore these paths should be considered in order to reduce the indicator value.

3. Total value added (TVA): This indicator describes the economic influence of a compound in a given path and is the difference between MVA and EWC. Negative values of this indicator show high potential for improvements in terms of decrease in the variable costs.

4. Accumulation factor (AF): This indicator determines the accumulative behavior of the compounds in the closed-paths. This corresponds to the amount that is recycled relative to the input to the process and not the inventory. High values of this indicator show high potentials for improvements.

It should be noted that having the indicator values is not enough. It is necessary to perform a sensitivity analysis to identify the operation requiring the least change to obtain the largest effect in the correct direction for the indicator. Also, the effect in the indicator needs to be evaluated against the improvement in the sustainability metrics. The result of this analysis is the identification of a target indicator with a target value for improvement that can be achieved through minimum changes in one or more operational variables related to the indicator. Based on this analysis, the indicators are ordered according to a scoring system [10]. That is, the indicators are ranked (high, medium, low) in terms of their potential to improve the process (or make it more sustainable).

Tools: The method developed by Carvalho et al. [10,11] and its corresponding software-tool, SustainPro [12].

2.2.3. Step 2.3: Calculate and Analyze Economic and Lifecycle Assessment (LCA) Factors

In this step the remaining flowsheet alternatives (M3) from Step 2.2 are further analyzed using an economic and LCA evaluation. The economic evaluation provides an estimate of the operational cost and the LCA provides the environmental impact of the reduced flowsheet alternatives, for example, the carbon footprint.

Tools: For the economic evaluation: ECON [20], which is based on the economic evaluation model given in [21]. For the LCA: LCSoft [22], which uses US-EPA and IPCC emission factors to calculate the environmental impact and assesses the environmental impact and carbon footprint of a process.

2.2.4. Step 2.4: Screen Process Alternatives to Identify the Base Case Design

In this step the results from Step 2.3 are used for further reduction of the remaining flowsheet alternatives from Step 2.2 based on the economic and LCA evaluation. The flowsheet alternative that matches the specified production constraints with the least bottlenecks is selected as the base case design.

2.2.5. Step 2.5: Analyze Base Case Design

The objective in this step is to identify the potential bottlenecks through a more detailed and rigorous simulation and indicator based analysis. That is, the input flowrates and some process design variables are re-evaluated because the condition of all binary variables equal to unity and all split streams being equal are no longer valid. Therefore, the measures by which the new alternatives would be compared to establish the more sustainable designs are re-calculated in this step for the base-case design. These measures are the sustainability metrics (obtained through SustainPro), economic measures (obtained through ECON) and LCA factors (obtained through LCSoft). These measures will be used to set the design targets for Part 3.

2.3. Part 3–Sustainable Design Options

Compared to the base-case design of step 2.5, to achieve a more sustainable process design, phenomena based synthesis/design intensification method (PBS) is applied. A brief explanation of the method is presented here and the reader can obtain further details from Lutze et al. [13]. The method operates at the phenomena level and the concept is similar to that of computer-aided molecular design (CAMD) where groups of atoms are used as building blocks to generate new as well as known molecules with desired (target) properties [13].

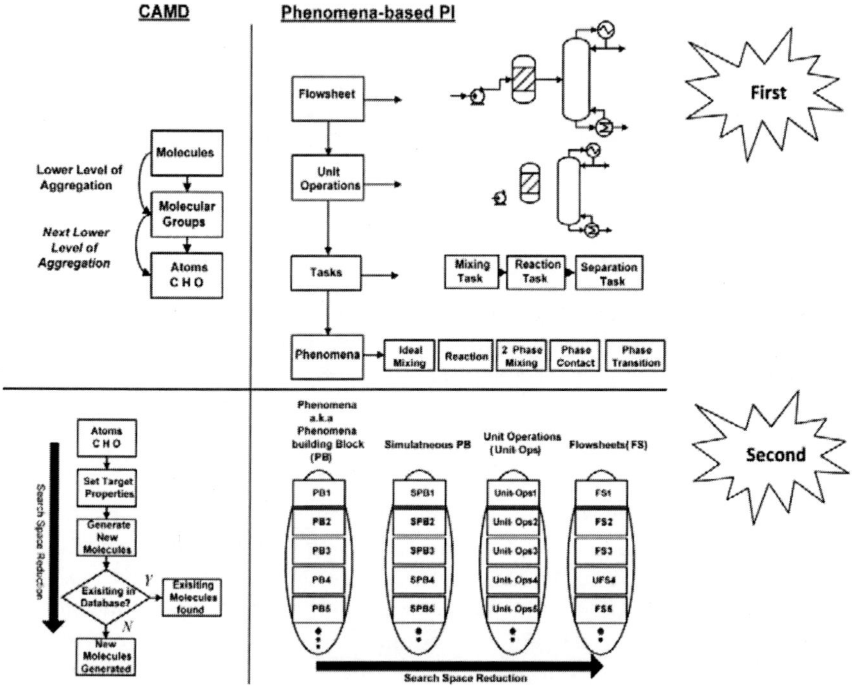

Figure 2. Phenomena-based Process intensification (PI) concepts using computer-aided molecular design (CAMD).

In the same way the PBS method also operates at different levels of aggregation where the flowsheet is represented in terms of unit operations from which the necessary tasks are identified. For each task, the involved phenomena are identified. Therefore, just as groups of atoms can be joined to form molecules, groups of phenomena can be joined to form existing as well as new tasks. In a similar way, groups of tasks can be joined to form operations and groups of operations can be joined to form process flowsheets. The number of alternatives that can be generated increases by going to a lower level of aggregation, just as in the case of molecular design. This concept is illustrated in Figure 2.

The phenomena currently considered are classified into eight different classes. They are: mixing-M (mass flow within one or several phases), phase contact-PC (the contact and resistances at phase boundaries of phases), phase transition-PT (mass transfer of compounds between two phases), phase change-2phM (state change of a complete stream at no phase transition), phase separation-PS (the degree of separation of two phases), reaction-R (change in mass of a compound or compounds generated or consumed between inlet and outlet), energy transfer-H/C (energy transfer between sources and sinks of energy) and stream dividing-D (division of a stream into two or more streams).

2.3.1. Step 3.1: Problem and Objective Function Definition

In this step the synthesis/design problem and objective function is defined together with the process boundaries, underlying assumptions, and the performance metrics for screening PI options.

2.3.2. Step 3.2: Analyze the Process

In this step, the base case design is analyzed in order to further identify/classify the process bottlenecks through a knowledge-based and a model-based approach. All mass and energy data are collected from the rigorous simulation of the base case design and the flowsheet is transformed into a task and phenomena based flowsheet in order to decouple the analysis from physical unit operations. The identified process bottlenecks are subsequently linked to corresponding phenomena and tasks inside and outside of the unit operation in which they occur. For analyzing the properties of the compounds in the system in general the method developed by Jaksland et al. [18] based on thermodynamic insights is used.

2.3.3. Step 3.3: Identification of Desirable Phenomena

In this step additional phenomena for synthesis of flowsheets, which match the targets defined in Step 3.1 and thereby improve the process performance of the necessary phenomena (identified in Step 3.2) are selected. That is, the identified process bottlenecks and the corresponding phenomena and tasks are used to identify tasks and phenomena that need to be improved to achieve the desired performance. For the selection of suitable (desirable) phenomena, property analysis of single compound and mixtures of compounds as well as reactions in the system are needed. From the set of suitable phenomena the most promising phenomena are selected. However, some phenomena cannot be used on their own; hence, additional accompanying phenomena are selected. Finally, the operating window of all identified phenomena in the search space is determined. These are needed as input for the next step.

2.3.4. Step 3.4: Generate Feasible Operation/Flowsheet Options

In this step all phenomena based process options are generated. They are screened according to logical and structural constraints. The input to this step is the initial search space of identified phenomena and its corresponding operating windows. Next, all phenomena are connected to form feasible simultaneous phenomena building blocks (SPBs). Note that SPBs are connected to form operations, which are translated into unit operations, which are then combined to form flowsheet alternatives. Not all SPBs when connected to form operations are feasible and structurally promising and therefore, screening based on logical and structural constraints, which were defined in Step 3.1 is performed again to obtain the feasible set.

2.3.5. Step 3.5: Fast Screening Using Process Constraints

In this step the feasible SPBs are connected to form the feasible flowsheet alternatives. The flowsheet alternatives are then screened in terms of operational constraints and performance criteria at the unit operation level using process simulation and analysis tools.

2.3.6. Step 3.6: Solve the Reduced Optimization Problem

In this step the most promising feasible flowsheet alternatives inclusive of PI are identified. This is achieved by optimization of the objective function with respect to the full mathematical model simultaneously for each remaining flowsheet alternatives and ranking all options by their corresponding performance.

3. CASE STUDY

The application of the framework is highlighted through a case study. Due to the emerging demand for sustainable fuels from renewable resources in order to have lower environmental impacts, production of biodiesel was selected as a case study highlighting the application of the computer-aided framework. Throughout the application of the framework on the biodiesel production process, several software and computer-aided methods and tools have been employed, as listed in Table 1.

Table 1. Type of software used in the case study.

Software	Application	References
PRO/II	Process simulation	[23]
Aspen Hysys	Process simulation	[24]
SustainPro	Sustainability analysis	[10–12]
ECON	Economic analysis	[20,21]
LCSoft	Life cycle assessment	[22]
ICAS-ProPred	Chemical and physical properties	[25]

Table 2. List of catalysts that may be used for production of biodiesel.

Raw material	Alcohol	Catalyst	Ratio	Temperature (°C)	Pressure (atm)	Residence time (h)	Conversion (%)	References
Palm oil	Methanol	KOH 1 wt.%	6:1	60	1	–	95	[26]
Waste oil	Methanol	KOH 1 wt.%	50:1	80	1	–	97	[26]
Waste oil	Methanol	KOH 1 wt.%	6:1	70	1	–	95	[26]
Palm oil	Methanol	KOH	3:1	–	1	–	99.8	[27]
Palm oil	Methanol	KOH	–	–	1	–	99.7	[27]
Palm oil	Methanol	KOH 1 wt.%	6:1	60	1	1	82.5	[28]
Waste cooking oil	Methanol	H_2SO_4 4 wt.%	20:1	95	1	10	90	[29]
Waste cooking oil	Methanol	Carbon based catalyst 10 wt.%	30:1	80	1	8	Yield = 92	[30]
Palm oil	Methanol	NaOH 0.2 wt.%	6:1	60	1	1.33	94.3	[31]
Palm oil	Methanol	κ-carrageenan 52.5 wt.%	7:1	30	1	72	99	[32]
Palm oil	Methanol	20% w/w DTPA/Clay, 5%	15:1	170	17	6	94	[33]
Palm oil	–	Immobilized lipase (*Candida rugosa*)	3.3% v/v	55	1	6	Yield = 82	[34]

3.1. Part-1: Superstructure Generation

3.1.1. Step 1.1: Collect Data/Information on Available Technologies

Available data for several biodiesel processing routes were collected from reported technologies in the literature. The different catalysts applied for biodiesel production can be classified in terms of alkali catalyst, acid catalyst, solid base catalyst, solid acid catalyst and enzyme catalyst [9].

Different types of catalysts were considered to design multiple processing routes and they are given in Table 2.

3.1.2. Step 1.2: Create Superstructure Configuration

By using the information from Step 1.1, a generic superstructure of processing routes was created (see Figure 3). It describes a network of configurations representing multiple flowsheets. In principle, hundreds of flowsheet alternatives can be generated for the biodiesel production from this superstructure (M1 = 14256). Five main stages/tasks are identified for biodiesel production from waste oil and vegetable oil: feed, reaction, separation 1, separation 2 and separation 3. A pretreatment section has been considered as a black-box, that is, given the raw material resource, the output is the liquid raw material needed for the conversion step (reaction section). For the reaction section, alkali- and acid-catalyzed (homogeneous and heterogeneous), enzymatic and supercritical methanol transesterification are considered and for the separation, methanol recovery and purification of fatty acid methyl ester (FAME) and glycerol are considered. The separation task is divided into three sub-tasks: Separation-1 for methanol recovery; Separation-2 for water washing and neutralization; and, Separation-3 for biodiesel purification and glycerol recovery. The throughput for each unit is determined for each section. For example, in the separation-1 section, the input is determined from the output from the reactor and the output is determined based on the input in the separation-2 section or if recycle exists, a methanol recovery recycle stream is needed. The type of separation task was selected based on available information in the literature.

3.1.3. Step 1.3: Screening of Alternatives

The superstructure is analyzed and the alternatives have been screened based on criteria that are also used by Simasatitkul et al. [9]. In this work, the feedstocks considered are palm oil and waste (cooking) palm oil. Besides, the productivity of the catalyst was also used as a part of the screening procedure where the processes that only give the highest productivity of biodiesel are retained. The reduction of the superstructure for the highest separation task is carried out as well as the limitations of the separation technique.

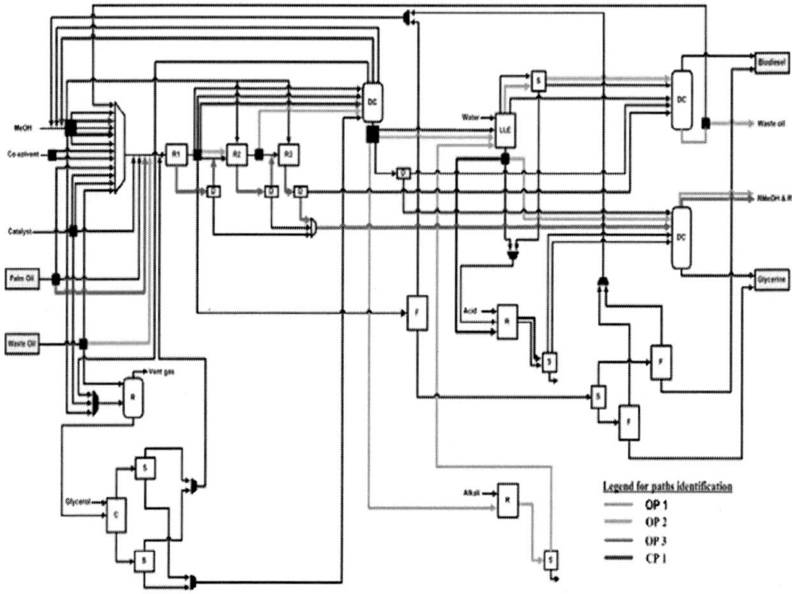

Figure 3. Closed- and open-paths that were identified as major bottlenecks in the superstructure.

In order to reduce the size of the combinatorial problem, the following assumptions were made:

1. The different types of the same group of catalysts were not considered, for example, for all alkaline catalyst options, only one option was considered: the alkali-catalyzed transesterification.
2. The triglycerides were only assumed to be Triolein.
3. The design was based on the use of palm oil and waste palm oil as feedstock with low free fatty acid content (5% FFA).
4. The total operation time is assumed to be 350 days per year.
5. The following reactions were considered to represent transesterification and esterification steps [35]:

   Reaction 1: Esterification

   Oleic acid + Methanol→Methyl oleate + Water

   Reaction 2: Transesterification

   Triolein + Methanol→Methyl oleate + Glycerol

Finally, the superstructure was reduced to six process configurations to be considered in the next step of the framework.

3.2. Part-2: Base Case Design

3.2.1. Step 2.1: Perform Process Simulation of Superstructure

In this step the superstructure of feasible flowsheet combinations (M2 = 6) for producing biodiesel was simulated for all processing routes in order to obtain steady state mass and energy balance data needed in the next step. The flowsheet alternatives were simulated simultaneously by setting all binary variables to unity, by equally dividing all split streams and by using different reactors for different catalyst types (modeled by their different conversion values). The interested reader can obtain the total simulated superstructure flowsheet, the corresponding computed stream summaries from the simulator and associated details from the corresponding author.

The feed to the transesterification reactor and the operating conditions for each reactor are different, but because different reactors have been used, the superstructure can indeed be simulated simultaneously. For example, the reaction in one of the alkali catalyzed processes was carried out with a 6:1 molar ratio of methanol to oil while in the enzymatic processes the molar ratio was 3:1. In addition, the enzymatic process used immobilized enzyme catalyst while alkali catalyzed process used sodium hydroxide as the catalyst.

All data needed for input to the simulation, for example, kinetic data and product recoveries were obtained from collected data (from Step 1.1). The chemical composition of biodiesel is dependent upon feedstock from which it is produced, as the feedstocks such as palm oil and waste oil of differing origin have dissimiliar fatty acid composition. The compounds taken into account in the simulation are taken based on the average of palm oil and waste palm oil composition available in the literature. Typical fatty acid composition for palm oil and waste oil are given in Table 3. Oleic acid is the major fatty acid present in palm oil and also in waste palm oil. To reduce the complexity of the simulation problem, glycerol trioleate (triolein) is selected to represent the palm oil feedstock for tryglycerides, while the FFA is represented by oleic acid. Based on the selected compound, the fatty acid methyl esters (FAME) are represented by methyl oleate (moleate) as the biodiesel product. The UNIQUAC model based phase equilibrium constant was employed for phase equilibrium calculations. Missing pure compound property data were generated through ICAS-ProPred [25]. The PRO/II commercial process simulation [23] was used to perform the steady state process simulation. The detailed process data for the biodiesel production were obtained through process simulation (using PRO/II) and formed the basis for design decisions to obtain a base case design. For the simulation of this superstructure, the total number of equipment considered was 99 units with 154 streams and 6 compounds.

The generated steady state simulation data included mass/molar flowrates, temperature, pressure and enthalpy values for all streams in the superstructure, which are then used for the indicator based sustainability analysis in the next step.

Table 3. Typical fatty acid composition.

Composition	Palm oil [36]	Waste oil [37]
Myristic acid	0.5–2.0	–
Palmitic acid	32–45	20.4
Stearic acid	2.0–7.0	4.8
Oleic acid	38.0–52.0	52.9
Linoleic acid	5.0–11.0	13.5
Others	–	7.6

3.2.2. Step 2.2: Indicator Calculation and Analysis of Superstructure

The steady state mass and energy data obtained from Step 2.1 are used here. An example of results for the sustainability analysis is shown in Table 4 and Table 5 for the open- and closed-paths, respectively.

Table 4. The main open-paths in the superstructure that were identified as bottlenecks in terms of material value added (MVA), energy and waste cost (EWC), total value added (TVA) indicators.

Path	Compound	Flowrate (kg/h)	MVA (10^3\$/y)	EWC (10^3\$/y)	TVA (10^3\$/y)
OP 1	H_2O	11.7	−57.0	2.7	−59.8
OP 2	Methanol	8.2	−55.3	3148.4	3093.1
OP 3	H_2O	6.9	−34.1	26.8	−60.9

Table 5. The main closed-paths in the superstructure that were identified as process bottlenecks in terms of EWC and accumulation factor (AF) indicators.

Path	Compound	Flowrate (kg/h)	EWC (10^3\$/y)	AF (10^3\$/y)	Score *
CP1	Methanol	3.2	2671.3	1.1	High

* a high score indicates a high potential for change.

It can be seen in Table 4 that there is a high material loss (– ve values) as well as energy consumption (+ ve values) of water and methanol in the given open paths. Furthermore, high energy consumption is also observed for methanol in the given closed-path in Table 5. In the same table, the accumulation factor (AF) is also high, which means that there is a high potential for improvement in this path. It should be noted that the paths having the higher values in terms of sustainability metrics have been considered as the bottleneck in the superstructure as presented in Figure 3.

Therefore, this information has been used to identify the path that leads to an optimal processing route. Carvalho et al. [11] explained that the lower positive value of the metric indicates that either the impact of the process is less or the output of process is more.

3.2.3. Step 2.3: Calculate and Analyze Economic and Lifecycle Assessment (LCA) Factors

The remaining flowsheet alternatives from Step 2.2 were further analyzed using an economic and LCA evaluation.

In the biodiesel production process, heating, cooling and electricity are the main drivers for operating cost. The main utilities for heating/cooling are steam and cooling water, respectively. These are used by heat exchangers, reboilers, and condensers that are present in the flowsheet. The units that have a high utility cost relative to other units are identified as process bottlenecks in terms of operating cost. In this analysis, it was found that the reboiler of the columns to recover methanol, and biodiesel as well as the heat exchangers have the largest utility costs in the process. From Figure 4 it can be noted that heat exchangers E-110 and E-111 and the reboiler of column rT-108 have the largest utility costs in the process that employs an acid-catalyst for converting a feedstock of waste cooking oil. Therefore, these units were identified as process bottlenecks in terms of high energy consumption.

Life cycle assessment (LCA) was used to evaluate the inputs and outputs of the processes with respect to their carbon footprint and the unit operations that had a large carbon footprint relative to other units were identified as process bottlenecks. The environment impacts in terms of carbon footprint (CO_2 equivalent) are presented in Figure 5. The process equipment with the highest carbon footprint is highlighted in black in this figure. For example, rT-T7 corresponds to the reboiler of the distillation column in the acid-catalyzed process for pretreatment of waste cooking oil.

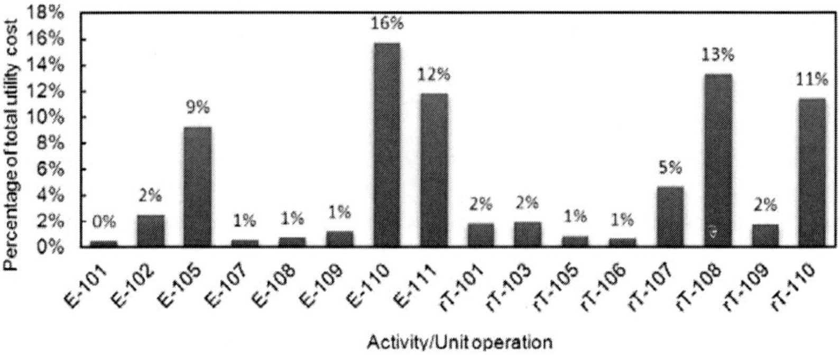

Figure 4. Percentage of the total utility costs with respect to each activity/unit operation.

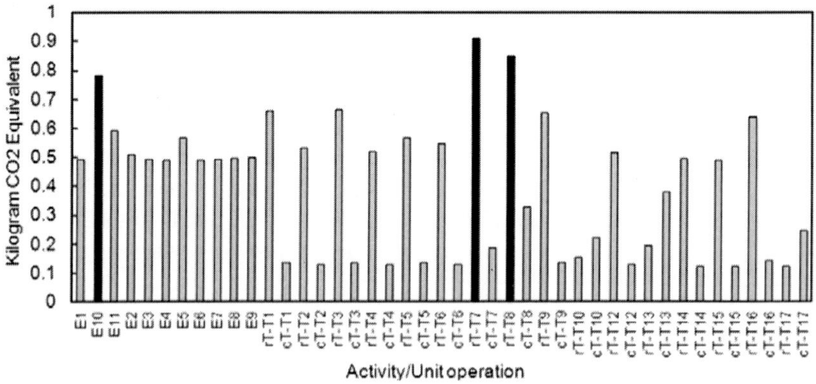

Figure 5. The carbon foot prints of different unit operations in the most promising processing routes.

3.2.4. Step 2.4: Screening of Alternatives to Identify Base Case Design

An example of comparison of the sustainability metrics in two processing routes obtained from the superstructure in order to reduce the search space is presented in Table 6. Here, the considered criteria for screening were the utility costs and carbon footprint for enzymatic catalyzed process (processing route 1) and supercritical methanol process (processing route 2). From Table 6, it can be seen that the processing route 1 is better in terms of carbon footprint and total utility cost. Processing route 2 operates at high temperature and pressure, resulting in higher utility consumption and the total amount of CO_2 emitted. Therefore, processing route 2 is eliminated and not considered further.

Table 6. An example of comparison of the sustainability metrics in two processing routes.

Sustainability metric	Processing route 1	Processing route 2
Total kg CO_2 Equivalent	1.0447	1.0755
Total utility costs ($)	2,350,000	2,680,000

The feasible alternative selected as the base case design is presented in Figure 6. The base case design was selected on the basis of having the least process bottlenecks (and likely operational costs) and carbon footprint. This alternative uses waste cooking palm oil as the feed over an enzyme as the catalyst [31]. The enzyme catalyzed process is recommended since it has several advantages over homogeneous catalysts, for example, high selectivity, lower energy consumption because of low temperature operation and high product yield.

Biodiesel is produced as the main product and glycerol as a value-added byproduct. There are 1 reactor, 4 distillation columns, 2 flash drums and 2 separators together with several heat exchangers and pumps in the process. Methanol is recovered throughout the process and is recycled to the feed stream by three recycle loops.

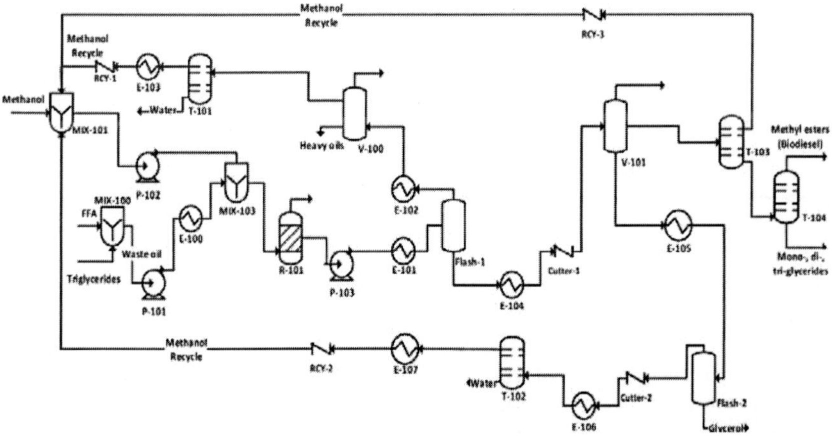

Figure 6. Biodiesel production process with glycerol as a value-added by product obtained from the superstructure which is the base case design.

3.2.5. Step 2.5: Analysis of Base Case Design

The feed in the base case design contains triglycerides and free fatty acids which correspond to a waste cooking oil feedstock. Therefore, having the base case flowsheet, a more detailed rigorous simulation was now performed. Note, in course of simultaneous simulation of the superstructure, less number of compounds and reactions were considered for the sake of simplicity of the simulations. Waste cooking palm oil consists of a mixture of triglyceride (e.g., trilaurin, tripalmitin, triolein, tristearin, trilinolein and trilinolenin) and free fatty acids (e.g., lauric acid, palmitic acid, stearic acid, oleic acid, linoleic and linolenic acid). The composition of the feedstock is given in Table 7.

Table 7. Mole fraction of the compounds that form the waste cooking oil in the base case design.

	Compound	Mole fraction in the feed
Triglycerides	Trilaurin	0.0010
	Trimyristin	0.0087
	Tripalmitin	0.3351
	Tristearin	0.0319
	Triolein	0.2892
	Trilinolein	0.0726
	Trilinolenin	0.0029
Free Fatty Acids	Lauric acid	0.0003
	Myristic acid	0.0030
	Palmitic acid	0.1132
	Stearic acid	0.0107
	Oleic acid	0.0972
	Linoleic acid	0.0244
	Linolenic acid	0.0097

The main reaction for biodiesel production consists of the three-step transesterification of triglyceride (trilaurin, tripalmitin, triolein, tristearin, trilinolein and trilinolenin), diglyceride (dilaurin, dipalmitin, diolein, distearin, dilinolein, dilinolenin) and monoglyceride (monolaurin, monopalmitin, monoolein, monolinolenin) with methanol as shown below. Note that the $RCOOCH_3$ is the biodiesel product which is a mixture of (methyl laurate, methyl myristate, methyl palmitate, methyl stearate, methyl oleate, methyl linoleate).

$$\text{Triglyceride} + CH_3OH \leftrightarrow \text{Diglycerides} + RCOOCH_3$$
$$\text{Diglyceride} + CH_3OH \leftrightarrow \text{Monoglyceride} + RCOOCH_3$$
$$\text{Monoglyceride} + CH_3OH \leftrightarrow \text{Glycerol} + RCOOCH_3$$

When an esterification is applied to reduce free fatty acids in waste cooking oil, the following reaction where water is produced as a by-product, takes place:

$$RCOOH + CH_3OH \leftrightarrow RCOOCH_3 + H_2O$$

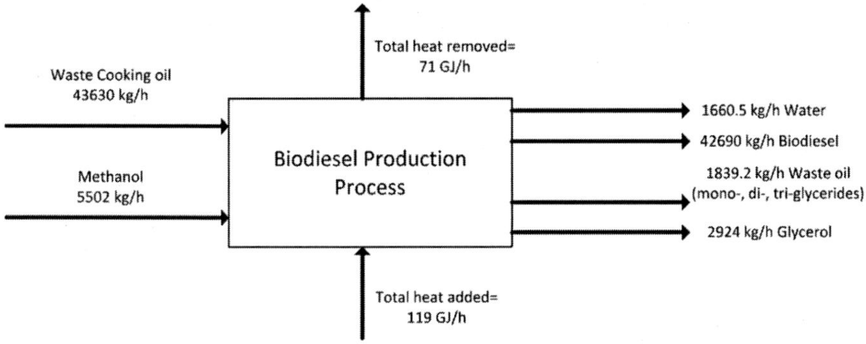

Figure 7. Calculated mass (in kg/h) and energy (GJ/h) balance for the base case design of biodiesel production process.

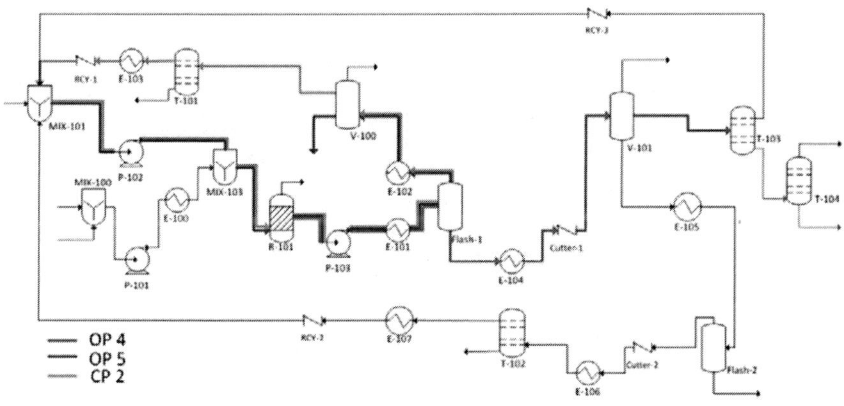

Figure 8. Presentation of closed- and open-paths that include process bottlenecks in the base design for production of biodiesel.

Simulation of biodiesel production is performed using the Aspen Hysys process simulator. All the triglycerides and free fatty acids mentioned above are defined using a "Hypo manager" tool. A waste cooking palm oil containing 10 wt.% of free fatty acids at the flow rate of 62.48 kmol/h is used as feedstock for detailed simulation of biodiesel production base case design that was obtained in Step 2.4. Fresh methanol was supplied at the flowrate of 171.1 kmol/h. The mass and energy balance are simply depicted in Figure 7. Note that the recycle of excess methanol to the reactor is not shown on this figure. This is shown in Figure 8 and also the suggested excess amounts are listed in Table 2.

The updated sustainability indicators, metrics and LCA on the more detailed base case calculations are given in Table 8 and Table 9 and the paths with which they are associated are given in Figure 8.

Table 8. List of selected open-path indicators.

Path	Compound	Flowrate (kg/h)	MVA (10^3S/y)	TVA (10^3S/y)
OP4	Methyl oleate	174.1	−2047.2	−2052.3
OP5	Methanol	3129.8	−14174.3	−14898.1

Table 9. List of the principal closed-path indicators (EWC and AF).

Path	Compound	Flowrate (kg/h)	EWC (10^3S/y)	AF (10^3 S/y)	Score *
CP2	Methanol	1402.6	496.7	0.246	High

It can be seen in the Table 8, that methanol has a very negative value in terms of MVA and TVA indicators; therefore, this indicates very high raw material loss and energy consumption in the OP4. The same also applies to methyl oleate in this path.

From Table 9, it can be observed that methanol has very high EWC value in a closed-path, which refers to a recycle stream. Thus, this path is identified as the one that contains the unit operations with potential bottlenecks. Table 10 gives a selected list of the sustainability metrics for the base case design.

Table 10. Sustainability metrics for the base case design.

MJ energy/kg product	kg product/kg raw material	kg waste/kg product	Total kg CO_2 Equivalent	Total utility costs (S)/year
2.5	0.94	0.034	0.183	7,790,000

The economic and life cycle assessment analysis were also updated for the more detailed base case calculations and these are highlighted in Figure 9 and Figure 10. From Figure 9 it can be seen that the reboilers of columns rT-103, rT-101 and rT-104 and heat exchanger E-101 have the largest utility costs in the process and therefore these are stored as process bottlenecks (high energy consumption) with respect to utility costs. This is not surprising, however, the analysis also points out the operation with the largest utility cost. From Figure 10, it can be seen that the same heat exchanger (E-101), reboilers of the same columns rT-103, rT-104 and rT-101, and the reactor (R-101) have the largest carbon footprint for biodiesel production because of their high utility (energy) consumption.

The characteristics of the unit operations, for example, the above energy consumption analysis were stored as the final set of process bottlenecks. This information is useful because if these process bottlenecks are improved/eliminated then the new process alternative will be more sustainable and offer more economic benefits. Note that removing these bottlenecks means improvements in the corresponding indicators (from sustainability analysis), which in turns means improvement in the sustainability metrics. Thereby, a more sustainable process design is obtained. A list of the equipment according to their identification as a major bottleneck in the three (sustainability, cost, LCA) analyses is given in Table 11. It can be seen that columns T-101, T-102, heat exchanger (E-101) and reactor (R-101) have appeared as a bottleneck in all three analyses. Therefore, from this analysis it can be concluded that the major bottleneck of the process is very high energy consumption, which can be overcome by reducing the utility consumption and consequently utility costs.

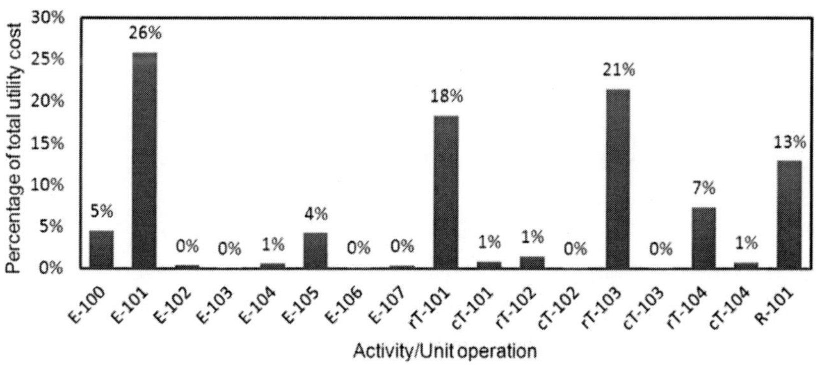

Figure 9. Percentage of the total utility costs with respect to each activity/unit operation.

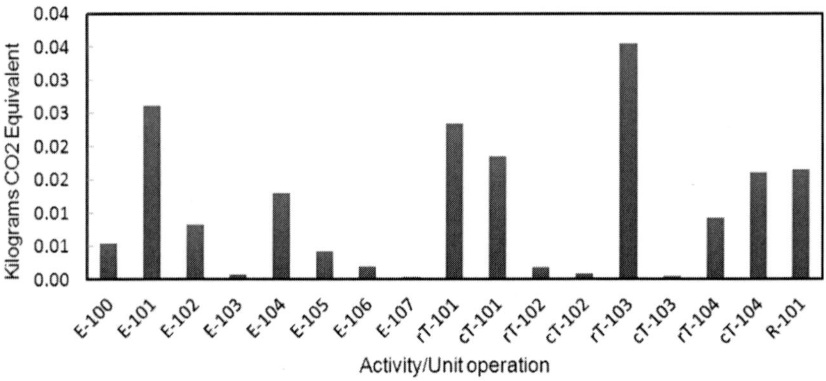

Figure 10. Carbon footprint of different activities/unit operations in the base case design for biodiesel production.

Table 11. Main identified bottlenecks with respect to sustainability, economic and life cycle assessment (LCA) analysis.

Unit	Label	Sustainability	LCA	Economic
Reactor	R-101	x	x	x
Column	T-101	x		x
Column	T-102	x	x	
Column	T-103	x	x	x
Column	T-104	x	x	x
Heat exchanger	E-100	x		
Heat exchanger	E-101	x	x	x

3.3. Part-3: Sustainable Design Options

3.3.1. Step 3.1: Problem and Objective Function Definition

The PI problem is defined as follows: to achieve and identify an intensified process flowsheet for the production of biodiesel and glycerol with maximum purity and conversion of raw materials that is >97% vegetable/waste oil while reducing the operating costs of the process.

Table 12. Performance metrics and their associated constraints at logical (L), Structural (S), Operational (O) and PI screening (PIs) levels.

Performance metric	L	S	O	PI_s	Constraint description
Reactor	x				Reaction is present
	x				Reaction is inside the first unit operation
	x				The reactor effluent is connected to a separation sequence according to the product purity defined in the base case design
Capital Costs		x			Identify possible units and allow only a number of units (defined in step 1 or by screening)
Efficiency		x			Do not integrate phenomena which inhibit each other's performance
				x	Product-yield
				x	Higher separation efficiency
		x			Do not connect phenomena in a series of co-current stages with decreasing efficiency/equilibrium
Energy		x			Do not connect phenomena to a series with alternating heat addition and heat removal
		x			Remove options in which phenomena are heated/cooled leading to a decrease of the efficiency
Operational Costs			x		Utility costs
				x	Utility costs
Simplification		x			Remove options with redundant stages
		x			Identify possible units and allow only a number of units (defined in step 1 or by screening)
Waste minimization		x			Ensure that phenomena are connected so that the best use of or recycle of raw materials is achieved
Raw materials			x		Raw materials are pure
			x		Yield ≥97%
Equipment				x	PI screening equipment: Novel equipment

The objective function is shown in Equation (9) where c_i represents the cost of raw materials and the c_j cost of energy, $\dot{m}_i$ represents the mass flowrate of the raw materials, $\dot{m}_k$ represents the mass flowrate of products and $\dot{e}$ represents the utility usage. Note that subscripts i, j, k represent raw material, energy and products, respectively. It should be mentioned that as a result of intensification the number and/or volume of unit operations may decrease which will consequently result in a reduction of capital costs.

$$\min F_{obj} = \left(\frac{\sum c_i \dot{m}_i + \sum c_j \dot{e}}{\sum \dot{m}_k} \right) \tag{9}$$

The performance metrics were selected and categorized in terms of logical (L), structural (S), operational (O) and PI screening constraints and are given in Table 12. The product specifications are defined to be the same as the base case design.

3.3.2. Step 3.2: Analyze the Process

In this step the mass and energy balance steady state data were collected from the rigorous simulation of the more detailed simulated base case design. The base case design flowsheet was then transformed into a task based flowsheet, which identifies different tasks in the process in terms of mixing, reaction and/or separation tasks. This task based flowsheet was then transformed into a phenomena based flowsheet using a PI knowledge base. The task-based and phenomena-based flowsheets are shown in Figure 11 and Figure 12, respectively.

Using the task and phenomena based flowsheets of the base case design; the previously identified process bottlenecks of the base case design are linked to the phenomena and the tasks inside and outside of the unit operation in which they occur. To further analyze the system to identify the bottlenecks, the needed pure compound properties are retrieved from ICAS database [24] for computing the binary ratio matrix of properties. Excerpts of these results are given in Table 13 (for pure compound properties) and Table 14 (for binary ratio matrix).

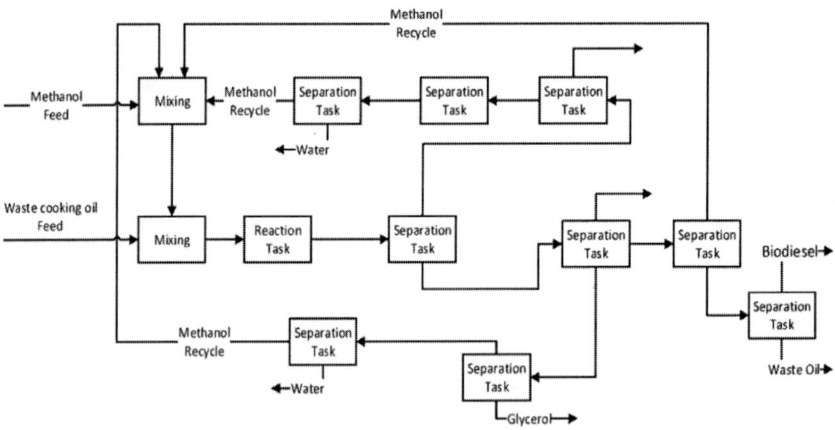

Figure 11. Task-based flowsheet of the base case design for biodiesel production process.

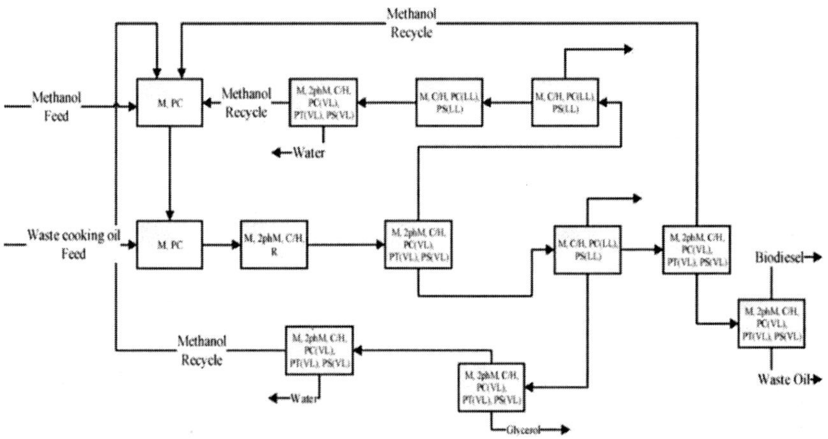

Figure 12. Phenomena-based flowsheet of the biodiesel production process according to the existing phenomena in the base case design.

Table 13. List of the pure compound properties of some of the compounds in the system.

Compound	MW (g/mol)	T_c (K)	P_c (bar)	T_b (K)	T_m (K)	Solubility parameter (MPa^0.5)	Radius of gyration (Å)	V_{vdW}	V_m
Methanol	32.0	512.4	72.8	337.6	175.3	29.61	1.5520	0.0217	4.08×10^{-2}
Water	18.0	747.1	218.4	373.0	273.0	47.79	0.6150	0.0123	1.79×10^{-2}
Glycerol	92.0	724.8	65.8	561.8	291.2	34.99	3.5199	0.0513	7.13×10^{-2}
Lauric acid	200.3	733.8	19.05	571.7	342.31	19.96	6.6009	0.13549	0.2262
Methyl laurate	214.3	711.8	19.9	539.2	278.1	19.74	18.6656	0.1448	0.2067
Monolaurin	274.4	857.4	17.3	625.6	351.3	18.81	7.0445	–	0.3076
Dilaurin	456.7	916.7	11.2	703.1	372.5	14.61	8.8250	–	0.5109
Trilaurin	639.0	998.7	9.5	767.2	386.8	13.97	9.6796	–	0.7151

Table 14. Computed binary ratio matrix for some of the compounds of the system.

Ratio of properties of binary pairs	Property ratio									
	MW	Critical temperature (T_c)	Critical pressure (P_c)	Boiling temperature (T_b)	Melting temperature (T_m)	Solubility parameter	Radius of gyration	V_{vdW}	V_m	Binary azeotrope at $P = 1$ atm
Methnol/Water	1.78	1.44	3.03	1.67	1.56	1.61	2.52	1.76	2.28	No
Methanol/Glycerol	2.85	1.41	1.11	1.59	1.67	1.17	2.27	2.38	1.75	No
Methanol/Methyl laurate	6.66	1.39	3.64	1.85	1.59	1.50	12.50	6.67	5.00	No
Methanol/Lauric acid	6.25	1.43	3.82	2.08	1.96	1.48	4.17	6.25	5.56	No
Methanol/Monolaurin	8.33	1.67	4.21	2.27	2.00	1.57	4.55	0.00	7.69	No
Methanol/Dilaurin	14.29	1.79	6.50	1.04	2.13	2.03	5.56	0.00	12.50	No
Methanol/Trilaurin	20.00	1.96	7.68	1.11	2.22	2.12	6.25	0.00	16.67	No
Water/Glycerol	5.00	1.03	3.32	1.25	1.06	1.37	5.88	4.17	4.00	No
Water/Methyl laurate	12.50	1.05	10.92	1.37	1.02	2.42	33.33	11.11	11.11	No
Water/Lauric acid	11.11	1.02	11.46	1.16	1.25	2.39	11.11	11.11	12.50	No
Water/Monolaurin	14.29	1.14	12.63	1.10	1.28	2.54	11.11	0.00	16.67	No
Water/Dilaurin	25.00	1.21	19.48	1.52	1.37	3.27	14.29	0.00	33.33	No

Ratio of properties of binary pairs	Property ratio									
	MW	Critical temperature (T_c)	Critical pressure (P_c)	Boiling temperature (T_b)	Melting temperature (T_m)	Solubility parameter	Radius of gyration	V_{vdw}	V_m	Binary azeotrope at $P = 1$ atm
Water/Trilaurin	33.33	1.33	23.03	1.30	1.41	3.42	16.67	0.00	33.33	No
Glycerol/Methyl laurate	2.33	1.02	3.29	1.43	1.05	1.77	5.26	2.86	2.94	No
Glycerol/Lauric acid	2.17	1.01	3.45	1.12	1.18	1.75	1.89	2.63	3.13	No
Glycerol/Monolaurin	2.94	1.18	3.81	1.22	1.20	1.86	2.00	0.00	4.35	No
Glycerol/Dilaurin	5.00	1.27	5.87	1.09	1.28	2.40	2.50	0.00	7.14	No
Glycerol/Trilaurin	7.14	1.37	6.94	1.67	1.33	2.51	2.78	0.00	10.00	No
Methyl laurate/Lauric acid	1.07	1.03	1.05	1.59	1.23	1.01	2.83	1.07	1.10	No
Methyl laurate/Monolaurin	1.28	1.20	1.16	1.85	1.27	1.05	2.65	0.00	1.49	No
Methyl laurate/Dilaurin	2.13	1.28	1.78	2.08	1.33	1.35	2.12	0.00	2.50	No
Methyl laurate/Trilaurin	2.94	1.41	2.11	2.27	1.39	1.41	1.93	0.00	3.45	No
Lauric acid/Monolaurin	1.37	1.16	1.10	1.04	1.03	1.06	1.06	0.00	1.35	No
Lauric acid/Dilaurin	2.27	1.25	1.70	1.11	1.09	1.37	1.33	0.00	2.27	No
Lauric acid/Trilaurin	3.23	1.37	2.01	1.25	1.14	1.43	1.47	0.00	3.13	No
Monolaurin/Dilaurin	1.67	1.06	1.54	1.37	1.06	1.29	1.25	0.00	1.67	No
Monolaurin/Trilaurin	2.33	1.16	1.82	1.16	1.10	1.35	1.37	0.00	2.33	No
Dilaurin/Trilaurin	1.41	1.09	1.18	1.10	1.04	1.05	1.10	0.00	1.41	No

Table 15. Decisions for finding the desirable task for overcoming the process bottlenecks caused by primary task.

Bottleneck	Primary task		Desirable task				
	Reaction	Separation	Reaction	Separation	Mixing	Cooling	Heating
Activation problem	x		x		x		x
Contact problems of raw materials/limited mass transfer	x	x			x		
Formation of undesirable side products	x	x	x	x			
High energy consumption/demand	x	x	x	x			
Highly endothermic reaction	x					x	x
Limiting equilibrium	x		x	x	x		
Difficult separation due to low *driving force*		x	x	x			

Besides the previously identified process bottlenecks, that is, high energy consumption (identified in Step 2.5), presence of limited equilibrium reaction is also identified as a further process bottleneck (see Table 15).

3.3.3. Step 3.3: Identification of Desirable Phenomena

In this step desirable tasks which are then translated into phenomena are identified for overcoming all the identified process bottlenecks. The identified desirable tasks are listed in Table 15.

In order to identify the feasible phenomena to overcome these bottlenecks, the tasks given in Table 15 need to be translated into phenomena using thermodynamic insights and the operating window of each phenomenon. The identified phenomena to overcome the bottlenecks are given in Table 16.

Table 16. Translation of the desirable tasks into phenomena.

Task	Feed condition	Separation type	Phenomena	Pure compound property	Solvent required?
Absorption separation	V	Gas-separation	PT(VL)	Solubility parameter	Yes
Supercritical extraction	L/V	VL	PT(VL)	Solubility parameter, critical pressure	Yes
V-L separation based on boiling points	V/VL/L	VL	PT(VL)	Vapor pressure *, heat of vaporization *, boiling point, solubility parameter	No
Vapor permeation separation	L	Vapor-separation	PT(P:VV)	Molar volume, solubility parameter, dipole moment	No
Pervaporation separation	V	Vapor-separation	PT(P:VL) **	Molar volume, solubility parameter, dipole moment *	No
Stripping separation	V/VL/L	VL	PT(VL)	Solubility parameter	Yes
Separation by Boiling points	V/VL/L	VL	PT(VL)	Vapor pressure *, heat of vaporization *, boiling point	No

* The parameter was not considered in calculating binary ratio matrix; **(P:) denotes permeation.

All accompanying phenomena are selected by consulting the knowledge base tool for each identified phenomena. In total 15 phenomena building blocks are identified: mixing; ideal (M_{id}), tubular flow (M_{tub}), rectangular flow (M_{rec}), 2-phase mixing (2phM): ideal; dividing (D); heating (H) and cooling (C); reaction (R); phase contact V-L, V-V (PC); phase transition V-L, P:VV, P:VL (PT); phase separation V-L, V-V (PS).

3.3.4. Step 3.4: Generate Feasible Operation/Flowsheet Options

In this step the phenomena building blocks are connected to form SPBs, for example, phenomena for mixing, reaction and heating can be combined to form an SPB if the reaction is endothermic, that is, M=R=H. The total number of SPBs that can be generated from the 15 identified phenomena building blocks are calculated using Equations (10) and (11) [13] and is found to be 16,278. In Equation (10), $n_{Ph,max}$ is the maximum number of phenomena that can be present in an SPB, Ph_{BB} is the total number of phenomena building blocks, 15, Ph_E is the total number of energy transfer phenomena, 2, Ph_M is the total number of mixing phenomena and Ph_D is the dividing phenomena, 1. In Equation (11) $NSPB_{max}$ is the total number of possible SPBs,

$$n_{Ph,Max} = Ph_{BB} - (Ph_E - 1) - (Ph_M - 1) - Ph_D = 11 \quad (10)$$

$$NSPB_{Max} = \sum_{k=1}^{n_{Ph,\max}} \left[\frac{(Ph_{BB}-1)!}{(Ph_{BB}-k-1)!k!} \right] + 1 = 16278 \tag{11}$$

Using connectivity rules, for example, heating and cooling cannot exist in the same SPB (see SPB 5 and 6 in Table 17) or simultaneous reaction and separation can exist in the same SPB (see SPB 16 and 18 in Table 17); and the information regarding the operating window of each phenomenon, a total number of 73 SPBs were found to be feasible. Considering the three different types of mixing that is M_{id}, M_{tub} and M_{rec}, the total number of SPBs are 24 × 3 +1 = 73, including dividing (D) as a separate phenomenon. An example of feasible SPBs is given in Table 17. In this table, mixing (M), 2-phase mixing (2phM), dividing (D); heating (H) and cooling (C); reaction (R); phase contact V-L, V-V (PC); phase transition V-L, P:VV (V-V permeation), P:VL (V-L permeation) (PT); phase separation V-L, V-V (PS); are connected as the phenomena in order to form the SPBs.

The SPB's were then connected to form an operation (which is a collection of SPBs) and these operations are then translated into unit operations which make up the flowsheet alternatives. An example of the generation of a flowsheet alternative is presented below.

In order to produce the products, a change of raw materials state must occur and in order to achieve this, the following task is identified: Reaction task. For overcoming the bottleneck caused by the limiting equilibrium, a task for the recovery and recycle of the unreacted raw materials to the reaction task would shift the equilibrium toward the product side. Therefore, the task identified is: Use of a recycle task. This is shown in Figure 13.

The effluent from the reactor consists of the reactants and products which must be separated. Using the binary ratio matrix (see Table 14) which is based on the pure compound properties (see Table 13) the values for ratios of the boiling points for Methanol/Glycerol, Water/Methanol, Methanol/Methyl-laurate, Methanol/Mono-, Di-, and Tri-laurin are 1.67, 1.10, 1.59, 1.85, 2.08 and 2.27 respectively, which makes separation by VL very feasible for separation of methanol and water from glycerol, methyl esters, mono-, di- and tri-glycerides [17]. Hence, two of the main involved phenomena are PT(VL)–phase transfer involving vapor-liquid and PS(VL)–phase separation involving vapor-liquid. Therefore the following task is identified for the recovery of methanol: Separation task. The task and phenomena based flowsheets are shown Figure 14.

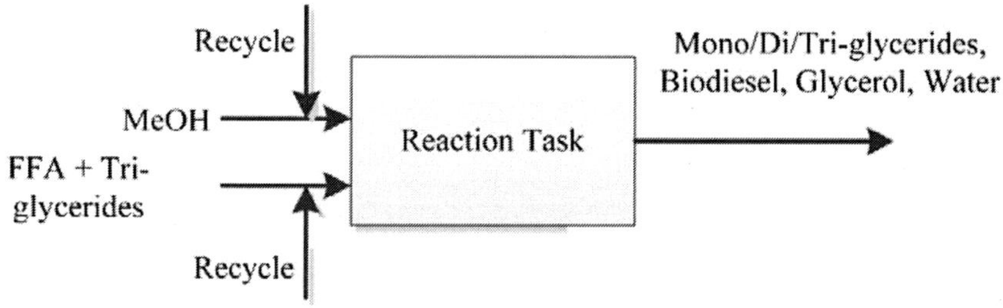

Figure 13. Task for product production.

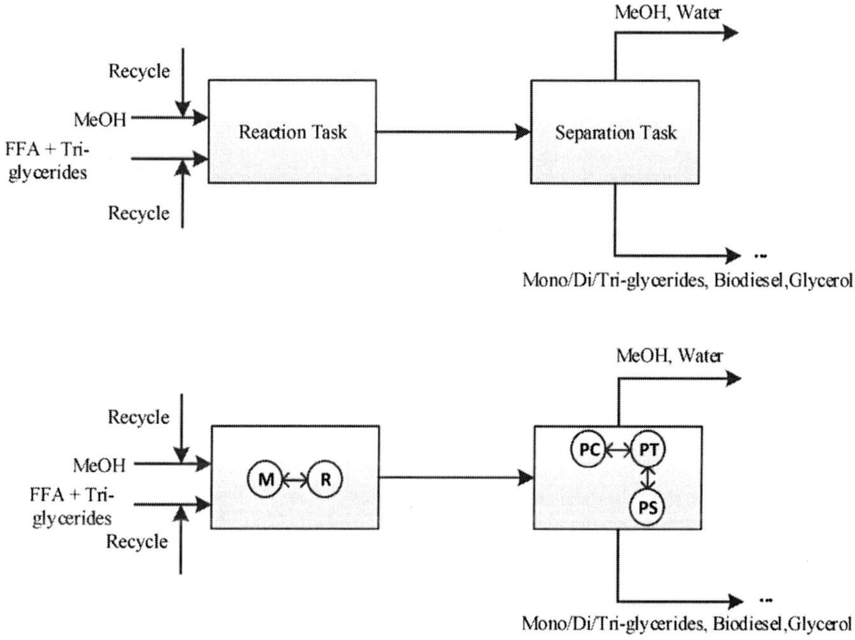

Figure 14. Task and phenomena based flowsheets for generation intensified operation.

From the system properties; transesterification reaction and limiting equilibrium reaction; and the available SPBs reaction and separation can occur simultaneously within an SPB, for example, SPB 18 (see Table 17) indicates M=H=R=2phM=PC(VL)=PT(VL)=PS(VL). Therefore, using this option provided by the SPB (at the phenomena level) and knowledge that simultaneous reaction and separation task (see Table 15, for example, for limiting equilibrium) increases the raw materials conversion, the possibility exists for combining the reaction and separation. Therefore, the following task is identified: Reaction-separation task. The task and phenomena-based flowsheets are shown in Figure 15.

Table 17. The identified feasible SPBs, interconnection phenomena, inlet and outlet conditions of an SPB.

SPB	Interconnection Phenomena	In	Out
1	M	1..n(L)	1(L)
2	M=R	1..n(L)	1(L)
3	M=H	1..n(L)	1(L)
4	M=C	1..n(L)	1(L)
5	M=R=H	1..n(L)	1(L)
6	M=R=C	1..n(L)	1(L)
7	M=2phM=PC(VL)=PT(VL)	1..n(L,VL)	1(V/L)
8	M=R=2phM=PC(VL)=PT(VL)	1..n(L,VL)	1(V/L)
9	M=R=2phM=PC(VL)=PT(P:VL)	1..n(L,VL)	1(V/L)
10	M=R=2phM=PC(VL)=PT(P:VL)=PS(VL)	1..n(L,VL)	2(V;L)
11	M=R=2phM=PC(VL)=PT(VL)=PS(VL)	1..n(L,VL)	2(V;L)
12	M=R=2phM=PC(VV)=PT(P:VV)=PS(VV)	1..n(V,VV)	2(V;V)
13	M=C=2phM=PC(VL)=PT(VL)	1..n(L,VL)	1(V/L)
14	M=H=2phM=PC(VL)=PT(VL)	1..n(L,VL)	1(V/L)
15	M=H=R=2phM=PC(VL)=PT(P:VL)	1..n(L,VL)	1(V/L)
16	M=C=R=2phM=PC(VL)=PT(P:VL)	1..n(L,VL)	1(V/L)
17	M=C=R=2phM=PC(VL)=PT(P:VL)=PS(VL)	1..n(L,VL)	2(V;L)
18	M=H=R=2phM=PC(VL)=PT(P:VL)=PS(VL)	1..n(L,VL)	2(V;L)
18	M=H=R=2phM=PC(VL)=PT(VL)=PS(VL)	1..n(L,VL)	2(V;L)
20	M=C=R=2phM=PC(VL)=PT(VL)=PS(VL)	1..n(L,VL)	2(V;L)
21	M=H=R=2phM=PC(VV)=PT(P:VV)=PS(VV)	1..n(L,VV)	2(V;V)
22	M=C=R=2phM=PC(VV)=PT(P:VV)=PS(VV)	1..n(V,VV)	2(V;V)
23	M=2phM=PC(VL)=PT(P:VL)=PS(VL)	1..n(L,VL)	2(V;L)
24	M=2phM=PC(VL)=PT(P:VV)=PS(VV)	1..n(L,VL)	2(V;L)
73	D	1(L;VL,V)	1..n(L/V/VL)

Note: V, vapor; L, liquid; "/" means "or"; ";" means "and".

However, SPB 18 cannot by itself fulfill process requirements because it does not include complete separation of the raw materials/products, which can then be recycled to increase the raw material conversion (which is desired). Therefore, more SPBs are added to the current SPB, which now includes non-reactive SPBs and this can now be termed as an operation and is shown in Figure 16.

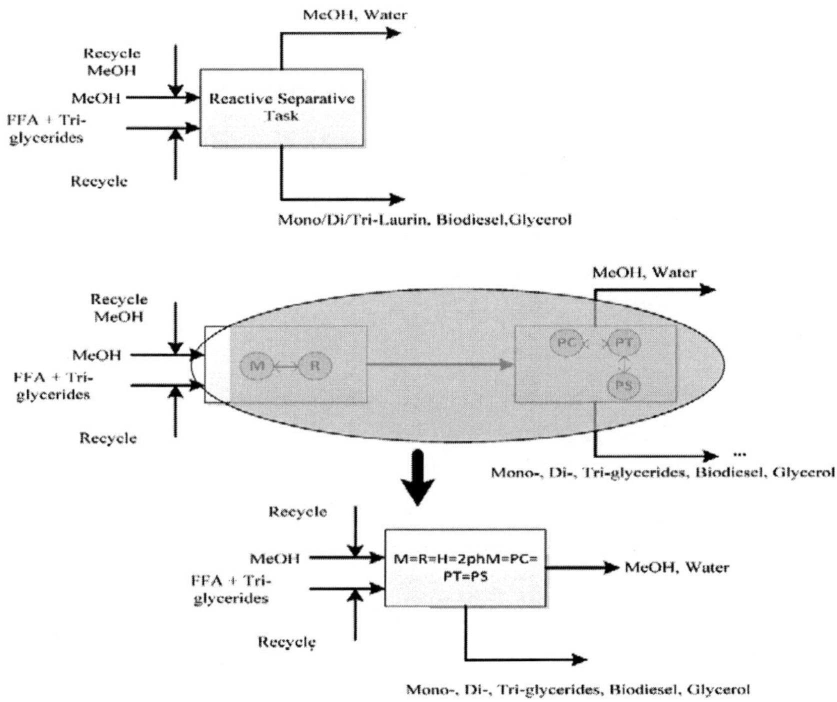

Figure 15. Task and phenomena based flowsheets for combining reaction and separation tasks.

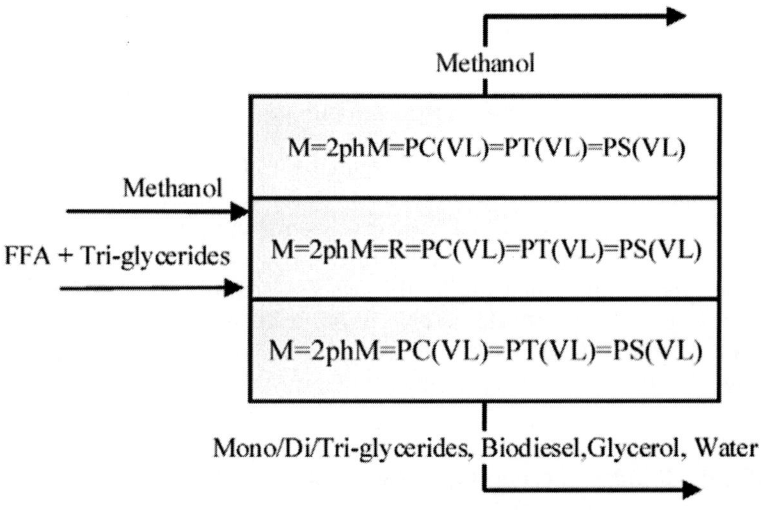

Figure 16. The phenomena building block as a result of connecting SPBs.

Figure 17 shows the operation, which is then transferred to the final flowsheet to replace the reaction and separation after it, in the base case design. The operation has the following characteristics: (a) simultaneous reaction and separation of raw materials and products, (b) the raw materials and products can now be purified for obtaining the raw materials for recycle and the product for collection. However in order to have the existence of VL due to the difference in boiling points of the raw materials and products, heating and cooling are added at the bottom and top of the operation, respectively.

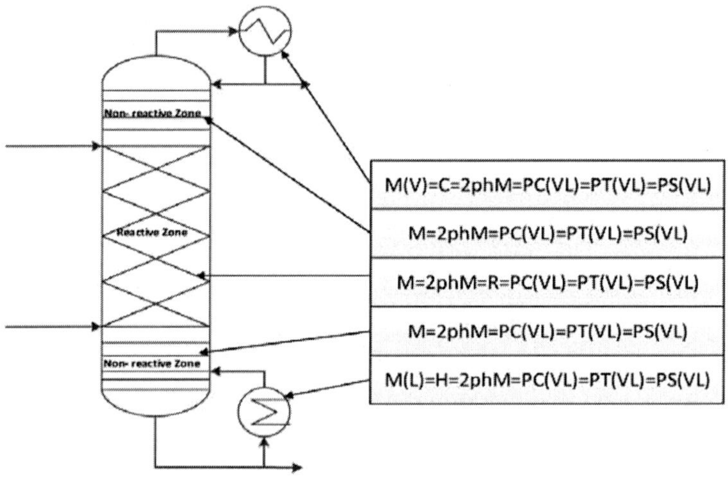

Figure 17. The generated intensified unit operations to combine reaction and separation task: Reactive distillation.

It was found that pure methanol could be obtained at the top of this operation but the biodiesel product at the bottom (besides having glycerol) also has some methanol, which has also been shown by Simasatitkul et al. [9] based on the analysis of the ternary reactive phase diagram. Therefore, the final operation, which fulfills the process requirements, must now be translated to a unit operation. This is done by using a knowledge base, which contains SPBs in terms of unit operations and the result found was a reactive distillation column presented in Figure 17.

3.3.5. Step 3.5: Fast Screening Using Process Constraints

This step is not applicable in the present application of the framework since an intensified unit operation was highlighted in the previous step that combines the reaction and separation tasks in the base case design. Generation of further alternatives and their screening has not been considered in this work.

3.3.6. Step 3.6: Solve the Reduced Optimization Problem

In order to check the feasibility of the proposed intensified operation (given in Step 3.4) for production of biodiesel and recovery of methanol; the reactive distillation column was simulated using rigorous models (Using Aspen Hysys process simulator). The total number of stages for this reactive distillation was found to be 6 including reboiler and condenser. Methanol and waste cooking oil are fed to the column from the second stage with a 10:1 mole ratio of methanol to triglycerides and FFA mixture after connecting all the other separation units in the flowsheet and recycle loops. The reflux ratio should be kept at a very low value since refluxing the water back to the column will consequently shift the equilibrium toward the reactants. Due to this, the separation section of the base case for the separation of methanol, biodiesel and glycerol was reused since the compounds within the stream entering this section is the same as the base case. The final flowsheet inclusive of an intensified unit operation (reactive distillation column) is given in Figure 18. Note that the controllability of the reactive distillation has not been investigated in this work and it is a subject for future studies. The optimality of this process alternative cannot be guaranteed unless all the possible alternatives are generated at the phenomena level and then screened to find the optimal solution. In this work, only the adjacent reaction and separation steps were targeted for intensification. However, it can be claimed that a more sustainable alternative has been found.

The most important performance criteria, including the sustainability metrics for the base case and the intensified alternative are given in Table 18. Note that, at this point of the work, the capital investment was not considered. However, we plan to include this in future studies where it will be shown that the payback time due to extra capital investments for the new intensified alternatives is very short due to the extra profits. Therefore, what are considered in this work are the performance metrics (inclusive of operational costs) and the LCA metrics. Looking at these metrics it can be seen that the objective function that was defined in Step 3.1 has been fulfilled.

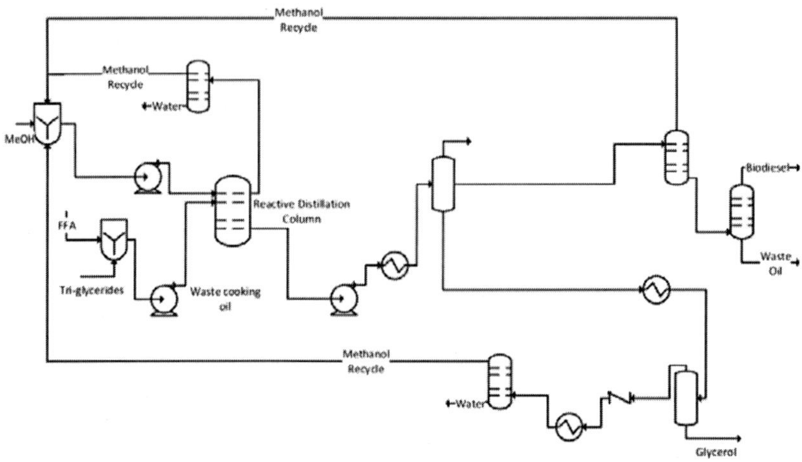

Figure 18. Final flowsheet inclusive of an intensified unit operation which is a reactive distillation column.

Comparing the base case design presented in Step 2.4, with the alternative design inclusive of PI given in Step 3.6 of this framework; it can be observed that the alternative design is more environmentally sustainable and economically profitable. It can be seen that in analyzing the base case design the reactor and the separation were identified as bottlenecks because they were in the path of the selected indicator (OP4; see Table 8). In terms of overcoming the bottlenecks that were identified through the intensification approach (see Table 15), the following process bottlenecks have been overcome as a result of combining the reaction and separation tasks: contact problem of raw materials in the reactor; high energy consumption (reduced utility costs as given in Table 18); limiting equilibrium was overcome due to the simultaneous removal of reactants and products; and difficult separation due to the low driving force has been to some extent overcome since pure methanol (raw materials) is recovered at the top of the reactive distillation column. It should also be noted, that all performance and LCA metrics have been either improved or remained neutral in the intensified process alternative.

Table 18. Comparison of the sustainability metrics in the base case design and the alternative design inclusive of PI.

	Sustainability Metrics	Base case design	Intensified alternative	%Improvement
Performance metrics	Total utility cost ($/year)	7,790,000	4,660,000	40.2
	Total energy consumption (GJ/h)	119.163	73.104	38.6
	product/raw material (kg/kg)	0.94	0.94	0
	Energy/ products (GJ/kg)	0.0025	0.0017	32
	Net water added to the system (m^3)	0	0	0
	Water for cooling/product (m^3/kg)	0.017	0.017	0
	Waste/raw material (kg/kg)	0.032	0.026	18.8
	Waste/products (kg/kg)	0.034	0.028	17.6
	Hazardous raw material/product (kg/kg)	0	0	0
	Number of unit operations	9	7	22
LCA	Total carbon footprint (kg CO_2 eq.)	0.183	0.143	21.8
	HTPI—Human Toxicity Potential by Ingestion (1/LD_{50})	0.51811	0.51111	0
	HTPE—Human Toxicity Potential by Exposure ($mg_{emission}/m^3$)	0.03558	0.03564	0
	GWP—Global Warming Potential (CO_2 eq.)	0.55214	0.55241	0
	ODP—Ozone Depletion Potential (CFC-11 eq.)	5.18×10^{-9}	5.18×10^{-9}	0
	PCOP—Photochemical Oxidation Potential (C_2H_2 eq.)	0.04968	0.04976	0
	AP—Acidification Potential (H^+ eq.)	0.00010	0.00010	0
	ATP—Aquatic Toxicity Potential (1/LC_{50})	0.00366	0.00366	0
	TTP—Terrestrial Toxicity Potential (1/LD_{50})	0.51811	0.51111	0
	HTC (Benzene eq.)—human toxicity (carcinogenic impacts)	2062.7	1794.5	13
	HTNC (Toluene eq.)—human toxicity (non-carcinogenic impacts)	1.3301	1.1795	11.3
	ET (2, 4-D eq.)—Fresh water ecotoxicity	0.00525	0.00490	6.7

LC50 is lethal concentration (mg emission/kg fathead minnow); LD50 is one kg body weight of rat administered in milligrams of toxic

The question of uncertainty of data and/or model parameters on the final more sustainable design that has been obtained has not been investigated. Steimel et al. [38] have proposed a framework for optimizing flowsheet superstructures under uncertainty. A natural next step of this work would be to extend our computer aided framework for more sustainable design to include also uncertainty issues.

4. CONCLUSIONS

In this work, a systematic framework for sustainable design and process intensification has been presented together with its application to a biodiesel production process. The framework is divided into 3 main parts: superstructure generation, base case design selection, and identification of sustainable design options. The framework is generic and can be applied to any chemical or biochemical process. In this work, it has been applied to biodiesel production, where the most promising process alternative in terms of economic, LCA and sustainability metrics was obtained in a systematic and efficient manner, taking into account the use of a number of different methods and tools and data from various sources. The final, intensified option was found by first setting a target for improvement on the selected base case and then by matching and evaluating alternatives to find the more sustainable design option. For the evaluation of alternatives, the measures used included sustainability metrics, economic potential and LCA factors. It was found that the alternative option for production of biodiesel is more economic and environmentally sustainable. Current and future work is extending the problem solution by considering all the chemicals present in the system and also by adding other likely feedstocks so that a truly optimal production process for biodiesel can be found. In this case, because the price and availability of the feedstock may vary in different geographical locations, this factor would also be added in the analysis. Therefore, using the framework and the collected data, different, more sustainable options for different geographical locations would be possible to generate without too much extra effort.

REFERENCES

1. Escobar, J.C.; Lora, E.S.; Venturini, O.J.; Yáñez, E.E.; Castillo, E.F.; Almazan, O. Biofuels: Environment, technology and food security. Renew. Sustain. Energ. Rev. **2009**, 13, 1275–1287.
2. Singh, A.; Pant, D.; Korres, N.E.; Nizami, A.S.; Prasad, S.; Murphy, J.D. Key issues in life cycle assessment of ethanol production from lignocellulosic biomass: Challenges and perspectives. Bioresour. Technol. **2010**, 101, 5003–5012.
3. Kiss, A.A.; Bildea, C.S. A review of biodiesel production by integrated reactive separation technologies. J. Chem. Technol. Biotechnol. **2012**, 87, 861–879.
4. International Energy Statistics. Available online: http://1.usa.gov/13OrOBn (accessed 6 March 2013).
5. Biomass Research & Development Initiative. Available online: http://www.esd.ornl.gov/eess/ (accessed 1 June 2013).
6. Lam, M.; Lee, K.; Mohamed, A. Homogenoues, heterogenous and enzymatic catalys for transesterification of high free fatty acid oil (waste cooking oil) to biodiesel: A review. Biotechnol. Adv. **2010**, 28, 500–518.

7. Talebian-Kiakalaieh, A.; Amin, N.A.S.; Mazaheri, H. A review on novel processes of biodiesel production from waste cooking oil. Appl. Energ. **2013**, 104, 683–710.
8. Leung, D.; Wu, X.; Leung, M. A review on biodiesel production using catalyzed transesterification. Appl. Energ. **2010**, 87, 1083–1095.
9. Simasatitkul, L.; Arpornwichanop, A.; Gani, R. Design methodology for bio-based processing: Biodiesel and fatty alcohol production. Comput. Chem. Eng. **2013**.
10. Carvalho, A.; Gani, R.; Matos, H. Design of sustainable chemical processes: Systematic retrofit analysis generation and evaluation of alternatives. Process Saf. Environ. Prot. **2008**, 86, 328–346.
11. Carvalho, A.; Matos, H.A.; Gani, R. Design of batch operations: Systematic methodology for generation and analysis of sustainable alternatives. Comput. Chem. Eng. **2009**, 33, 2075–2090.
12. Carvalho, A.; Matos, H.A.; Gani, R. SustainPro—A tool for systematic process analysis, generation and evaluation of sustainable design alternatives. Comput. Chem. Eng. **2013**, 50, 8–27.
13. Lutze, P.; Babi, D.K.; Woodley, J.M.; Gani, R. A phenomena based methodology for process synthesis incorporating process intensification. Ind. Eng. Chem. Res. **2013**, 52, 7127–7144.
14. Yuan, Z.; Chen, B.; Gani, R. Applications of process synthesis: Moving from conventional chemical processes towards biorefinery processes. Comput. Chem. Eng. **2013**, 49, 217–229.
15. West, A.H.; Posarac, D.; Ellis, N. Assessment of four biodiesel production processes using HYSYS. Plant. Bioresour. Technol. **2008**, 99, 6587–6601.
16. Martín, M.; Grossmann, I.E. Simultaneous optimization and heat integration for biodiesel production from cooking oil and algae. Ind. Eng. Chem. Res. **2012**, 23, 7998–8014.
17. Karunanithi, A.T.; Achenie, L.E.K.; Gani, R. A new decomposition-based computer-aided molecular/mixture design methodology for the design of optimal solvents and solvent mixtures. Ind. Eng. Chem. Res. **2005**, 44, 4785–4797.
18. Jaksland, C.A.; Gani, R.; Lien, K.M. Separation process design and synthesis based on thermodynamic insights. Chem.Eng.Sci. **1995**, 50, 511–530.
19. Gani, R.; Bek-Pedersen, E. Simple new algorithm for distillation column design. AIChE J. **2000**, 46, 1271–1274.

20. Saengwirun, P. ECON: A Software for Cost Calculation and Economic Analysis. M.S. Thesis, Chulalongkorn University, Bangkok, Thailand, 2011.
21. Peters, M.S.; Timmerhaus, K.; West, R. Plant Design and Economics for Chemical Engineers; McGraw-Hill: Singapore, 2004.
22. Kalakul, S. Development of Software for Life Cycle Assessment. M.S. Thesis, Chulalongkorn University, Bangkok, Thailand, April 2013.
23. PRO/II User's Guide; Simulation Sciences, Inc: Brea, CA, USA, 2011.
24. Aspen HYSYS User's Guide; Aspen Technology, Inc.: Burlington, MA, USA, 2009.
25. Marrero, J.; Gani, R. Group-contribution based estimation of pure component properties. Fluid Phase Equilbria **2001**, 183, 183–208.
26. Zhang, Y.; Dubé, M.A.; McLean, D.D.; Kates, M. Biodiesel production from waste cooking oil: 1. Process design and technological assessment. Bioresour. Technol. **2003**, 89, 1–16.
27. Sotoft, L.F.; Rong, B.; Christensen, K.V.; Norddahl, B. Process simulation and economical evaluation of enzymatic biodiesel production plant. Bioresour. Technol. **2010**, 101, 5266–5274.
28. Darnoko, D.; Cheryan, M. Kinetic of palm oil transesterification in a batch reactor. J. Am. Oil Chem. Soc. **2000**, 77, 1263–1267.
29. Wang, Y.; Ou, S.; Liu, P.; Xue, F.; Tang, S. Comparison of two different processes to synthesize biodiesel by waste cooking oil. J. Mol. Catal. A **2006**, 252, 107–112.
30. Lou, W.; Zong, M.; Duan, Z. Efficient production of biodiesel from high free fatty acid-containing waste oils using various carbohydrate-derived solid acid catalysts. Bioresour. Technol. **2008**, 99, 8752–8758.
31. Nárvaez, P.; Rincón, S.; Sánchez, F. Kinetic of palm oil methanolysis. J. Am. Oil Chem. Soc. **2007**, 87, 971–977.
32. Jegannathan, K.; Seng, C.; Ravindra, P. Economic assessment of biodiesel production: Comparison of alkali and biocatalyst processes. Renew. Sustain. Energ. Rev. **2011**, 15, 745–751.
33. Bokade, V.; Yadav, G. Transesterification of edible and nonedible vegetable oils with alcohols over heteropolyacids supported on acid-treated clay. Ind. Eng. Chem. Res. **2009**, 48, 9408–9415.
34. Chew, Y.; Chua, L.; Cheng, K.; Sarmidi, M.; Aziz, R.; Lee, C. Kinetic study on the hydrolysis of palm olein using immobilized lipase. Biochem. Eng. J. **2008**, 39, 516–520.

35. Edgar, L.; Yijun, L.; Dora, L.; Kaewta, S.; David, B.; Goodwin, J.G. Synthesis of biodiesel via acid catalysis. Ind. Eng. Chem. Res. **2005**, 44, 5353–5363.
36. Ma, F.; Hanna, M.A. Biodiesel production. Bioresour. Technol. **1999**, 70, 1–15.
37. Karmakar, A.; Karmakar, S.; Mukherjee, S. Properties of various plant and animals feedstocks for biodiesel production. Bioresour. Tecnol. **2010**, 101, 7201–7210.
38. Steimel, J.; Harrmann, M.; Schembecker, G.; Engell, S. A framework for modeling and optimization of process superstructures under uncertainty. Chem.Eng.Sci. **2013**.

2
CHAPTER

ALGAL BIOREFINERY FOR BIODIESEL PRODUCTION

Didem Özçimen[1], M. Ömer Gülyurt[1] and Benan İnan[1]

[1] YıldızTechnical University, Faculty of Chemical and Metallurgical Engineering, Bioengineering Department, Istanbul, Turkey

1. INTRODUCTION

In recent years, the rapid depletion of fossil fuels, increase in energy demand, global warming, increase in price of fossil fuels depends on economic and political behaviors increased orientation to alternative energy sources. In this context, biodiesel that is one of the renewable alternative energy sources draws attention because of its useful features such as easily biodegradable and environmentally friendly. However, biodiesel production from oil crops does not meet the required demand of vehicle fuel, and recently it is not economic and feasible. It needs to be improved to produce more economically to be able to compete with diesel in the market. Vegetable oils and crops which biodiesel produced from are a kind of human food sources and the shortage on food source cause to go up prices and make the biodiesel high-priced. To meet the requirements, the interest on algae is increased day by day since this technology has potential to meet global demand [1]. Microalgae have higher productivity per area and no need for farm field to grow as opposed to oil crops and animal fat. Microalgae use sunlight to reduce CO_2 to biofuels, foods, fertilizers, and valuable products. Furthermore, microalgae can be used to get different types of biofuels. Using microalgae as fuel source is not a novel idea but recently the prices of diesel and global warming hit this solution to the top [2].

Microalgae have lots of advantages for biodiesel production over other raw materials such as crops, waste cooking oils, and so on. Microalgae have short doubling time which is around 12-24 h since they have a simple structure and capable to high photosynthetic efficiency and they contain much more amount of oil than other oil crops that can be used as oil source for biodiesel production. Compared with the oil yields from various oil crops such as corn (172 L/ha), soybean (446 L/ha), canola (1190 L/ha), jatropha (1892 L/ha), coconut (2689 L/ha) and oil palm (5959 L/ha), oil yield from microalgae is very high as 136900 L/ha and 58700 L/ha for 70% oil in biomass and 30% oil in biomass, respectively [2-4].

The other significant feature is that algae can grow everywhere and every season in a year since there are thousands of algae species that have different adaptations and different properties. They can grow in saltwater, freshwater, lakes, deserts, marginal lands, etc. In addition to biodiesel production, algae can be also used as feedstock to produce different valuable products such as fertilizer, energy, neutraceuticals, protein, animal feed etc. The other significant property is that microalgae can remove some heavy metals, phosphorous, and nitrogen from water during its growth. Algae also clean up the water. Moreover, microalgae sequester lots of carbon by photosynthesis. Utilization of carbon dioxide by algae is significantly lowering the risk for greenhouse gas effects. Lastly, usage of microalgae for biodiesel almost cancels out the carbon dioxide and sulfur release to atmosphere [5]. These reasons mentioned above are enough to believe that microalgae can take the place of fossil fuels completely.

There are many of microalgae studies for biodiesel production. Because the most of the scientists believe that microalgae will take the place of the petroleum diesel, however, algal biodiesel production is not feasible yet since there is no much commercial or large scale production of microalgae for biodiesel. That is why most of the works are focused on decreasing the cost of biodiesel production or make it competitive versus petroleum diesel. Surely, until these improvements are achieved, algal biodiesel can not be an accurate alternative. The current problems making biodiesel expensive can be improved with some innovations. The first of all is about the algae strain which is also first step of algal biodiesel production. The algae strain should be better than recent ones. There are natural many kinds of algae strains and isolation of new natural algae strain may help procedure to be cost effective. The algae strain has to have high lipid productivity and adaptability to new environments. These features let it produce more and obtain more oil content [6, 7]. As an example, if the flue gas is used as carbon dioxide source, microalgae have to be adapted for this situation so that it can tolerate the high concentration of SO_x, NO_x, and other gases [8]. That will reduce the cost and increase the biomass growth rate. The other important innovation should focus on cultivation of algae. The large-scale production is one of the most cost-intense parts. The innovative thinking should show a tendency to lower the cost of operation and capital for cultivation systems. As it is explained below, open ponds are the cheapest way but the efficiency of them has to be worked on. Moreover, the closed photobioreactors (PBR) are also being improved for a cheaper way to control and lighten the system. Furthermore, microalgae can be fixed in a cultivation system with an immobilization technique to get higher biomass. The last way to lower the cost is to produce sub-products from microalgae beyond biodiesel. There are lots of high value products and sub-products produced from microalgae such as biogas [9, 10], biobutanol, acetone [11], Omega 3 oil [12], eicosapentaenoic acid [13], livestock feed [14], pharmaceuticals and cosmetics [15, 16]. Especially sub-products can be preferred for economic support of main process.

For example, recovery of methane from microalgae pulp after biodiesel production develops renewability of conversion of microalgae biomass to biodiesel process as much as it makes the cost of process and environmental effects less. The microalgae pulps after oil removed contain significant amounts of protein and carbohydrate that can convert to biogas by anaerobic fermentation. Conversion of algal waste to biogas by anaerobic fermentation will play a dual role for renewable energy production and also sustainable development of microalgal biodiesel industry [17, 18].

Algae can be also used in bioethanol production. Algae are more uniform and continuous than terrestrial plant, due to lack of functional parts such as root and leaf composition. Their cell walls made of polysaccharides that can hydrolyze to the sugar. For this reason, microalgae can be used as carbon source in fermentation process. Ethanol produced by fermentation can be purified for using as a fuel, CO_2 as a nutrient may also be recycled to algae culture to grow microalgae [19, 20].

In this chapter, algae production methods that cover the algae strain and location selection, algae cultivation, harvesting, oil extraction, and algal biodiesel production processes are presented in detail with alternatives. New progresses in this area are also explained.

2. ALGAE STRAINS AND PROPERTIES

Algae are simple organisms including chlorophyll. They can be found in seas, soils and lakes wherever they can use the light for their photosynthesis. There are two types of main algae groups. The first group is macro algae, which includes green, brown and red algae. The second group is microalgae as phytoplankton in the coasts, lakes and oceans, which includes diatoms, dynoflagellates, green and brownish flagellate, and blue-green algae [21].

The classification of algae can be done in many ways since there is a millions of kind. Also there is no standard on classification so you can see different types of classification. The taxonomic group of algae can be given as follow: *Archaeplastida, Chlorophyta*(green algae), *Rhodophyta*(red algae), *Glaucophyta, Chlorarachniophytes, Euglenids, Heterokonts, Bacillariophyceae* (diatoms), *Axodine, Bolidomonas, Eustigmatophyceae, Phaeophyceae*(brown algae), *Chrysophyceae*(golden algae), *Raphidophyceae, Synurophyceae, Xanthophyceae*(yellow-green algae), *Cryptophyta, Dinoflagellates, Haptophyta*[22].

Algae are the most common wide photosynthetic bacteria ecologically. To grow algae some parameters such as amount and quality of ingredients, light, pH, turbulence, salinity, and temperature become prominent. Macro (nitrate, phosphate, silicate) and micro (some metals, B1, B12 and biotin vitamins) elements are required in the growth of algae. Light intensity has also an important role, the light demand changes up to microalgae density and type of microalgae. The other parameter pH is mostly between 7 and 9 for most of algae strains and mostly the optimum range is 8. 2-8. 7. The last parameter salinity should be between 20-24 ppt. Moreover, nitrogen also affects the growth of some algae strains as such as green algae [22-25].

2.1. Macroalgae

Macroalgae are adapted to life in ocean and it is a plant mostly seen on the costal strips. There are plenty of macro algae types. Algae can be classified as brown, red, and green based on type of pigments. Recently, several brown algae types have been used in the industry and energy production as an alternative source to fossil fuels, and green algae is also studied to produce biodiesel [26].

Brown algae have xanthophyll pigments and fucoxanthin, which results the colour of brown algae. These substances mask the other pigments [27]. Polysaccharides and higher alcohols are nutrition reserves of brown algae but the main carbohydrate reserve is laminarin. The cell walls of brown algae are made of cellulose and alginic acid. Brown algae have a lot of features such as: Cytotoxic and antitumor activity, Antifungal activity, Anti-inflammatory activity, Antiviral activity, Protection against herbivorous animals (fish, sea urchins), Antioxidant activity [21, 28, 29]. Composition of brown algae can vary according to species, their location, salinity and season. According to analysis, brown algae contain about 85% high moisture and 25 % high sodium carbonate [26].

Green algae contain chlorophyll a and b. Presence of these pigments makes green color of the green algae. There are a few reports about second metabolites of green algae. [21]. Moisture content of green algae is higher than brown algae but they have similar sodium carbonate content. Green algae species can access higher sugar levels and this makes them useful energy sources. They also have high cellulose content [26]. Green algae have a lot of features such as: Anti-inflammatory substances, Cytotoxic and immunosuppressive activities, Antibacterial activity, Antiviral activity, Antifungal activity [30].

Red Algae have phycoerythrin and phycothcyanin pigments that make red color of these algae. These pigments mask the other pigments. The cell walls of red algae made of cellulose, agar and carrageenan [27]. There are approximately 8000 red algae species. In comparison of the other algae species, red algae are considered as the most important active metabolite resource. They have a lot of features such as: Cytotoxic activities, Antiviral activity, Anti-inflammatory activity, Antimicrobial activity, Free radical scavenger activity [21, 31].

2.2. Microalgae

There are at least 30000 microalgae species in the world. Microalgae are mostly defined as unicellular photosynthetic cells but some complex associations create larger colonies. This is a heterogenic group, which contains prokaryotic organisms similar to bacteria and eukaryotic cells [26, 32]. Microalgae production is concentrated on particular species, which have special tolerance for extreme conditions in their growth. This situation enables the production in open ponds and canals. In future, microalgae production will focus on more advanced species for the demand of energy and pure monocultures which have specific capabilities like production of carbohydrate, lipid or hydrogen will be cultivated [33]. According to use of algae, biomass of microalgae has variable chemical composition. They can be rich or balanced composition of protein, lipid and sugar. Microalgae selection should be made according to desired biofuels. Microalgae have important lipid content even in the extreme conditions they reach higher lipid content [26].

Green algae or diatoms are the most used microalgae species for production of alternative energy derives. Just a handful of these species has commercial importance. This group contains *Chlorella, Spirulina, Dunaliella* and

Haematococcus. Only *Dunaliella* is a dominant sea species. These are usually cultivated for extraction of high value component like pigments or proteins [26].

Blue-green algae (cyanobacteria) have a lot of common structural features with bacteria. They are classified as algae because they contain chlorophyll and other components. They have also nitrogenic components because all of the prokaryote species convert atmospheric nitrogen to ammonium [21, 34]. Morphologically blue green algae can have filamentous, conical or unicellular shape. They have a lot of features such as: anticancer and cytotoxic activities, antibacterial activity, antifungal activity, immunosuppressive activity [21, 35, 36].

Pyrrhophyta (Dinoflagellates) are unicellular organisms, which are classified as primitive algae. Large amount concentrations of these organisms exist in ocean surface and they cause fish deaths. Also because of their pigments, dinoflagellates give the water brown to red coloration in the sea [34, 37]. Particular dinoflagellate species produce toxin in case of consumed by species such as shellfish. Consumption of contaminated shellfish by humans can cause a lot of health problems including death [21].

Bacillariophyceae (Diatoms) are the most versatile and frequent family. They are more feasible for large-scale productions due to short doubling time and easy to grow. Unlike *Dinoflagellates* they create less second metabolites [38].

Microalgae are investigated as biodiesel feedstock because of their high photosynthetic efficiency, their ability to produce lipids. Macroalgae usually don't contain lipids too much and they are taken into consideration for the natural sugars and other carbohydrates that they contain. These contents can be fermented to produce alcohol-based fuels or biogas.

2.3. Lipid Content of Microalgae Species

As the structure of many microalgae species can accumulate significant amounts of lipid and provide high oil yield. Their average lipid contents can be reached to 77% of dry biomass under some certain conditions [39]. Table 1 shows lipid content of some microalgae species.

Also high productivity is very important beside high oil content. As shown in table 1, microalgal lipid content can reach 77% by weight of dry biomass but it is observed that there can be low productivity of *Botryococcusbraunii*, however, *Chlorella* appears to be a good choice in biodiesel production, since it has high productivity though lower oil content [39].

Table 1.**Lipid content of some microalgae species [15, 39, 40-45].**

Microalgae	Oil content (dry weight %)
Botryococcus braunii	25-75
Chlorella protothecoides	14-57
Crypthecodinium cohnii	20-51
Dunaliella tertiolecta	16-71
Nannochloris sp.	20-56
Neochloris oleoabundans	29-65
Phaeodactylum tricornutum	18-57
Schizochytrium sp.	50-77
Skeletonema coastatum	13-51

Lipid content can be affected by several parameters such as nutrition, environment, cultivation phases and conditions growth can affect fatty acid composition [32], Fatty acid composition is important in microalgae selection because it has a significant effect on biodiesel properties. For example, if unsaturated fatty acid content is high in algal oils and their presence reduces the efficiency of esterification to produce biodiesel [39].

Value chain stages of biodiesel production from microalgae can be given as algae and site selection, algae cultivation, harvesting, filtration, dewatering, oil extraction and biodiesel production [39].

3. BIODIESEL PRODUCTION FROM MICROALGAE

The selection of species depends on some factors like ability to usage of nutrition or grow under specific environment conditions. All these parameters should be evaluated for biodiesel production.

3.1. Selection of algae strain and location

To make algal biodiesel cost effective lots of researchers keep going on algae culturing. The criteria to select location and sources are mentioned below [46]:

- Water sources and demand, salinity, content
- The region information such as topography, geology
- Weather conditions, isolation, evaporation
- Availability of carbon and food resources

The next decision should be on the algae culturing process type. It can be either batch or continuous process. Depending on microalgae strain, environmental conditions, availability of nutrition and moreover industrial pollutions the process type has to be selected. The devices and apparatuses also have to be adjusted for these conditions and nutrients [39].

Algae strains have different contents, different doubling time (the total biomass per time and volume) and resistance to change in environmental conditions. Biodiesel production directly depends on the oil content of microalgae and its efficiency. So that, even the process and culturing systems are selected perfectly, time and other related factors plays an important role [39].

3.2. Methods Used for Algae Growth

Not only the microalgae strain is important for efficiency of oil but also growing conditions are important. There are different ways to grow algae. Each type of microalgae has a different mechanism which let them to respond different weather and environmental conditions [39, 47]. Different growing conditions affect the microalgae doubling time. There are 4 growing type basically: phototrophic, heterotrophic, mixotrophic, and photo heterotrophic. All of them will be explained in detail.

3.2.1. Phototrophic growth

Microalgae are mostly thought to be phototrophic since it requires light [48]. Phototrophic growing method is based on using light and carbon dioxide to produce chemical energy during photosynthesis. This is the most common way used to grow microalgae. The best advantage of the process is using carbon dioxide as a carbon source to grow or produce fatty acid. Since carbon dioxide is only the carbon source, locations close to fabrics and companies could be selected to procure carbon dioxide. If it is compared to other growing types, phototrophic method has the lowest contamination risk [49].

3.2.2. Heterototrophic growth

Some microalgae are not able to grow phototrophic conditions but they can grow in dark using organic carbon as a carbon source like bacteria. If microalgae is using organic carbon these microalgae are heterotrophic growing algae. Heterotrophic growth has advantages over phototrophic growth because light is not required. The biggest problem with the phototrophic is the light penetration when the density of the culture gets higher. In that way one of the biggest problems is solved with heterotrophic growth. Heterotrophic growth will be more cost effective compared to phototrophic growth [48]. And this method is said the most practical and promising way to increase the productivity [50-52]. Also higher oil rates and efficiency can be obtained when the algae grow heterotrophic, but the contamination risk is much higher compared to phototrophic [49].

Microalgae uses different organic carbon sources such as glucose, acetate, glycerol, fructose, sucrose, lactose, galactose, and mannose, especially growth with sugar is more efficient [49].

Mostly the organism growing heterotrophic should have adaptation property to new habitat as soon as possible since when culturing to new media the lag phase should be too short, and durability during processing in fermenters and other machines [48].

3.2.3. Mixotrophic growth

Mixotrophic growth is a combination of phototrophic and heterotrophic growth. Mixotrophic growth is using organic and inorganic carbon and the process requires light because of photosynthesis. Thus the microalgae have ability to live in both conditions. Microalgae uses organic compounds and carbon dioxide as a carbon source and the released carbon dioxide are also captured with the photosynthesis. Although mixotrophic-growing method mostly is not preferred compared to heterotrophic and phototrophic growth [49], because of other advantages even so mixotrophic method is applied in some studies. For example; Park *et al.* found that biomass and lipid productivities were boosted by mixotrophic cultivation [53]. Bhatnagar et al. found the mixotrophic growth of some microalgae strains resulted in 3–10 times more biomass production compared to that obtained under phototrophic growth conditions [54].

3.2.4. Photoheterototrophic growth

When microalgae use organic compounds as carbon sources, sometimes it requires light. The main difference between mixothrophic and photoheterotrophic is that mixotrophic growth using organic compounds as energy sources, as photoheterotrophic growth requires light as energy source. This method is mostly used for production of some beneficial metabolites; however, it is rarely used for biodiesel production [49]. Metabolisms can split into groups due to pH changes. *Chlorella vulgaris, Haematococcus pluvialis, Arthrospira (Spirulina)* platensis strains are the examples for the growth by mixothrophic, phototrophic and heterotrophic methods. *Selenastrum capricornutum* and *Scenedesmus acutus* are able to grow in phototrophic, heterotrophic, photoheterotrophic conditions [47].

Algae require more than organic carbon, sugar, protein, oil or any carbon sources. Algae cannot grow without vitamins, salts, or some other nutrients (nitrogen and phosphor). Moreover, there are lots of parameters has to be controlled during algae growth to maximize and stabilize the production. Some of these parameters are oxygen rate, carbon dioxide rate, pH, heat, light intensity and so on. When appropriate weather conditions and enough nutrients are provided microalgae grow faster. Mostly doubling time is between 3. 5 h and 24 h [39].

As a result, if we compare different methods mentioned above for microalgae growth;Heterotrophic growth is much better than the others for the application of biodiesel. These methods can produce more oil than other growing types. However, heterotrophic cultures may contaminate especially in open pond systems and result in big problems in large-scale production. Moreover, organic carbon as a carbon source is an expensive raw material and makes the process cost higher. Phototrophic growth is an easily scalable and mostly uses the carbon dioxide from exhaust gas for the production of oil. However, the efficiency of the oil is lower than heterotrophic growth because the biomass doubling time is higher and total biomass rate is lower at the end. Phototrophic method mostly preferred to set a cost effective system [49].

3.2.5. Conditions for growth of algae

3.2.5.1. Light

The microalgae growing photosynthetically needs light and the light intensity is the most significant limiting factor. Algae culture systems mostly use both sun and lamp light. Mostly lamp-lightened algae culture systems uses wider screens to be able to absorb more light from the system. For photosynthetic production, at least 50 % of the volume of PBR has to get enough light [55]. Open raceway ponds, plate, plate PBR, Vertical-column PBRs, Internally-illuminated PBRs, inclined tubular type, horizontal/continuous type, bubble column and air-lift PBRs are the systems used for photosynthetic algae growth. Plate photo bioreactor is more efficient than tubular photo bioreactor because the light can penetrate to bottom more in plate design. Recent works are on closed system photobioreactors to improve the capacity. Some works are done to increase the capacity; however the light penetration becomes a major problem. Light source for open ponds is only Sun. That is why the alteration is not possible for raceway ponds. The depth of the pond that the only thing can be changed. Thus mostly researches are going on closed systems to optimize light emission. Mostly photobioreactors in lab scale are lightened by fluorescence lights from inside and outside [56]. The light wavelength should be between 600-700 nm to maximize the photosynthesis. Light intensity depends on microalgae density. Higher algae density requires higher light intensity. Light also affects the lipid content. Yeesang and Cheirsilp reported that the lipid contents in all strains increased with increasing light intensity in their study [57].

Changes in light intensity and quality can alter biofuel quality [58]. Each type of microalgae has its own optimal light absorbing point. If this point exceeds the optimum point, microalgae light absorption ratio decreases. After a specific point, light decreases the biomass production and this is called photoinhibition. Photoinhibition processes depend on time and after stress of light for a few minute biomass loss starts. 10-20 min later more than 50 % damage can be seen. Cheirsilp and Torpee investigated the effect of light intensity on growth and lipid content of marine *Chlorella sp.* and *Nannochloropsis sp.* The growth of marine *Chlorella sp.* increased when the light intensity was increased from 2000 to 8000 lux. But up to 10000 lux its growth decreased. They reported that this could be some extent of effect from photoinhibition. The growth of *Nannochloropsis sp.* continuously increased up to the maximum level when increasing light intensity up to a maximum light intensity of 10000 lux. [59]. High light intensity limited algal growth, but gave the benefit of higher lipid content and yield. It can be seen in Ruangsomboon's study whose cultures exposed to low light intensity showed a higher biomass compared to others [60].

To increase the microalgae production, photoinhibition should be cut off or exceed to high light intense. In addition, photorespiration decreases the photosynthetic efficiency. Therefore the process has to avoid photorespiration. Photorespiration occurs when the oxygen concentration increases depending on carbon dioxide [56].

Sara et al. investigated the light effects on microalgae. The research was done by using red and blue lasers as light source for photosynthetic growth of green algae. The results showed that the both blue and red lasers increased the algae cell count [61].

Allen and Arnon tested the effect of light on green algae growth. The light intensity was around 16000 lux. There were two samples. One of the samples was analyzed under 11 h darkness and 13 h light. The other sample was analyzed under light for 24 h and the results showed that the growth rate was same. However after 5 days the growth rate for the sample with 24 h light was declined [62].

The effects of light on *Parietochloris incisa* was analyzed by Solovchenko et al. The results showed that best growth was seen on high light (400 μmol photons m^{-2} s^{-1}). With high light condition, total fatty acid and arachidonic amount was increased due to increase in biomass [63].

Another study (Yeh et al.) was focused on effects of different light sources on microalgae (*C. vulgaris)* growth. In the sutdy, three different light sources was used which are tungsten lamp, fluorescent lamp (TL5), fluorescent lamp (helix lamp). The results showed that fluorescence lamps were much better for algae growth. In an other study by Floreto et al., it was mentioned that high light intensity increased the palmitic acid and most fatty acids ratio [64].

3.2.5.2. Carbon dioxide

Carbon dioxide is the natural carbon source of the microalgae culture. Oxygen is releasing depending on decreasing carbon amount and it is delivered to the medium. Carbon dioxide is an general carbon source for photosynthetic microalgae. When the carbon amounts get low, oxygen is produced by photolysis of water and released to media. Since algae lives in high carbon dioxide concentration, greenhouse gases, nitrogen dioxide and atmospheric pollutants came from different sources became a food for algae. The exhausted gases can feed algae production facilities from fossil fuels and also its efficiency would be increased. Works on usage of stack gases as carbon source were done but the toxicity of the stack gas components couldn't be documented well. The amount of carbon dioxide required for the growth relates to type of microalgae and photo bioreactor. Some types of algae strains are able to keep growing in high carbon dioxide conditions, in contrast for faster growth lower carbon dioxide concentration is required [56]. Widjaja studied the effect of CO_2 on growth and it was seen that this effect correlates directly to the lipid productivity since growth was enhanced tremendously by increasing the CO_2 concentration [65]. CO_2 requirement can change up to strains. VirthieBhola et al. reported in their studies that at 15% CO_2 concentration there is a 3-fold decline in biomass yield when compared to the yield produced at a 4% CO_2 concentration. This suggests that the strain under study could not endure CO_2 concentrations greater than 4% [66]. Also Ebrahimzadeh et al. reported that increasing CO_2 injection had a significant effect on microalgae growth [67]. CO_2 input is also important. Sonnekus reported that the CO_2 should make up 0. 2 -5% of the total gas flow and being careful about the CO_2 input does not lower the pH of the culture [68].

3.2.5.3. Heat

Algal growth is also dependent on temperature. For maximum growth there is a need to know the optimal temperature. The temperature changes also lipid production and composition [69]. The degree of unsaturation of algal membrane lipids increases if cultures are maintained at temperatures below their optimum [70]. Other than this temperature is significant for dissolubility of carbon particles, which helps carbon to be used for photosynthesis. Heat effects respiration and photorespiration more than photosynthesis. However, if carbon dioxide and light are the limiting factor, the effect of heat is not significant anymore. Optimal temperature for microalgae cultures is between 20-24 °C. This can be different according to media composition, type of culture and strain. The most general cultured microalgae can tolerate the temperature between 16- 27 °C. The temperatures lower than 16 °C will increase the duplication time and higher than 35 °C will have a fatal effect on algae [56]. However, these ranges can be changed by environmental factors such as salinity, pH, carbon dioxide etc.

In the study of Floreta etal., the factors affecting algae growth were determined. Temperature effect was determined with salinity simultaneously. The results showed that low temperature (15 °C) with high salinity is the best choice. Low temperature increases the level of oleic and linoleic fatty acids. Moreover, high salinity increases the amount of C_{16} and C_{18} poly-unsaturated fatty acids [71].

3.2.5.4. pH

Microalgae require different pH values according to the media. During high pH concentration, the carbon dioxide might be limiting factor for growth and photosynthesis. The most used pH range for algal growth is around 7-9. The optimal pH for algae is between 8. 2- 8. 7. But it can change with different strains. For example, Weissel and Stadler studied with *Cryptomonas sp.* which showed positive population growth rates over a wide pH range, from 4. 4 to 9.65 [72]. Appropriate pH can be adjusted by ventilation or gassing. There is a complex relationship between CO_2 concentration and pH in microalgal bioreactor systems, owing to the underlying chemical equilibrium among such chemical species as CO_2, H_2CO_3, HCO_3 and CO_3. Increasing CO_2 concentrations can increase biomass productivity, but will also decrease pH and this causes important effect upon microalgal physiology [73]. Water contaminated with a high pH has negative effects on algal abundance [74]. If there is not enough CO_2 gas supply, algae will utilize carbonate to maintain its growth [75].

Although high concentration of carbon dioxide provides high biomass efficiency, on the other side higher contamination risk and effect of low pH on microalgae physiology occurs [56].

Except the parameters mentioned above; there are also some parameters which affect on algal growth or lipid accumulation. Nitrogen, phosphorus and salinity can be examples for these parameters [76]. Widjaja et al. studied about nitrogen starvation effect on lipid accumulation. They reported that longer time of nitrogen starvation obviously resulted in higher accumulation of lipid inside the cells. Under all CO_2 concentrations, the lipid content tend to increase when the algae was exposed to nitrogen starvation condition that total lipid content was higher than lipid obtained during normal nutrition [75]. Ruangsomboon found the highest biomass concentration was found under the highest phosphorus concentration [60]. Li Xinet all.have reported in their study that lipid productivity was not at its highest when the lipid content was highest under nitrogen or phosphorus limitation [77]. Yeesang and Cheirsilp also studied about nitrogen and salinity effect. They found an increase in algal biomass under nitrogen-rich condition for all strains and in the absence of a nitrogen source, no growth was observed. They reported that although some loss in algal biomass was found, the lipid contents of four strains increased. They also noticed that growth and lipid accumulation by these microalgae could be affected by salinity. Under nitrogen-rich condition, all strains survived at high salinity but growth of some strains decreased [57, 78].

3.3. Microalgae Cultivation Systems

Cultivating microalgae can be achieved in open systems like lakes and ponds and in high controlled closed systems called photobioreactor. A bioreactor is defined as a system, which carries out biological conversion. Photobioreactors are reactors, which used for prototroph to grow inside or photo biological reactions to occur [79].

3.3.1. Open ponds

Generally open ponds are used in microalgae cultivation. Open ponds have various shapes and forms and certain advantages and disadvantages. In the scientific investigations and industrial applications, raceway ponds, shallow big ponds, circular ponds tanks and closed ponds are used [80]. Area where pool exist is critical factor for selection of pond type. Ponds become local climate function due to lack of control in open ponds [80, 81]. Therefore, area contributes to the success. Open ponds are limited by key growth parameters, which include light intensity, temperature, pH and dissolved O_2 concentration. Another problem seen in open ponds is contamination. It limits cultivation system of algae, which can grow under certain conditions [79].

Cost of cultivation systems is an important factor for comparison of open and closed systems. Construction, operation and maintaining costs are less than photobioreactors in ponds and these systems are simpler than the others [79, 82].

3.3.2. Photobioreactors

Nowadays researches are made for designing photobioreactors due to cultivating microalgae. Photobioreactors offer better control than open systems [2]. Their controlled environment allows high yield for cultivating.

Productivity is the most important indicator for bioreactor technology. It is very difficult to compare productivity of bioreactors due to various strains and scale of microalgae [80].

Photobioreactors basically can be tubular and flat type. When it is compared with the other bioreactors, tubular reactors considered as more suitable for open cultivating. Large illumination surface of reactor, which made of transparent tubes, is the main factor to being suitable for cultivation. Tubes can be adjusted in various types, adjustments convenience is depend to the specification of system.

A general configuration includes straight line and coiling tubes [83]. Reactor geometry is also important, tubular reactors can be vertical, horizontal or inclined shape. There are important differences between configurations of vertical and horizontal. Vertical designs provide more mass transfer and reduce energy consumption; horizontal designs can be scaled but needs more space. There are more studies about tubular photobioreactors but usually flat type photobioreactors is preferred because it can offer high cell density [84]. In addition, this type of reactors is advantageous due to low energy consumption and high mass transfer capacity, reduction of oxygen increases, high photosynthetic efficiency, no dark volumes when compared with the other photobioreactors. Suitable reactor design should be provided with maximum cell mass. Various flat-plate photobioreactor designs are made of glass, thick transparent PVC materials and V-shape and inclined. Although the other designs are cheap and easy to construct, glass and PVC is more transparent for maximum light penetration [80, 84-86].

3.3.2.1. Flat-plate photobioreactors

These systems have large illuminated surfaces. Generally these photobioreactors are made of transparent materials to utilize the solar light with maximum degree. Dissolved oxygen concentration is low compared to the horizontal tubular photobioreactors. In this system high photosynthetic activity can achieve. Although it is very suitable for culturing algae but it has some limitations [83].

3.3.2.2. Tubular photobioreactors

Most of tubular photobioreactors are made of glass or plastic tubes. They can be horizontal, serpentine, vertical, near horizontal, conical and inclined photobioreactors. Ventilation and mixing is generally performed by pump or ventilation systems. Tubular photobioreactor is suitable with their illuminated surfaces. But one of the important limitations of this system is poor mass transfer. It is a problem when photobioreactor is scaled. Also photoinhibition is seen in photobioreactors [83, 87].

If there is not sufficient mixing system cells don't have enough light for their growth. Developing mixing systems can provide effective light distribution.

Also controlling culture temperature is very difficult in these systems. Thermostat can be used but it is expensive and hard to control. Also cells can attach the walls of tubes. Long tubular photobioreactors are characterized with transfer of oxygen and CO_2 [83, 88].

Vertical column photobioreactors are low cost, easily constructed and compact systems. They are promising for large scale of algae production. Bubble column and airlift photobioreactors can reach specific growth rate [56].

3.3.2.3. Internally illuminated photobioreactors

Florescent lamps can illuminate some photobioreactors internally. Photobioreactor is equipped with wheels for mixing algal cultures. Sprayer provides air and CO2 to culture. This type of photobioreactors can utilize solar light and artificial light [90]. When solar light intensity is low (night or cloudy day) artificial light is used. Also in some researches, it is told that solar light can be collected and distributed with optic fibers in cylindrical photobioreactors [91]. Another advantages of this system are can be sterilized with heat under pressure and minimizing the contamination [56, 83].

3.3.2.4. Pyramid photobioreactor

The Pyramid photobioreactor is using fully controlled and automatic system that increases the production rate. With this system, it is easy to grow any microalgae at any climate conditions. The design is in pyramid shape to absorb light more effectively. As mentioned above, light is one of the significant parameters affecting algae growth rate and with this recent system algae can be supplied with optimal light intensity. That is why the shape of the system is the last innovation for production step. So, having optimal light intensity during high microalgae production decreases the energy consumption. The body design is angled to reduce to pump costs by using air-lifting method and decrease the deformation on cell walls. Thermo-isolated and high technologic materials are used to avoid energy lost and over heating [92].

3.4. Biocoil Microalgae Production System

Biocoil is a holozoic tubular photobioreactor which made of plastic tubes with small diameter (between 2.4-5 cm), centrifuges, diaphragm pumps or peristaltic pumping are utilized in this system. Biocoil design provides equal mixing and reduces the attachment of algae to the walls. It automates the production process. It is not suitable for all algae species. Some of algae species damages by circulation system and some of them attach to the internal surface of tubes and affects algae production negative. In this system, when the level of algae increases maximum degree, because of the light limitation photosynthesis can slow. Biocoil systems with utilizing solar light in or outsides can executable. Light is given with an angle so algal cell can utilize better and photosynthesis occurs easily [89, 93, 94].

3.4.1. Design of culture growth systems

Depends of local conditions and suitable materials various culture systems can be designed by various sizes, shapes of construction material, slope and mixing type. These factors affect performance, cost and resistance. To construct suitable photobioreactor material has main importance. Materials like plastic or glass relax and rigid shouldn't be toxic, they should have mechanical power, resistance, chemical stability and low cost. Tubular photobioreactors are the most suitable ones for open culture systems. They have big illumination surface, good biomass productivity and they aren't expensive because they are made of glass or plastic tubes. Flat-type photobioreactors are made of transparent materials to utilize solar light energy in maximum degree. This type of photobioreactors allows good immobilization of algae and they are cleaned easily [56]. Pond walls and deep side can made of simple sand, clay, brick or cement even PVC, glass fiber or polyurethane. For coating mostly long lasting plastic membrane is used. (e. g., 1-2 mm thick, UV-resistant, PVC or polyethylene sheets) sometimes to lower the cost uncoating ponds are used but that time some problems occur like contamination, a layer of mud and sand [39].

3.4.2. Mixing

Mixing is a process for increasing the productivity of biomass in photobioreactors. Mixing provides distribution of light intensity, sufficient CO_2 transfer and maintains uniform pH. Mixing is necessary for preventing algae sedimentation and avoiding cell attachment to the reactor wall. Mixing is also provides equal light and nutrients to all cells and increases the gas transfer between culture medium and air [95]. The second of priority measures is carbon supply for using in photosynthesis. In very dense cultures, CO_2 from air (includes 0.035 % of CO_2) and bubbles during the culture can be limited for algal growth. CO_2 addition creates a buffer for the result of changing pH in the water [56].

Poor mixing allows cells to clumping like different size of aggregates; therefore it leads 3 phase (solid-liquid-gas) system in reactor. This situation tends to reduce the mass transfer. But all algae cannot tolerate agitation. Because they are sensitive to hydrodynamic stress. High mixing rate can cause the damaging of cells. Mixing in bubble column and air lift reactors can characterize with axial dispersion coefficient, mixing time, circulation time and Bodenstein number [96]. Analysis of mixing in bubble column shows it has shorter time than airlift reactors. Bubbles beyond the suction pipe provide less blurry area and causes better exposure to the light. In addition, existence of suction pipe in airlift reactors causes more effective mixing because internal loop provides a circulation. Airlift reactor gives information about fluid flow and high gas-liquid mass transfer rate. Bubble column causes unbalance cell density and these causes to death of algae [56, 97].

3.4.3. Light penetration

Another key of successfully scale up is light penetration. Ilumination in the photobioreactor affects biomass composition, growth rate and products. Microalgae need light for their photosynthesis [98]. Photosynthetic active radiation wave changes about 400-700 nm and this is equal to the visible light [99]. In intense cultures, light gradient changes over the photobioreactor radius due to the weakening of the light. Reduction of light intensity related to wave length, cell concentration, photobioreactor geometry and distance of the light transmittance. Light intensity in photobioreactor related to light way, cell concentration and light which emits by microalgae [56].

3.4.4. Gas injection

Supplement of CO_2 by bubbles is an important factor to be considered in designs. Injection of CO_2 bases on giving CO_2 to photobioreactor artificially. Researches show that rich ventilation of CO_2 provides CO_2 to algae, supports deooxygenation of suspension, to improve cycling provides mixing and limits the light inhibition [100]. But high ventilation rate leads to higher cost that is why in large scale of microalgae production it is not recommended. These researches results for microalgae production necessary optimum aeration rate of CO_2 gas. Includes about 5% or 10% of CO_2 (v/v), rate of 0.025-1 vvm [100]. Volume of air/medium/time is found cost effective for air mass culture [56].

3.4.5. Comparison of open and closed culture systems

Open and closed culture systems have advantages and disadvantages. Construction and operation of open culture systems are cheaper and they are more resistant than closed reactors and have large production capacity [101]. Ponds use more energy to homogenize to nutrients and to utilize the solar energy for growth their water level cannot be less than 15 cm [41]. Ponds are exposing to air conditions because water temperature evaporation and illumination cannot be controlled. They produce large amounts of microalgae but they need larger areas than closed systems and they are open to other contaminations from the other microalgae and bacteria. Also when atmosphere has only 0. 03-0.06 % of CO_2, mass transfer limitation slows the growth of microalgae cell.

Photobioreactors are flexible systems, which can operate for biological and physiological characteristics of cultured microalgae. It can be possible to produce microalgae, which cannot produce in ponds. Exchange of gas and contaminants between atmosphere and cultured cells in photobioreactor is limited or blocked by reactor walls [39]. Depends on the shape and design, photobioreactors have more advantages than open ponds. Culture conditions and growth parameters can be controlled better, it prevents evaporation, reduces loss of CO_2, provides high microalgae density or cell concentration, high yield, creates more safe and preserved environment, prevents contamination. Despite the advantages, photobioreactors have problems to be solved. Over heating, biological pollution, accumulation of oxygen, difficulty of scale-up, high cost of construction and operation and cell damage because of shear stress and degradation of material in photo phase are main problems in photobioreactors [39].

Comparing photobioreactors and open ponds is not easy because growth of algae related to al lot of different factors. Three parameters are considered in algae production units for yield [41]:

- Volumetric productivity (VP): productivity per unit reactor volume (expressed as g/L. d).
- Areal productivity (AP): productivity per unit of ground area occupied by the reactor (expressed as g/m^2d).
- Illuminated surface productivity (ISP): productivity per unit of reactor illuminated surface area (expressed as g/m^2d).

According to researches closed systems don't provide advantage for areal productivity but provide volumetric productivity (8 times) and cell concentration (16 times) more than open ponds [39, 41].

3.4.6. Comparison of batch and continuous process

Photobioreactors can be operated in batch or continuous process. There are a lot of advantages for using continuous bioreactors than batch bioreactors. Continuous bioreactors provide more control than batch bioreactors. Growth rates can be regulating in long time periods, can be saved and with variable dilution rates biomass concentration can be controlled. With steady state continuous bioreactors results is more dependable, products can be easily produced and can be reached desired product quality. Continuous reactions offer many opportunities for system research and analysis [102].

But some type of bioreactors is not suitable for continuous process. For some productions, cell aggregation and wall growth can inhibit the steady state growth. Another problem is loss of original product strain in time.

Mixtures viscosity and heterogenic nature make difficult for maintaining filamentous organisms. Long growth periods increase the contamination risks [83].

3.5. Harvesting Alternatives

There are several ways to harvest microalgae and dry them. Some main harvesting methods are sedimentation, flocculation and filtration.

Sedimentation: When a particle moves continuously in a phase, the velocity is affected by two factors. First of them is increasing the velocity because the density gradient between particle and fluid create buoyant force. At the end, buoyant force gets equal to dragging force and particle starts moving with a constant velocity. The same idea is applied to collect microalgae from the ponds. Gravity force is used for settling of suspended particles in fluid. This method is cheap and easy. However, the particles suspended in the fluid have to be incompressible. The problem with the *Scenedesmus sp.* and *Chlorella sp.* is that they are compressible. That is why sedimentation cannot be used for these types [103]. For low value products, sedimentation might be used if it is improved with flocculation [104].

Flocculation: is also used for harvesting microalgae. The general idea is microalgae carries negative charge on it and if the flocculants disappear the negative charge, algae starts coagulation. Some used flocculants are $Al_2(SO_4)_3$, $FeCl_3$, $Fe_2(SO_4)_3$ [105].

Filtration: This is one of the most competitive methods for the collection of algae. There are different types of filtrations, for example, dead end, microfiltration, ultrafiltration, pressure filter and vacuum filter. Mostly filtrations require the liquid media with algae to come through filtration. Filter can be fed until a thick layer of microalgae is collected on the screen. This method is very expensive for especially microalgae. The pore sizes of the filters are the most important part. If the pore size is bigger than algae you cannot collect it. In contrast, if the pore size is too small it might result in decrease of the flow rate and block the pores [106].

3.6. Extraction of Lipid From Microalgae

There are a lot of methods for extraction of lipid from microalgae but the most common techniques are oil presses, liquid-liquid extraction (solvent extraction), supercritical fluid extraction (SFE) and ultrasonic techniques. Oil presses are usually used for extracting of lipids from nuts and seeds. The same process and devices can be used for lipid extraction from microalgae.

For the purpose of this process to be effective, firstly microalgae must be dried. Presses use pressure for breaking cells and removing oil [107]. This method can extract 75% of oil but in longer extraction times it is less effective [80].

Solvent extraction is more successful for extracting lipids from microalgae. In this method organic solvents such as hexane, acetone, and chloroform are added in the algae paste. Solubility of oil is higher in organic solvents than water. Therefore solvent breaks the cell wall and extracts oil easily. Solvent extraction continues with distillation process for separating oil from the solvent [108]. Hexane is cheap and has high extraction capacity. For this reason it is reported to be the most effective solvent in extractions.

In addition to this studies, 2 stage process using ethanol improves lipid extraction. The yield of recovery of oil reaches about 80%. Butanol is also effective in extraction of lysophospholipids. But evaporation of butanol is difficult and there are some impurities because of its high polarity [80].

Supercritical extraction uses high pressure and temperature for breaking cells. This method is widely used and efficient for extraction time. Studies reported that temperature and pressure don't affect the yield of components but it affects extraction rate. Similar effects are seen in SFE system and solvent extraction [109].

Another method is using ultrasonic techniques. In this method microalgae is exposed to high intensity ultrasonic waves and these waves creates bubbles around the cell. Shock waves are emitted by collapsing bubbles. It breaks cell wall and desired components release to the solution. This method is also improves the extraction rate with the same way. This technique is widely used in laboratory scale but in commercially scale there is not enough information about cost and applicability [110, 80].

3.7. Biodiesel Production from Oil

After extraction there are 4 main methods for producing biodiesel: direct used and mixing with raw oils; microemulsion; pyrolysis and transesterification.

3.7.1. Dilution

This is a dilution method that certain proportion of vegetable and waste oils blended with diesel fuel and another solvent. The most used oils for producing biodiesel with this way are waste oils and vegetable oils like sunflower and rapeseed.

Direct use or blending generally considered being unsatisfactory and impractical for both direct and indirect diesel engines. There are specific problems such as high viscosity, acid composition, free fatty-acid content, gum formation because of oxidation, polymerization during storage and combustion, carbon deposits and also lubricating-oil thickening [111].

Dilution of vegetable oils with solvents lowers the viscosity. The viscosity of oil can be lowered by blending with pure ethanol [112]. The low viscosity is good for better performance of engine, which decreases with increasing the percentage of diesel [33]. In this method there is no chemical process and viscosity can be lower but there are also carbon deposits and lube pollution problems to be solved. To solve problems caused by high viscosity, micro-emulsion, pyrolysis and transesterification methods are used [113].

3.7.2. Micro-emulsion

It is defined that the size of 1-150 nm, the two immiscible liquid organic mixtures with ionic or non-ionic, self-formed stable colloidal distribution. With this method it is possible to form alternative diesel fuels except petroleum [28]. In this method vegetable oils with an ester and dispersant (co-solvent), or of vegetable oils, an alcohol and a surfactant, with or without diesel fuels can be used to make a microemulsion. Due to their alcohol contents, microemulsions have lower volumetric heating values than diesel fuels. But these alcohols have high latent heats of vaporization and also tend to cool the combustion chamber, which cause a reduction of nozzle coking. A microemulsion made of methanol and vegetable oils can perform like diesel fuels [111]. To solve the problem of the high viscosity of vegetable oils, microemulsions with solvents and immiscible liquids, such as methanol, ethanol, 1-butanol and ionic or non-ionic amphiphiles have been studied [114].

3.7.3. Pyrolysis

Pyrolysis is the conversion of organic substance into another by means of heat or by heat in the presence of a catalyst. Vegetable oil, animal fat, algae oil, natural fatty acids or methyl esters of fatty acids can be pyrolyzed [111]. Although this method is not very cheap, however, fuel can be produced without extraction of lipids or hydrocarbons. More uniform product can be obtained and ideally increases yields over transesterification with this method [115]. Products are chemically similar derived from petroleum products, which are to gasoline and diesel fuel derived [28]. Also with pyrolysis some low value materials and sometimes more gasoline than diesel fuel are produced [116]. In comparison between pyrolysis and the other cracking processes, pyrolysis is seen more simple, pollution free and effective [33]. Sharma et al. reported that pyrolysis of the vegetable oil can produce a product which has high cetane number, low viscosity, acceptable amounts of sulfur, water and sediments contents, acceptable copper corrosion values [117].

3.7 4. Transesterification

Transesterification of the oil is the most promising solution to the high viscosity problem [114]. In this process, triglycerides are converted to diglycerides, then the diglycerides are converted to monoglycerides, and the monoglycerides are converted to esters (biodiesel) and glycerol (by-products) [118]. There are three common kinds of catalysts used in transesterification process such as lipase catalysts, acid catalysts and alkali catalysts. Each catalyst has advantages and disadvantages [113].

In the acid-catalytic transesterification, the reaction can be catalyzed by sulfuric, phosphoric, hydrochloric and organic sulfonic acids. Very high yields can be obtained by using this catalyst. These reactions need the use of high alcohol-to-oil molar ratios in order to obtain good product yields in practical reaction times. But ester yields do not proportionally increase with molar ratio and the reaction time is very long (3–48 h) [114, 119, 120]. Xu et al. studied the acidic transesterification of microalgae (Heterotrophic C. Protothecoides) oil. They used methanol for alcohol and they achieved 80% of FAME yield [121].

Johnson made a study on *Schizochytrium limacinum* microalgae species. He converted this algal oil to biodiesel with acidic transesterification and he achieved 82.6% of biodiesel yield [122].

In the alkali-catalytic transesterification, the reaction can be catalyzed by alkaline metal alkoxides, and hydroxides, as well as sodium or potassium carbonates. Sodium methoxide is the most widely used biodiesel catalyst. This reaction is faster than acid-catalytic transesterification and reactions can occur in low temperatures with a small amount for catalyst and with little or no darkening of colour of the oil [114]. High quality can be obtained however this process is very sensitive to the presence of water and free fatty acids and needs lots of methanol. If the raw materials have a high percentage of free fatty acids or water, the alkali catalyst reacts with the free fatty acids to form soaps [113]. There are some studies on microalgae oil to produce biodiesel by using alkali transesterification. Velasquez-Orta et al. studied on biodiesel production from *Chlorella vulgaris*. In that study, alkali transesterification was used for conversion and they achieved 71% of FAME yield [123]. Ferrentino et al. studied on biodiesel production from microalgae too. They used Chlorella sp. oil and their production method was alkali transesterification. They have obtained high yield from their experiment [124]. In another study, Carvalho et al. used alkali transesterification for biodiesel production from algae oil. In their study, they used *Chlorella emersonii* oil and they have obtained 93% conversion yield [124].

It can be seen that there are some problems such as recovery of glycerol or removing catalysts from product and need of wastewater treatment in acid or alkali-catalytic transesterification. Enzymatic catalysts like lipases are able to catalyze the transesterification of triglycerides effectively. With this process glycerol can be easily recovered however enzymatic catalysts are often more expensive than chemical catalysts. The high cost of enzyme production is the main obstacle to the commercialization of enzyme-catalyzed processes. But using solvent-tolerant lipases and immobilized lipases can be a solution for this. Lipase-catalyzed transesterification is considered to be one of the most effective reactions for production of biodiesel [114]. In another study Tran et al. used microalgae oil (*Chlorella vulgaris* ESP-31) for producing biodiesel. Their method was enzyme-catalyzed transesterification and they used lipase in this process. In the result, they reported that they achieved 94.78 % of FAME yield [126]. Table 2 presents the transesterification studies for biodiesel production from microalgae oil.

Supercritical process, microwave-assisted method and ultrasonic-assisted process are novel methods used in biodiesel production area. Since these methods are novel methods and also algae are new materials for biofuel area, there is a few studies biodiesel production from algae oil with these novel methods, these studies were reviewed and presented below.

With *supercritical process* biodiesel production can be easily achieved without catalysts. Supercritical fluid is a substance whose temperature and pressure is above the critical point. These fluids are environmentally friendly and economic. Usually water, carbon dioxide and alcohol is used for supercritical fluid. In biodiesel production generally supercritical methanol and supercritical ethanol is used. Advantages of this process are being easier for purification, shorter the reaction time and more effective reaction [130]. In the study of Patil et al., using supercritical methanol produced biodiesel. The wet algae were used and the ratio of alcohol/ oil was chosen as 9:1. The temperature of the reaction occurred at 255 and 1200 psi and resulted in 90% of FAME yield [131].

Microwaves activate differences in small degrees of polar molecules and ions, because the molecular friction and chemical reactions start. Molecules have not the enough time to relax and heat generation occurs in a short

Table 2. The transesterification studies for biodiesel production from microalgae oil

Algae strain	Method	Alcohol	Alcohol / oil molar ratio	Temp.	Time	Results	Ref.
Heterotrophic C. Protothecoides (microalga)	Acidic trasesterification	Methanol	56:1	30 °C	4 h	80% (FAME Yield	[121]
Chlorella vulgaris ESP-31 (microalga)	Enzymatic transesterification (Lipase)	Methanol	98.81	25-40 °C	48 h	94.78% (FAME Yield)	[126]
Chlorella vulgaris (microalga)	in situ alkaline transesterification	Methanol	600:1	60 °C	75 min	71% (FAME Yield)	[123]
Nannochloropsis oculata (microalga)	heterogeneous transesterification	Methanol	30:1	50 °C	4 h	97.5% (FAME Yield)	[127]
Chlorella (microalga)	In-situ acidic transesterification	Methanol	315:1	23 and 30 °C	15 min-2 h	70-92% (FAME Yield)	[17]
Chlorella sp. (microalga)	Alkali Transesterification	Methanol	-	100 °C	25 h	90 (Fluorometric Reading)	[124]
Schizochytrium limacinum (microalga)	Acidic Transesterification	Methanol	-	90 °C	40 min.	82.6% (biodiesel Yield)	[122]
Chlorella emersonii	Alkali trasesterification	Methanol	5:1	60 °C	2 h	93% conversion	[125]
Fucus spiralis (macroalga)	Alkali Transesterification	Methanol	6:1	60 °C	4 h	1.6-11.5% (Process Yield)	[128]
Commercially refined macroalga (Kelp)	McGyan process	Methanol	32:1	360 °C	30 s	94.7% (FAME Yield)	[129]

time because energy interacts with molecules very quickly. Transesterification reaction is carried out with microwave in a short time and microwave results in an efficient manner. As a result in a short time separation and pure products with high yield is obtained. Thus, production costs and the formation of by-product are reduced [130]. Patil et al., made a study on biodiesel production from dry microalgae by using microwave-assisted process. KOH was used as catalyst in the study and microwave condition is set to 800 W. The performance of the study is around 80% [132]. The other study with macroalgae for microwave-assisted algal biodiesel was showed that methanol to macroalgae ratio of 1:15 was the best condition. In the study, sodium hydroxide concentration was 2 wt % and reaction time of 3 min for the best condition [133]. Koberg et al. was reported the study used *Nannochloropsis* for algal biodiesel production with microwave-assisted method. The higher biodiesel yield was observed which was around 37.1% with microwave technique. The same conditions for sonication technique resulted in lower yield [134].

Recent years, *ultrasonic-assisted process* is widely used in biodiesel production. Mixing is very important factor for biodiesel yield in transesterification reactions. It is an effective mixing method in liquid-liquid mass transfer to provide better mixing. Powerful mixing creates smaller droplets than the conventional mixing and increases the contact areas between the oil phases. Also it provides the activation energy, which needs for initiating transesterification reactions [130]. In the study of Eihaze et al., they are focused on the in situ transesterification of microalgae by ultrasound technique. The reaction takes 1 h with the use of methanol/oil ratio to 315:1. The result was 0.295 ± 0.003 g biodiesel/g dry *Chlorella* which shows that this is higher than mechanically stirred in situ technique [135].

3.8. Design of Algae And Biodiesel Production

In this section of study, algae production stages that cover the algae strain and location selection, algae cultivation, harvesting, oil extraction, and biodiesel production process from microalgae are presented by using ChemCad design program. All stages are given in this process flow diagram (pfd) and equipment table in detail. As it is seen in a process flow diagram (pfd), the streams between 1-8 are the area of the process where algae growth occurs. The algae bodies contain a lipid, which can be extracted and converted into a type of biofuel. The area where between stream1-8 has several large ponds to grow algae containing large amounts of lipid in preparation for lipid extraction. Once a pond is harvested, it is re-inoculated for another crop of algae (stream 11-13). Once the algae reach maturity in the growth ponds and have the desired lipid content, the cells are harvested in the area where stream 9-10. This area at a concentration of 1g-algae/L water. The algae collected will be dewatered, and the usable lipid is extracted for the reaction process where stream 9,10,14-16. The remaining algal biomass will be sent to algal pulp tank, it may be evaluated for biogas production in digesters. Lipids, catalysts and alcohol are sent for fuel conversion to heat-jacketed transesterification reactor. Once the lipid is harvested from the algae cells, the usable triglycerides are converted to biofuel in streams 16-18. Then products sent to the separator to separate biodiesel and byproduct glycerol in stream 21-27. The byproduct of this reaction is glycerol, which is removed and treated as waste. The biofuel is then ready to be used in modern farm equipment, or as a fuel supplement for diesel. All the equipments, tanks and ponds are labeled in the Figure 1.

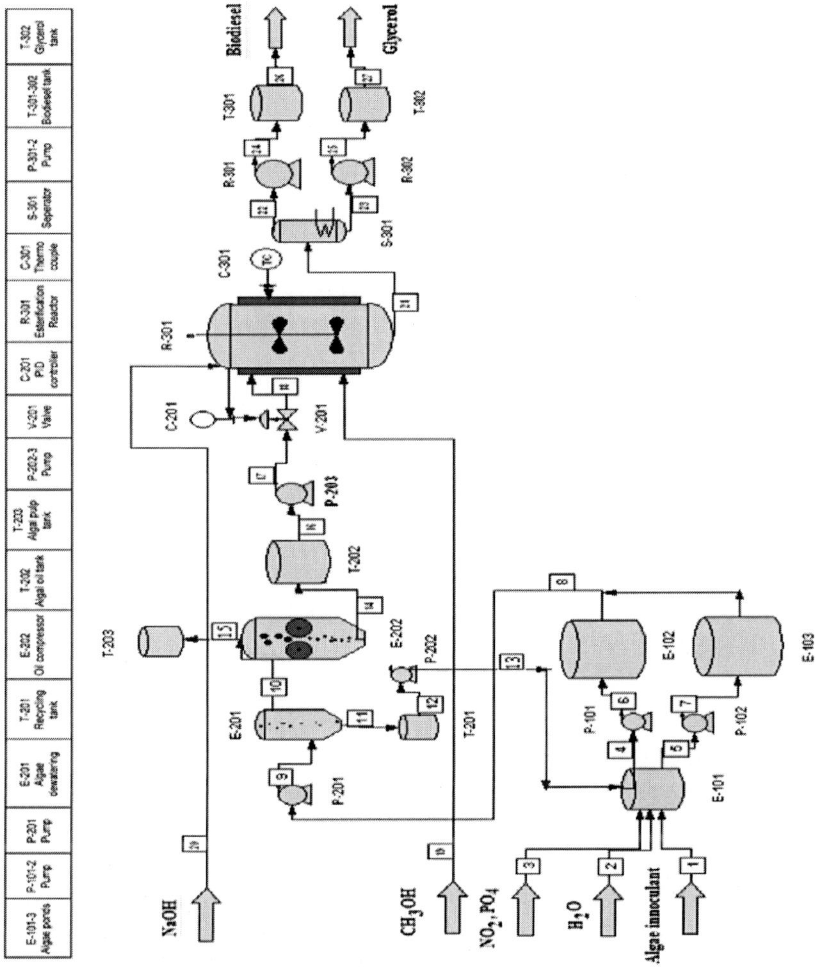

Figure 1. The process flow diagram of biodiesel production process from microalgae by ChemCAD.

4. CONCLUSION

Nowadays, demands on energy are caused to reduction of sources and environmental problems let the world to use alternative fuels. Microalgae have important potential as an alternative energy source. A lot of valuable products can be produced from microalgae such as biodiesel, biogas, bioethanol, medicines and nutraceuticals. Biodiesel is one of the most important alternative fuels. Microalgal biodiesel production is very new technology. In this study, microalgae and their classifications, important steps of biodiesel production from microalgae have been mentioned. In production sections, steps are explained briefly and easily understandable. Also advantages and disadvantages in the production are mainly discussed. At the end of this chapter, a biodiesel production from microalgae is designed by ChemCadprogram, which shows a simple process flow diagram for who desires to produce biodiesel from microalgae. Recently, microalgae are not economically viable. The main problems are the cost of capital cost. The rate of return is not short as it is expected. The operation cost is also affecting the total cost significantly. The main part, which makes the process expensive due to operation and capital costs, are algae growth, harvesting, dewatering, and fuel conversion. Beyond these, oil extraction step significantly increases the cost. If the oil could be extracted easily and at higher rates, the cost would be much lower. However, there are needs to innovate new ways to make the process economically feasible. Regardless, microalgae are seen as important resources for the future and there will be a lot of improvements on recent technology.

REFERENCES

1. Szulczyk KR, McCarl BA. Market penetration of biodiesel. Renewable and Sustainable Energy Reviews 2010;14: 2426–2433.
2. Chisti Y. Biodiesel from microalgae. Biotechnology Advances 2007; 25 (3): 294-306.
3. Khan SA, Rashmi, Hussain MZ, Prasad S, Banerjee UC. Prospects of biodiesel production from microalgae in India. Renewable Sustainable Energy Reviews 2009; 13: 2361–2372.
4. Teixeira CM, Morales ME. Microalga como mate´ ria-prima para a produc¸a˜o de biodiesel. Revista: Biodiesel o Novo combustı´vel do Brasil 2007; 91–96.
5. Wu X, Ruan R, Du Z, Liu Y. Current Status and Prospects of Biodiesel Production from Microalgae. Energies 2012; 5: 2667-2682.
6. Yoo C, Jun SY, Lee JY, Ahn CY, Oh HM. Selection of microalgae for lipid production under high levels carbon dioxide. Bioresource Technology 2010; 101: 571–574.
7. Jiang LL, Luo SJ, Fan XL, Yang ZM, Guo RB. Biomass and lipid production of marine microalgae using municipal wastewater and high concentration of CO2. Applied Energy 2011; 88: 3336–3341.

8. Lee JN, Lee JS, Shin CS, Park S. C, Kim SW. Methods to enhance tolerances of *Chlorella* KR-1 to toxic compounds in flue gas. Applied Biochemistry and Biotechnology 2000; 84–86: 329–342.
9. Vergara-Fernandez A, Vargas G, Alarcon N, Velasco A. Evaluation of marine algae as a source of biogas in a two-stage anaerobic reactor system. Biomass and Bioenergy 2008; 32:338–44.
10. Yen HW, Brune DE. Anaerobic co-digestion of algal sludge and waste paper to produce methane. Bioresource Technology 2007; 98: 130–134.
11. Nakas JPS, Parkinson CM, Coonley CE, Tanenbaum SW. System development for linked-fermentation production of solvents from algal biomass. Applied and Environmental Microbiology 1983; 46:1017–23.
12. Belarbi EH, Molina EY. A process for high yield and scalable recovery of high purity eicosapentaenoic acid esters from microalgae and fish oil. Process Biochemistry 2000; 35:951–69.
13. Cheng-Wu Z, Zmora O, Kopel R, Richmond A. An industrial-size flat plate glass reactor for mass production of *Nannochloropsis* sp. (Eustigmatophyceae). Aquaculture 2001; 195:35–49.
14. Besada V, Andrade JM, Schultze F, Gonzalez JJ. Heavy metals in edible seaweeds commercialised for human consumption. Journal of Marine Systems 2009; 75:305–13.
15. Spolaore P, Joannis-Cassan C, Duran E, Isambert A. Commercial applications of microalgae. Journal of Bioscience and Bioengineering 2006; 101: 87-96.
16. Olaizola M. Commercial development of microalgal biotechnology: from the test tube to the marketplace. Biomolecular Engineering 2003; 20: 459-466.
17. Ehimen EA, Sun ZF, Carrington CG. Variables affecting the in situ transesterification of microalgae lipids. Fuel 2010; 89(3): 677–684.
18. Yang Z, Guo R, Xu X, Fan X, Luo S. Hydrogen and methane production from lipid-extracted microalgal biomass residues. İnternational Journal of Hydrogen Energy 2011;36: 3465-3470.
19. Hirayama S, Ueda R, Ogushi Y, Hirano A, Samejima Y, Hon-Nami K, Kunito S. Ethanol production from carbon dioxide by fermentative microalgae. Studies in Surface Science and Catalysis 1998; 114:657–60.

20. Bush RA, Hall KM. Process for the production of ethanol from algae. U. S. Patent 7,135,308; 2006.
21. Gamal A. Biological importance of marine algae. Saudi Pharmaceutical Journal 2010; 18: 1–25.
22. Hasan MR. Use of algae and aquatic macrophytes as feed insmall-scale aquaculture. FAO Fisheries and aquaculture technical paper, 2009; 531.
23. Hall J, Payne G. Factors controlling the growth of field population of Hydrodictyon reticulatum in New Zealand. Journal of Applied Phycology 1997; 9: 229-236.
24. Rafiqul IM, Jalal KCA, Alam MZ. Environmental factors for optimisation of *Spirulina* biomass in laboratory culture. Biotechnology 2005; 4: 19-22.
25. McHugh DJ. A guide to the seaweed industry. Fisheries Technical Paper 2003; No. 441. Rome, FAO, 118.
26. Bruton T, Lyons H, Lerat Y, Stanley M. A Review of the Potential of Marine Algae as a Source of Biofuel in Ireland. Sustainable Energy Ireland 2009; 1-88.
27. Bold HC, Wynne MJ. Introduction to the algae structure and reproduction, second ed., Prentice-Hall Inc., Englewood Cliffs, NJ, 07632,1985; 1–33.
28. Guardia SD, Valls R, Mesguiche V, Brunei JM, Gulioli G. Enantioselective synthesis of (-)-bifuracadiol: a natural antitumor marine product. Tetrahedron Letters 1999; 40: 8359–8360.
29. Bennamara A, Abourrichi A, Berrada M, Charrouf M, Chaib N, Boudouma M, Garneau XF. Methoxybifur-carenone: an antifungal and antibacterial meroditerpenoid from the brown alga *Cystoseira tamariscifolia*. Phytochemistry 1999; 52: 37–40.
30. Awad NE. Biologically active steroid from the green alga *Ulva lactuca*. Phytotherapy Research 2000; 14: 641–643.
31. Choi JS, Park HJ, Jung HA, Chung HY, Jung JH, Choi WC. A cyclohexanonyl bromophenol from the red alga *Symphyocladia latiuscula*. Journal of Natural Products 2000; 63: 1705–1706.
32. Sheehan J, Dunahay T, Benemann J, Roessler P. A Look Back at the U. S. Department of Energy's Aquatic Species Program: Biodiesel from Algae 1998.

33. Singh SP and Singh D. Biodiesel production through the use of different sources and characterization of oils and their esters as the substitute of diesel: A review. Renewable and Sustainable Energy Reviews 2010; 14: 200–216.
34. Garson J. Marine natural products. Natural Product Reports 1989; 6, 143–170.
35. Koehn FE, Longley RE, Reed JK. Microcolins A and B, new immunosuppressive peptides from the blue–green alga *Lyngbya majuscule*. Journal of Natural Products 1992; 55: 613–619.
36. Gerwick WH, Roberts MA, Proteau PJ, Chen JL. Screening cultured marine microalgae for anticancer-type activity. Journal of Applied. Phycology 1994; 6: 143–149.
37. Trease GE, Evanes WC. Pharmacognosy. 14th ed., W. B Saunders Company Ltd., London, Philadelphia, Toronto, Sydney, Tokyo, 1996. 18–27.
38. Shimizu Y. Dinoflagellates as sources of bioactive molecules. Attaway, D. H., Zaborsky, O. R. (Eds.), Marine Biotechnology, vol. 1: Pharmaceutical and Bioactive Natural Products, New York, Plenum, 1993; 391–310.
39. Mata MT, Martins AA, Caetano SN. Microalgae for biodiesel production and other applications: A review. Renewable and Sustainable Energy Reviews 2010; 14: 217–232.
40. Li Y, Horsman M, Wu N, Lan CQ, Dubois-Calero N. Biofuels from microalgae. Biotechnology Progress 2008; 24(4): 815–20.
41. Richmond A. Handbook of microalgal culture: biotechnology and applied phycology. Blackwell Science Ltd; 2004.
42. Renaud SM, Thinh LV, Parry DL. The gross chemical composition and fatty acid composition of 18 species of tropical Australian microalgae for possible use in mariculture. Aquaculture 1999; 170:147–59.
43. Rodolfi L, Zittelli GC, Bassi N, Padovani G, Biondi N, Bonini G, et al. Microalgae for oil: strain selection, induction of lipid synthesis and outdoor mass cultivation in a low-cost photobioreactor. Biotechnology and Bioengineering 2009; 102(1):100–12.
44. Barclay B. Microalgae Culture Collection 1984–1985. Microalgal Technology Research Group (MTRG) 1984; SERI/SP-231-2486.
45. Zhu CJ, Lee YK. Determination of biomass dry weight of marine microalgae. Journal of Applied Phycology 1997; 9:189–94.

46. Maxwell EL, Folger AG, Hogg SE. Resource evaluation and site selection for microalgae production systems May 1985; SERI/TR-215-2484.
47. Chojnacka K, Marquez-Rocha FJ, Kinetic. Stoichiometric relationships of the energy and carbon metabolism in the culture of microalgae. Biotechnology 2004; 3(1): 21–34.
48. Wen Z, Chen F. Heterotrophic production of eicosapentaenoic acid by microalgae. Biotechnology Advances 2003; 21: 273–294.
49. Chen C, Yeh K, Aisyah R. Cultivation, photobioreactor design and harvesting of microalgae for biodiesel production: A critical review. Bioresource Technology 2011; 102: 71–81.
50. Chen F. High cell density culture of microalgae in heterotrophic growth. Trends in Biotechnology 1996; 14: 421–426.
51. Li XF, Xu H, Wu QY. Large-scale biodiesel production from microalga *Chlorella protothecoides* through heterotrophic cultivation in bioreactors. Biotechnology and Bioengineering 2007; 98: 764–771.
52. Doucha J, Lívansky K. Production of high-density *Chlorella* culture grown in fermenters. Journal of Applied Phycology 2012; 24: 35–43.
53. Park KC., Whitney C, McNichol, JC, Dickinson KE, MacQuarrie S, Skrupski BP. Zou, J. ; Wilson, K. E. ; O'Leary, S. J. B. ; McGinn, P. J. Mixotrophic and Photoautotrophic cultivation of 14 microalgae isolates from Saskatchewan, Canada: Potential applications for wastewater remediation for biofuel production. Journal of Applied Phycology 2012; 24: 339–348.
54. Bhatnagar A, Chinnasamy S, Singh M, Das KC. Renewable biomass production by mixotrophic algae in the presence of various carbon sources and wastewaters. Applied Energy 2011; 88: 3425–3431.
55. Ogbonna JC, Yada H, Tanaka H. Light supply coefficient: A new engineering parameter for photobioreactor design, Journal of Fermentation and Bioengineering 1995; 80 (4): 369-376.
56. Bitoga J. P, Lee I. -B, Lee C. -G, Kim K. -S, Hwang H-S, Hong S. -W, Seo I. -H, Kwon K. -S, Mostafa E. Application of computational fluid dynamics for modeling and designing photobioreactors for microalgae production, Computers and Electronics in Agriculture 2011; 76: 131–147.

57. Yeesang C, Cheirsilp B. Effect of nitrogen, salt, and iron content in the growth medium and light intensity on lipid production by microalgae isolated from freshwater sources in Thailand. Bioresource Technology 2011; 102 (3): 3034-3040.
58. Converti A, Casazza AA, Ortiz EY, Perego P, Del Borghi M. Effect of temperature and nitrogen concentration on the growth and lipid content of *Nannochloropsis oculata* and *Chlorella vulgaris* for biodiesel production. Chemical Engineering and Processing: Process Intensification 2009; 48 (6): 1146-1151.
59. Cheirsilp B, Torpee S. Enhanced growth and lipid production of microalgae under mixotrophic culture condition: Effect of light intensity, glucose concentration and fed-batch cultivation, Bioresource Technology, 2012; 110: 510-516.
60. Ruangsomboon S. Effect of light, nutrient, cultivation time and salinity on lipid production of newly isolated strain of the green microalga, *Botryococcus braunii KMITL 2*. Bioresource Technology 2012; 109: 261-265.
61. Sara S. Kuwaharaa, Joel L. Cuelloa, Graham Myhreb, Stanley Pau. Growth of the green algae Chlamydomonas reinhardtii under red and blue lasers. Optics and Lasers in Engineering, 2010; 49: 434-438.
62. Allan MB, Arnon DI. Studies on nitrogen-fixing blue-green algae. I. Growth and nitrogen fixation by Anabaena cylindrica Lemm. Plant Physiology 1955; 30:366–372.
63. Solovchenko AE, Khozin-Goldberg I, Didi-Cohen S, Cohen Z, Merzlyak MN Effects of light intensity and nitrogen starvation on growth, total fatty acids and arachidonic acid in the green microalga *Parietochloris incise*. Journal of applied Phycology 2008; 20:245–251.
64. Yeh KL, Chang JS. and Chen WM. Effect of light supply and carbon source on cell growth and cellular composition of a newly isolated microalga *Chlorella vulgaris* ESP-31. Engineering in Life Scince 2010; 10: 201–208.
65. Widjaja A, Lipid production from microalgae as a promising candidate for biodiesel production. Makara Teknologi 2009, 13(1): 47-51.

66. Bhola V, Desikan R, Santosh SK, Subburamu K, Sanniyasi E, Bux F. Effects of parameters affecting biomass yield and thermal behaviour of *Chlorella vulgaris*, Journal of Bioscience and Bioengineering 2011; 111 (3): 377-382.

67. Ebrahimzadeh MM. Torkian Boldaji SA. Hosseini Boldaji, Sh. Ettehad. Effects of varying sodium nitrate and carbon dioxide quantities on microalgae growth performance of a designed tubular photobioreactor and a channel photobioreactor. Iranian Journal of Plant Physiology 2011; 1(4): 257-263.

68. Sonnekus MJ, Effects of Salinity on the Growth and Lipid Production of Ten Species of Microalgae from the Swartkops Saltworks: A Biodiesel Perspective Nelson Mandela Metropolitan University, 2010.

69. Sheng J, Kim H. W, Badalamenti J. P, Zhou C, Sridharakrishnan S, Krajmalnik-Brown R, Rittmann BE, Vannela R. Effects of temperature shifts on growth rate and lipid characteristics of *Synechocystis* sp. PCC6803 in a bench-top photobioreactor. Bioresource Technology 2011; 102 (24): 11218-11225.

70. Sivakumara G, Xua J, Thompsonb RW, Yangc Y, Randol-Smithd P, Weathersc PJ. Integrated green algal technology for bioremediation and biofuel, Bioresource Technology 2012; 107: 1-9.

71. Floreto EAT, Hirata H, Ando S, Yamasaki S. Effects of Temperature, Light Intensity, Salinity and Source of Nitrogen on the Growth, Total Lipid and Fatty Acid Composition of Ulva pertusa Kjellman (*Chlorophyta)*" Botanica Marina, 2012; 36 (2): 149-158.

72. Weisse T, Stadler P. Effect of pH on growth, cell volume, and production of freshwater ciliates, and implications for their distribution. Limnology and Oceanography 2006; 51: 1708-1715.

73. Kumar A, Ergas S, Yuan X, Sahu A, Zhang Q, Dewulf J, Malcata F. X, Langenhove H. Enhanced CO 2 fixation and biofuel production via microalgae: Recent developments and future directions. Trends in Biotechnology 2010; 28(7): 371-380.

74. Bergstrom C, McKeel C, Patel S. Effects of pH on algal abundance: a model of Bay Harbor, Michigan, Biological Station of University of Michigan, 2007.

75. Widjaja A, Chien CC, Ju YH. Study of increasing lipid production from fresh water microalgae Chlorella vulgaris, Journal of the Taiwan Institute of Chemical Engineers 2009; 40(1):13-20.

76. Moisander PH, McClinton E. and Paerl HW. Salinity Effects on Growth, Photosynthetic Parameters, and Nitrogenase Activity in Estuarine Planktonic Cyanobacteria 2002; 43(4): 432-442.
77. Xin L, Hong-ying H, Ke G, Ying-Xue S. Effects of different nitrogen and phosphorus concentrations on the growth, nutrient uptake, and lipid accumulation of a freshwater microalga *Scenedesmus* sp. Bioresource Technology 2010; 101 (14): 5494-5500.
78. Lin Q, Lin J. Effects of nitrogen source and concentration on biomass and oil production of a *Scenedesmus rubescens* like microalga, Bioresource Technology 2011; 102 (2):1615-1621.
79. Harun R, Singh M, Forde GM, Danquah MK. Bioprocess engineering of microalgae to produce a variety of consumer products, Renewable and Sustainable Energy Reviews 2010; 14: 1037–1047.
80. Borowitzka MA. Commercial production of microalgae: ponds, tanks, and fermenters, Progress in Industrial Microbiology 1999; 35 (C): 313-321.
81. Masojidek J, Torzillo G. Mass cultivation of freshwater microalgae. Ecological Engineering. Encyclopedia of Ecolology 2008; 3: 2226. 2235.
82. "Biodiesel from Algae Oil", www. oilgae. com", Access date: May 10, 2012.
83. Ugwu CU, Aoyagi H, Uchiyama H. Photobioreactors for mass cultivation of algae. Bioresource Technology 2008; 99: 4021–4028.
84. Hu Q, Guterman H, Richmond A. A flat inclined modular photobioreactor for outdoor mass cultivation of photoautotrophs. Biotechnology and Bioengineering 1996; 51:51–60.
85. Ramos de Ortega A, Roux JC. Production of Chlorella biomass in different types of flat bioreactors in temperate zones. Biomass 1986; 10:141–56.
86. Iqbal M, Grey D, Stepan-Sarkissian F, Fowler MW. A flat-sided photobioreactor for culturing microalgae. Aquacultural Engineering 1993 ;12: 183–90.
87. Vonshak A, Torzillo G. Environmental stress physiology. In: Richmond, A. (Ed.), Handbook of Microalgal Culture. Blackwell Publishers, Oxford, 2004; 57–82.

88. Camacho Rubio F, Acie ´n Ferna ´ndez F. G, Sa ´nchez Pe ´rez J. A, Garcı ´a Camacho F, Molina Grima, E. Prediction of dissolved oxygen and carbon dioxide concentration profiles in tubular photobioreactors for microalgal culture. Biotechnology and Bioengineering 1999; 62: 71–86.
89. Yılmaz HK. Mikroalg Üretimi İçin Fotobiyoreaktör Tasarımları, MEU. Su Ürünleri Fakültesi, (2006).
90. Ogbonna JC, Soejima T, Tanaka H. An integrated solar and artificial light system for internal illumination of photobioreactors. Journal of Biotechnology 1999; 70: 289–297.
91. Mori K. Photoautotrophic bioreactor using visible solar rays condensed by fresnel lenses and transmitted through optical fibers. Biotechnology and Bioengineering Symposium 1985; 15: 331–345.
92. "Microalgae production", SOLEY INSTITUTE, www. algaeinstitute. com
93. Harun R, Davidson M, Doyle M, Technoeconomic analysis of an integrated microalgae photobioreactor, biodiesel and biogas production facility. Biomass and Bioenergy 2011; 35: 741-747.
94. Kennedy CA, In Home Photosynthetic Bioreactor. Advanced Biology 1998.
95. Carvalho AP, Meireles LA, Malcata FX. Microalgal reactors: a review of enclosed system designs and performances. Biotechnology Progress 2006; 22: 1490–1506.
96. Miron SA, Garcia MC, Camacho GF, Molina GEY. Mixing in bubble column and airlift reactors. Chemical Engineering Research and Design 2004; 82 (A10): 1367–1374.
97. Fan L, Zhang Y, Cheng L, Zhang, L, Tang D, Chen H. Optimization of carbon. 2007.
98. Fernandes BD, Dragoner GM, Teixiera JA, Vicente AA. Light regime characterization in an airlift. Photobioreactor for production of microalgae with high starch content. Applied Biochemical Biotechnology 2010; 161: 218–226.
99. Kommareddy AR, Anderson GA. Study of Light as a Parameter in the Growth of Algae in a Photo-Bio Reactor (PBR). ASAE Paper No. 034057. ASAE, St. Joseph, Michigan. 2003.

100. Zhang K, Kurano N, Miyachi S. Optimized aeration by carbon dioxide gas for microalgal production and mass transfer characterization in a vertical flat-plate photobioreactor. Bioprocess Biosystems Bioengineering 2002; 25: 97–101.

101. Scott AS, Davey MP, Dennis SJ. Biodiesel from algae: challenges and prospects. Current Opinion in Biotechnology 2010; 21: 277–288.

102. Williams JA. Keys to Bioreactor selection. CEP Magazine 2002; 34–41.

103. Becker EW. Microalgae: Biotechnology and Microbiology. ISBN 0521350204. 1995.

104. Demirel Z. Eğirdir Gölünden İzole Edilen Yeşil Mikroalg: *ChlorophytaScenedesmus protuberans fris.* 'in Antimikrobiyal ve Antioksidan Özelliğinin Araştırılması, MSc. Thesis, Ege University, İzmir. 2006.

105. Molina Grima E, Belarbi E, Acie ́n Ferna ́ndez F, Robles Medina A, Chisti Y. Recovery of microalgal biomass and metabolites: process options and economics. Biotechnology Advances 2003; 20:491–515.

106. Danquah M, Ang L, Uduman N, Moheimani N, Forde G. Dewatering of microalgal culture for biodiesel production: exploring polymer flocculation and tangential flow filtration. Journal of Chemical Technology and Biotechnology 2009; 84:1078–83.

107. Popoola TOS, Yangomodou OD. Extraction, properties and utilization potentials of cassava seed oil. Biotechnology 2006; 5:38–41.

108. Serrato AG. Extraction of oil from soybeans. Journal of the American Oil Chemistry 1981; 58-3:157-159.

109. Macias-Sanchez MD, Mantell C, Rodriguez M, Martinez De La Ossa E, Lubian LM, Montero O. Supercritical fluid extraction of carotenoids and chlorophyll a from *Nannochloropsis gaditana.* Journal of Food Engineering 2005; 66:245–51.

110. Wiltshire KH, Boersma M, Moller A, Buhtz H. Extraction of pigments and fatty acids from the green alga *Scenedesmus obliquus* (*Chlorophyceae*). Aquatic Ecology 2000; 34:119–26.

111. Yusuf N, Kamarudin SK, Yaakub Z. Overview on the current trends in biodiesel production. Energy Conversion and Management 2011; 52: 2741–2751.

112. Demirbas A. Progress and recent trends in biodiesel fuels. Energy Conversion and Management 2009; 50: 14–34.

113. Lin L, Cunshan Zhou, Vittayapadung S, Xiangqian S, Mingdong D. Opportunities and challenges for biodiesel fuel. Applied Energy 2011; 88: 1020–1031.

114. Balat M, Balat H. Progress in biodiesel processing. Applied Energy 2010; 87: 1815–1835.

115. D'Elia L, Keyser A, Young C. Algae Biodiesel, An Interactive Qualifying Project Report. 2010.

116. Maa F, Hanna M. Biodiesel production: a review. Bioresource Technology 1999; 70: 1-15.

117. Sharma YC, Singh B, Upadhyay SN. Advancements in development and characterization of biodiesel: A review. Fuel 2008; 87(12): 2355–73.

118. Ahmad AL, Yasin NH, Derek JK. Microalgae as a sustainable energy source for biodiesel production: A review. Renewable and Sustainable Energy Reviews 2011; 15: 584–593.

119. Atabani AE, Silitongaa AS, Badruddina IA, Mahliaa TMI, Masjukia HH, Mekhilef S. A comprehensive review on biodiesel as an alternative energy resource and its characteristics. Renewable and Sustainable Energy Reviews 2012; 16: 2070– 2093.

120. Rachmaniah O, Ju YH, Vali SR, Tjondronegoro I, Musfil AS. A study on acid-catalyzed transesterification of crude rice bran oil for biodiesel production. In: youth energy symposium, 19th World energy congress and exhibition, Sydney (Australia), September 5–9; 2004.

121. Xu H, Miao X, Wu Q. High quality biodiesel production from a microalga *Chlorella protothecoides* by heterotrophic growth in fermenters. Journal of Biotechnology 2006; 126: 499–507.

122. Johnson MB. Microalgal Biodiesel Production through a Novel Attached Culture System and Conversion Parameters. MS thesis Virginia Polytechnic Institute and State University; 2009.

123. Velasquez-Orta SB, Lee JGM, Harvey A. Alkaline in situ transesterification of *Chlorella vulgaris*. Fuel 2012; 94: 544–550.

124. Ferrentino JM, Farag IH, Jahnke LS. Microalgal Oil Extraction and In-situ Transesterification. Chemical Engineering, University of New Hampshire (UNH) Durham, NH http://www. nt. ntnu. no/users/skoge/prost/proceedings/aiche-2006/data/papers/P69332. pdf (accessed 02. 08. 12).
125. Carvalho J, Ribiero A, Castro J, Vilarinho C, Castro F. Biodiesel Production by Microalgae and Macroalgae from North Littoral Portuguese Coast. WASTES: Solutions, Treatments and Opportunities, 1st International Conference September 12th – 14th 2011.
126. Tran D, Yeh K, Chen C, Chang J. Enzymatic transesterification of microalgal oil from *Chlorella vulgaris* ESP-31 for biodiesel synthesis using immobilized Burkholderia lipase. Bioresource Technology 2012; 108: 119–127.
127. Umdu ES, Tuncer M, Seker E. Transesterification of Nannochloropsis oculata microalga's lipid to biodiesel on Al2O3 supported CaO and MgO catalysts. Bioresource Technology 2009; 100: 2828–2831.
128. Maceiras R, Rodrı ́guez M, Cancela A, Urréjola S, Sánchez A. Macroalgae: Raw material for biodiesel production. Applied Energy 2011; 88: 3318–3323.
129. Krohn JB, McNeff VC, Yan B, Nowlan D. Production of algae-based biodiesel using the continuous catalytic Mcgyan process. Bioresource Technology 2011; 102: 94–100.
130. Özçimen D, Yücel S. Novel Methods in Biodiesel Production, In: Marco Aurélio dos Santos Bernardes (Ed.) Biofuel's Engineering Process Technology. InTech; 2011. p353-384.
131. Patil P, Deng S, Isaac Rhodes J, Lammers PJ. Conversion of waste cooking oil to biodiesel using ferric sulfate and supercritical methanol processes. Fuel 2009; 89 (2): 360-364.
132. Patil PD, Gude VG, Mannarswamy A, Cooke P, Munson-McGee S, Nirmalakhandan N, Lammers P, Deng S. Optimization of microwave-assisted transesterification of dry algal biomass using response surface methodology. Bioresource Technology 2011;102: 1399-1405.
133. Cancela A, Maceiras R, Urrejola S, Sanchez A. Microwave-Assisted Transesterification of Macroalgae. Energies 2012, 5: 862-871.

134. Koberg M, Cohen M, Ben-Amotz A, Gedanken A. Bio-diesel production directly from the microalgae biomass of *Nannochloropsis* by microwave and ultrasound radiation. Bioresource Technology, 2011; 5 (102): 4265-4269.

135. Ehimen EA, Sun ZF, Carrington CG. Use of ultrasound and co-solvents to improve the in-situ transesterification of microalgae biomass. Procedia Environmental Sciences 2012; 15: 47-55.

3 CHAPTER

ISOLATION AND SCREENING OF FILAMENTOUS FUNGI PRODUCING EXTRACELLULAR LIPASE WITH POTENTIAL IN BIODIESEL PRODUCTION

Sabrina Moro Villela Pacheco[1], Américo Cruz Júnior[1], Ayres Ferreira Morgado[1], Agenor Furigo Júnior[1], Onyetugo Chioma Amadi[2], José Manuel Guisán[3], Benevides Pessela[4,5*]

[1]*Department of Food and Chemistry Engineering, University Campus UFSC, Florianópolis, Brazil*
[2]*Department of Microbiology, Faculty of Biological Science, University of Nigeria, Enugu, Nigeria*
[3]*Departamento de Biocatálises Enzimática, Instituto de Catalisis y Petroleoquimica, ICP (CSIC-UAM), C/Marie Curie 2, Cantoblanco, Campus de la Univerisdad Autónoma de Madrid, Madrid, España*
[4]*Instituto de Investigación en Ciencias de la Alimentación CIAL (CSIC-UAM), C/Nicolás Cabrera 9, Cantoblanco, Campus de la Universidad Autónoma de Madrid, Madrid, España*
[5]*Departamento de Engenharia e Tecnologías, Instituto Superior Politécnico de Tecnología e de Ciências, ISPTEC, Luanda Sul, Talatona, República de Angola*

ABSTRACT

Nineteen fungal strains were isolated from a chicken slaughterhouse effluent and within those, only one showed high values of lipolytic activity in submerged cultures. This fungus was identified as Trametes hirsuta. The crude extract was immobilized in chitosan/clay beads, with an immobilization yield of 80.9%. The analyses of the crude extract and the immobilized derivative at different temperatures, pH (s), solvents, metallic ions and storage showed that the immobilization process increased the enzyme life span. Ethyl esters were obtained in solvent free systems using chicken viscera oil and the enzyme crude extract. For effective comparison, a reaction using viscera oil and commercial lipase Novozym 435 was carried out. The result revealed 35% and 28% esters conversion in the reactions containing chicken viscera oil, using Novozym 435 and the crude extract respectively. The extract was also used in a reaction with soybean oil, traditionally used as starting substrate for biodiesel production.

Keywords: Biodiesel Production, Trametes Hirsuta, Lipase

1. INTRODUCTION

The growing concern with the global environmental crisis associated with the use of fossil fuels, and spurred the search for new sources of energy. An interesting alternative to replace fossil fuels is biofuel. The term biofuel refers to liquid or gaseous fuel which is predominantly produced from biomass. Thus, biodiesel is a biofuel comprising a mixture of esters of fatty acids which can be obtained from vegetable oils or animal fats. The production of this biofuel may be performed through a reaction or a transesterification reaction of esterification. The reaction most widely used for the production of biodiesel is the transesterification reaction because it provides one coproduct that has many applications in various industrial segments: glicerol. The enzymes generally used as catalysts of the transterification reactions are lipases [1] .

Lipases (triacylglycerol acylhydrolase, EC 3.1.1.3) are hydrolytic enzymes that catalyze the hydrolysis of fats and oils to glycerol and free fatty acids at the oil-water interface. An important characteristic of lipases is their ability not only to hydrolyze ester bonds, transesterify triglycerides, resolve racemic mixtures but also to synthesize ester bonds in nonaqueous media. These enzymes are widely used in hydrolysis, esterification, alcoholysis and transterification catalysts, and are produced by micro-organisms, animals and plants [2] . Commercial lipases are, generally, from microbial sources, with fungi being able to produce the enzyme in a more intense and diversified way, with regard to enzyme properties and substrate specificity, facilitating its use in industrial process. Due to these characteristics, the search for new lipase sources has increased. However, like any other biocatalyst, lipases are subject to inactivation by several chemical, physical and biological factors, which makes its application in a larger scale difficult. One way to reduce this behavior is to immobilize the enzyme [3] . Immobilized enzymes are defined as enzymes that are "physically confined or localized in a certain region with retention of their catalytic activity, enzymes and that can be used several times". Depending on the immobilization technique used, the enzyme properties such as stability, selectivity, values of Vmax, Km, enzymatic activity profile in different pHs and temperatures can be significantly altered [4] . There are several methods for enzyme immobilization based on physical and chemical mechanisms. Among the chemical immobilization methods there are: a) enzyme/support bond through ionic bond; b) enzyme/support bond through covalent bond; c) enzyme/support cross bond and d) enzyme/support bond achieved through multifunctional reactants. The multifunctional reactants are molecules capable of bonding the enzyme in the support, for example, glutaraldehyde [5] . Several kinds of supports with both hydrophilic and hydrophobic characteristics have been studied for enzyme immobilization, including aerogels [6] , resin [7] , synthetic polymers [7] - [10] , silica [11] , nanoparticles [12] and biopolymers such as alginate, agarose and chitosan [12] - [14] .

Chitosan is obtained through alkaline hydrolysis reaction of chitin, resulting in the biopolymer constituting predominantly of β (1 - 4)-linked *N*-acetylglucosamine) units, presenting in its chain 70% - 90% degree of deacetylation [15] . The use of chitosan in the form of beads is required since it has the advantage of a better superficial characterization, which allows the establishment of geometrical parameters for the reproducibility of the process and further comparisons, also the reactors packing optimization aspects and filtrating devices. Nevertheless, there are disadvantages, such as operational problems regarding the fact that the enzymes density value is very similar to water which can cause flotation. These characteristics limit several industrial processes but can be reduced when chitosan is mixed with solid materials that increase its density and mechanical resistance, this way makes its applications broadened [16] . Among the solid materials that can be used, clay displays outstanding potential, according to the International Association of the Study of Clay and the Clay Minerals Society, it is readily present in nature and constitutes small dimensions minerals which generally exhibit plastic behavior with certain amounts of water and hardens after drying [17] .

Hence, in this study, we reported the isolation of a wild fungal strain from chicken slaughterhouse with lipolytic activity, the work further investigated the crude extract potential for posterior immobilization on chitosan/clay beads, characterization of its biochemical properties as well as analysis for application in biodiesel production.

2. MATERIAL AND METHODS

2.1. Source of Microorganisms

Microorganisms were isolated from a crude effluent of a chicken slaughterhouse located in the region of Florianópolis-Brazil. The effluent used had the following characteristics physico-chemical: pH 7.3, conductivity of 887.9 μS/cm, orthophosphate of 13.2 mg/l, ammonia of 94.7 mg/l, chemical oxygen demand of 2799.4 mg/l, biochemical oxygen demand of 4567.7 mg/l and oils and grease of 901.8 mg/l.

2.2. Screening

Microorganisms were isolated using spread plate method. To this, 0.1 ml of the effluent was spread over the surface of the solid culture medium with a Drigalski handle. The culture medium used, Sabouraud agar with Chloramphenicol at a pH of 5.5, was autoclaved at 121°C for 15 min. The Sabouraud agar with Chloramphenicol was obtained from ACUMEDIA. The isolation steps were performed by successive inoculations of fungi in culture media to obtain more axenic selective cultures [18] . The screening of microorganisms with lipolytic activity was carried out in a culture media constituted of (w/v): 2% bacteriological agar (VETEC), 0.05% anhydrous calcium chloride (DIFCO), 0.5% sodium chloride (DIFCO), 1% bacteriological peptone (DIFCO) and 1% Tween 80 (VETEC). This method allows detecting the strain that exhibit lipolytic activity by measuring the height of precipitate calcium laurate in test tubes [19] . The height precipitate produced by isolates was comparable with the height produced by strains obtained from the Culture Collections of the Department of Food and Chemistry Engineering (Brazil) that are recognized for producing lipase.

2.3. Identification of Fungi

Based on the sequencing and phylogenetic analysis from gene fragments of the operon ribosomal, the fungus was identified [20] . The extracted genomic DNA was performed and amplification of the region ITS1-5.8S-ITS4 was based on PCR technology. The primers used for amplification of the ITS region were ITS-1 and ITS-4. The fragments were purified and submitted directly sequencing in automatic sequencer ABI3500 XL Series (Applied Biosystem). The phylogenetic analysis of partial sequences of the ITS region, was built on a combination of different obtained fragments and compared with the sequences from organisms represented in the Genbank data base. The sequences were aligned using the CLUSTAL X program and phylogenetic analysis was performed using the MEGA 4.0 software [20] - [22] . The evolutionary distance matrices were calculated using the Kimura model [23] and the construction of the phylogenetic tree based on the evolutionary distances were made by Neighbor-Joining method [24] with bootstrap values calculated from 1000 re-sampling, using MEGA 4.0 software.

2.4. Preparation of the Crude Extract

The fungal isolates were inoculated in Roux bottles of 1000 ml with 200 ml of potato dextrose agar medium for 14 days at 25°C. A 5 ml inoculum was used which was obtained by adding 20 ml of 0.2% (v/v) Tween 80, followed by scraping the surface of the mold with the aid of a platinum loop and filtering gauze. The extracts were obtained by submerged cultivation in flasks of 250 ml, using a medium (previously optimized-data not shown) constituting of (w/v in wastewater): 3.41% of chicken viscera oil (obtained of a chicken slaughterhouse located in the region of Florianópolis-Brazil); 2% of Tween 80; 2% yeast extract (VETEC); 1% of NH_4Cl (NUCLEAR); 0.1% of $NaNO_3$ (DIFCO); 0.1% of KH_2PO_4 (NUCLEAR); 0.5% of peptone, pH 5.5, 0.25% olive oil (BUNGE), 0.05% of $MgSO_4$ (NUCLEAR). The flasks were incubated at 25°C, 100 rpm for 10 days, period of time for which it is possible to obtain high values of enzyme activity. The contents of the flask were filtered through gauze to remove the mycelium and centrifuged at 5000 rpm for 5 min. Ammonium sulfate was added to the supernatant at 40% of saturation. The extract was then maintained by stirring gently at 4°C for 240 min. The supernatant was removed and the precipitate was resuspended in a minimal volume (approximately 2 ml) of 100 mM phosphate buffer pH 7.0 [25] .

2.5. Immobilization of Enzyme

Immobilization of the concentrated crude extract was made in beads made of chitosan/clay. The beads were produced through the drip solution of 1 g of chitosan, 1 g of clay and 50 ml of acetic acid 1000 mM in 100 mM NaOH solution. The chitosan used had the following characteristics: average molar mass of 83.7 kDa, deacetylation degree of 72.4% and surface area of 4.9 m^2/g. The smectite clay used was collected from city Palmeira- Brazil and activated with hydrochloric acid to pH 4.2 and humidity of 14.5% and surface area of 181.4 m^2/g. The beads were activated with 3% glutaraldehyde (v/v in phosphate buffer pH 8.0) for 24 h.

2.6. Assay of Enzyme Activity

The activity of crude extract and immobilized derivative was measured using 0.5 g of p-nitrophenyl palmitate (p-NPP) (SIGMA) dissolved in 100 ml of ethanol (VETEC) as substrate. The increase in absorbance at 410 nm caused by the release of p-nitrophenol in the hydrolysis of p-NPP was measured using a spectrophotometer. Crude extract of 0.1 ml (or 0.07 g immobilized extract) was added to a mixture of 1 ml 0.5% (w/v) p-NPP solution and 1 ml buffer Tris-HCl pH 7.0 (SOLARBIO). The system was incubated for 5 min at 30°C. The reaction was terminated by adding 2 ml of 0.5 N Na_2CO_3 (VETEC) to the mixture followed by centrifugation (10,000 rpm for 10 min). The supernatant (0.5 ml was diluted 10 times with deionized water and the absorbance at 410 nm) was measured (UV-VIS Varian Cary 50 spectrophotometer). A molar extinction coefficient (ε 410) for p-nitrophenol of 15,000 $M^{-1} \cdot cm^{-1}$ was used. One unit (U) of lipase activity was defined as the amount of enzyme which catalyzed the production of 1 μmol p-nitrophenol per minute under the experimental conditions [26] . The total protein originally taken for immobilization and protein present in the supernatants (unbound protein) after immobilization were estimated by the method of Bradford using BSA as a standard [27] . To obtain the coupled protein percentage value and immobilization yield, calculus described in [28] was used.

2.7. Effects of Temperature on Stability

For thermal stability study, the immobilized enzyme and free enzyme was incubated at temperatures between 30°C and 70°C for 60 min and relative activity was measured. The enzyme activity was analyzed as described above. Controls were performed with boiled enzyme. The relative enzyme activity was calculated as the ratio of enzyme activity after the incubation period and enzyme activity before the incubation period [29] .

2.8. Effect of pH on Activity

Activity experiments were done using the p-NPP method at 30°C in experimental mixtures buffered to various pH values. The following buffers were used: tartrate pH 3.0; 4.0 and 5.0; phosphate pH 6.0, 7.0 and 8.0; Tris-HCl pH 9.0, 10.0 e 11.0 all at 100 mM. Controls were performed with boiled enzyme. The final results were presented in terms of relative enzyme activity determined between final enzyme activity (after 5 min incubation with buffer solution) and initial enzyme activity (enzyme activity analyzed immediately after the addition of buffer solution).

2.9. Stability in Organic Solvents

A crude extract and immobilized extract was added to a mixture of 50% (v/v in buffer phosphate pH 7.0). The solvents tested were heptane, hexane, acetone, ethyl ether, methanol, ethanol and chloroform, all purchased from VETEC. The reactions occurring in Erlenmeyer flasks of 250 ml on an orbital shaker conditioned for 6 days at 30°C and 100 rpm. Aliquots of 1 ml were removed from each flask for enzyme activity analysis after 0, 2, 4 and 6 days of incubation. Furthermore, the effect of 3% glutaraldehyde solution (v/v) on the enzymatic activity of the lipase was also verified under the same conditions described earlier. The results were expressed as relative enzyme activity, using as reference the initial enzyme activity without addition of solvent, in phosphate buffer [30] .

2.10. Effect of Various Substances on Lipase Stability

The immobilized enzymes were incubated with phosphate buffer pH 7.0, containing 1 mM of different substances: FeSO4, ZnSO4, NaCl, CaCl2, MnCl2, CuSO4 and EDTA. The time of incubation of systems constituted by enzyme and the substances cited above were for 30 min at 30°C [31] .

2.11. Effect of Storage on Lipase Activity

The stability of the crude extract and immobilized extract enzymatic solution was measured by calculating the relative activity after ten days of storage. 0.1 ml free enzyme and 0.07 g immobilized enzyme was stored at 12°C. Enzyme activity analysis was carried out after ten days. The results were presented in terms of relative enzyme activity determined between final enzyme activity, value obtained for enzyme activity after a 10 days incubation at 12°C, and the initial enzyme activity (enzymatic activity analyzed from the moment that the temperature reached 12°C) [32] .

2.12. Lipolytic Activity with Different Substrates

The reaction media constitutes 1 g of substrate (Sunflower, Soybean, Castor, Olive and Chicken viscera oil), 0.3 ml of Tween 80; 4 ml of phosphate buffer (100 mM), pH 7.0, 0.07 g of immobilized extract. Then, the system was stirred in orbital shaker for 30 min at 30°C and 150 rpm. Aliquots of 8 ml were taken from the reaction media and added to a 50 ml Erlenmeyer flask containing 16 ml of acetone:ethanol 1:1. The activity was measured by titration of the free fatty acids with NaOH solution (50 mM) using 3 drops of phenolphthalein 1% as indicator. One unit (U) of enzyme activity was defined as the amount of enzyme which catalyzed the production of 1 μmol of fatty acid per minute under the experimental conditions [31] .

2.13. Reusability of Immobilized Lipase

The immobilized extract was placed in a test-tube to evaluate the enzyme activity using p-nitrophenyl palmitate. After 5 minutes at 30°C, the beads were removed through a plastic sieve, washed with a phosphate buffer (100 mM), pH 7.0 and put back into the reaction media using the same method. This procedure was repeated five times.

2.14. Biodiesel Production

The synthesis of ethylic esters was performed with the use of the crude extract from isolated fungus, Novozym 435, soy oil and chicken viscera oil in 250 ml Erlenmeyers flasks. The chicken viscera oil has the following characteristics: saponification index of 126.42 mg KOH/g, density of 0.91 g/ml and acid index of 2.34%.

Taking into account the average molecular weight of the oils used, it was possible to calculate the mass to be used in reactions to satisfy the molar ratio of 3:1 (alcohol/oil), 1.5 ml of the enzyme solution (Novozym 435 and crude extract) with enzymatic activity of 88.0 U/ml [33] .

The experiments were carried out in an orbital shaker at 30°C, 200 rpm for 24 h. The contents of the Erlenmeyers flasks were added to a separation funnel [34] . A sample of approximately 1 g of the superior phase, that has the ethylic esters were analyzed through an Gas chromatography coupled mass spectrometry (CG-MS) Shimadzu, Model 2010, column-Restek-RTX Model-5MS. The analysis were performed using a heating ramp was started at 100°C, with an increase of 8°C per minute to 300°C, obtaining a 10 minutes isotherm. The injector temperature was 280°C and with a source interface ion 300°C, flow divider (split) 1/50. The injection of 1 ml sample was taken in manual mode and the carrier gas used was nitrogen. Were used the following chromatographic standards: ethyl laurate 99%, myristate acetate 99%, ethyl oleate 98% (ALDRICH), ethyl palmitate 99%, ethyl stearate 99% and ethyl linoleate 99% (SIGMA). The identification of chromatographic peaks was performed by comparing with retention times of standards mentioned. The quantification of peaks was performed by normalizing areas. The concentration of each compound was obtained by the chromatograms, which were used to calculate the conversion reaction according to the chemical equation: 1 mol oil + 3 mols of ethanol → 3 mols of ester + 1 mol of glycerol. To obtain the value of the maximum mass that can be obtained ester was used in Equation (1), where ne is the value related to the maximum number of moles of esters which can be obtained from the reaction (mol/l), no is the number of moles oil (mol/l).

$$ne = 3 * no \quad (1)$$

To obtain the percentage conversion reaction in Equation 2 was used. Where X is the conversion reaction of ethyl esters, C is the mass ethyl ester (g), Me is the maximum mass of esters that can be obtained in the reaction between the oil and the ethanol (g) [35] .

$$X = \frac{C}{Me} \quad (2)$$

3. RESULTS AND DISCUSSION

3.1. Screening and Preparation of the Crude Extract

From the slaughterhouse crude effluent was possible to isolate nineteen filamentous fungi strains which were classified according to the height of calcium laurate precipitated that is related lipolytic activity from fungal isolates. Through this method was possible to verify that, between the nineteen isolates, only one stood out for lipase activity as illustrated in Table 1. This strain was subjected to cultivation in liquid substrates for analysis of enzyme activity. After a10 day period, high values of enzyme activity of 43.5 U/ml was obtained. It was observed that the highest enzyme activity values were obtained using a higher concentration of viscera oil, compared to olive oil in the culture medium. This can be attributed to the fact that viscera oil has higher saturated fatty acids. The viscera oil has 30.3% of saturated fatty acids while olive oil has 20.0%, which indicates that the fungus utilized saturated fatty acids. This same behavior was observed by another author, who studied the production of lipase produced by Geotrichum candidum and found that higher yields were obtained when used to induce the oil that has the greatest percentage of saturated fatty acids [36] .

The numbers 1, 2 and 3 that following the parasentes represent proportions of precipitate formed. Based on the tubes with solid culture medium straight, was possible measured with a caliper the formation of the precipitate. Therefore, the number 1 represents the formation of 1 to 2 cm precipitate. The number 2 represents the formation of 0.7 to 1 cm precipitate and the number 3 represents is the formation of 0 to 0.7 cm.

Table 1. Screening of fungal strains lipase producers. The symbols (+++) (++) (+) and (−) represent very intense formation of precipitate, mean precipitate formation, little formation of precipitate and no precipitate formation, respectively.

Strains tested	Results for lipase production
Monascus purpureus (CCT 3802)	(+) 3
Rhizopus oryzae (ATCC 9092)	(+) 3
Aspergillus niger	(+ + +) 1
Penicillium simplicissimum (CCT 6686)	(+ +) 2
Isolated 1	(−)
Isolated 2	(−)
Isolated 3	(−)
Isolated 4	(−)
Isolated 5	(−)
Isolated 6	(−)
Isolated 7	(+ +) 2
Isolated 8	(+) 3
Isolated 9	(−)
Isolated 10	(−)
Isolated 11	(−)
Isolated 12	(−)
Isolated 13	(−)
Isolated 14	(−)
Isolated 15	(−)
Isolated 16	(−)
Isolated 17	(+ +) 2
Isolated 18	(−)
Isolated 19	(−)

3.2. Identification of Fungi

Fragments of the region ITS from sample was amplified with success. The DNA sequence obtained was analyzed with BLAST from GenBank and CBS (Centraalbureau voor Schimmecultures, Fungal Biodiversity Centre). The phylogenetic tree was built from sample and is illustrated in Figure 1. The sequence had 100% identity to Trametes hirsuta. Furthermore, a group formed was supported by a bootstrap of 76% with the sequence of the species mentioned in Figure 1. Thus, the sample was identified as Trametes hirsuta, which is a basidiomycete white-rot fungus (WRF) that promotes wood decay. The term "white-rot" refers to the wood material after selective degradation of lignin, which leaves the white cellulose exposed. This fungus is traditionally known to produce enzymes that degrade lignin, also known as lacases [36] - [38] . Although this fungus has not been associated with lipase production, it is known that fungi in general, can secrete certain enzymes when stimulated. The fact that the strain was grown in a medium containing high concentrations of fatty acids must have induced lipase production.

3.3. Enzyme Immobilization

In this step of the process, the lipase contained in the concentrated crude extract obtained by submerged culture of Trametes hirsuta was immobilized in the chitosan/clay solid support. The support beads were activated using glutaraldehyde and immobilization was performed by direct contact of the enzyme solution with the activated chitosan/clay beads [39] . The percentage of glutaraldehyde solution used to activate the beads is very important as it promotes modification in the spheres geometrical parameters, and is also associated with structural modifications in the enzyme after its activation. The glutaraldehyde molecules bond to the chitosan amine groups, creating an imine group (CH=N) and, through these groups, the enzyme bonds covalently to the support. However, excess glutaraldehyde molecules on the support's surface may increase the numbers of bonds with the enzyme molecule, leading to a stiffening of the enzyme's tridimensional structure and further loss of biological activity. It is also important to point out that activation of the chitosan/clay beads with glutaraldehyde was carried out in a media with alkaline pH to favor the imine groups' formation [40] . The formation of such bonds occurs preferentially in alkaline pH, for, in this pH, the amine groups are deprotonated. For the immobilization process, besides the beads activation, their contact with the enzyme solution, enzyme activity analysis and total proteins, both the beads and the supernatant were analyzed to verify if there were any modifications and if, in fact, the crude extract lipase was immobilized. The results revealed that, after 5 hours, 77.4% of coupled proteins and an immobilization yield of 80.9% were obtained.

3.4. Effects of Temperature on Stability

The thermal stability of enzymes is one of the important criteria for long-term and commercial application. The activity of immobilized enzyme is known to be more resistant against heat than the enzyme in its free state [41] . According to the results obtained as shown in Figure 2, the enzyme in the immobilized extract acquired a higher thermal stability when compared with the enzyme without a support. This difference was observed at temperature of 64°C in which the immobilized enzyme was 20% more active than the unsupported enzyme at the same

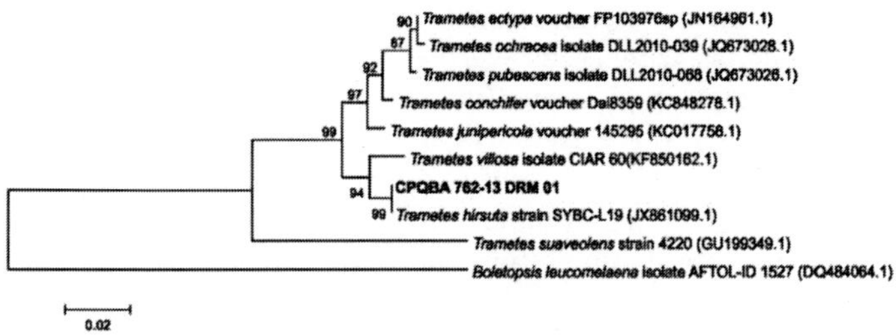

Figure 1. Phylogenetic tree illustrating relationships between the partial sequences of ribosomal genes from isolated fungi and sequences from strains contained in data base GenBank and CBS.

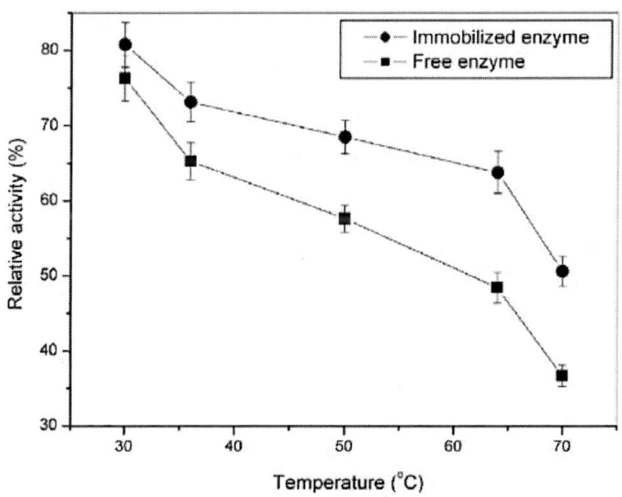

Figure 2. Influence of temperature on lipase activity. Activity was determined at pH 7.0 using p-NNP as substrate. The immobilized enzyme and free enzyme were incubated for 60 min at the indicated temperatures. Relative activity was assayed at 30°C as described above. Relative activity is expressed as percentage of the ratio of the activity after the incubation period and activity before the incubation period determined at 30°C under standard conditions.

temperature. At 70°C the unsupported enzyme activity was 35% whereas the immobilized enzyme activity was 50%. Similar behavior of thermal stability was observed in a study that illustrated the biochemical characteristics of Rhizopus oryzae lipase before and after immobilization process [5] . It was also observed when pancreatic lipase was immobilized on the chitosan beads. In this case, the authors observed that the immobilized lipase acquired a much higher thermostability compared to the free enzyme, as when they were incubated for two hours at 50°C the immobilized enzyme had a relative activity of 80%, while with the free enzyme there was no activity [42] . This behavior is attributed to interactions between enzyme and support, which often results in a more rigid structure of the protein molecule, giving the globular conformation of the active biocatalyst higher stability to temperature [41] .

3.5. Effect of pH on Activity

Results obtained as shown in Figure 3, revealed that the immobilization process did not cause modifications in the enzymes optimum pH, since both enzyme forms presented relatively higher enzyme activity values at pH 7.0. The different enzyme forms exhibited distinctive values of enzyme activity from pH 4.0 to 9.0; however, in the pH zone of 3.0 and 10.0, the enzyme activity values of the two forms were much alike. The fact that pH 7.0 promotes the maximum observed relative activity corroborates with many works quoted in literature that illustrate lipases to present their maximum values of enzyme activity at similar pHs [12] [29] [42] .

3.6. Stability in Organic Solvents

The stability in organic solvents is an important lipase characteristic and might determine the form in which the enzyme will be utilized, for example, as organic synthesis reaction catalyst. From the results illustrated in Figure 4(a), it was possible to observe that the solvents which led to higher decay of the enzyme activity were ethyl ether, ethanol and acetone. In fact, for these solvents, the enzyme immobilization meant no increase in its life span. A similar behavior was observed in other study, where the authors analyzed the enzyme activity from Rhizopus oryzae lipase in solvents with different polarities and noticed that Acetone led to a higher denaturation rate. This behavior occur due to the lipases low compatibility with hydrophilic solvents, though presenting high compatibility with hydrophobic solvents, since the hydrophobic solvents retain the enzymatic activity by not removing the water molecules of salvation from the enzyme's surface [43] .

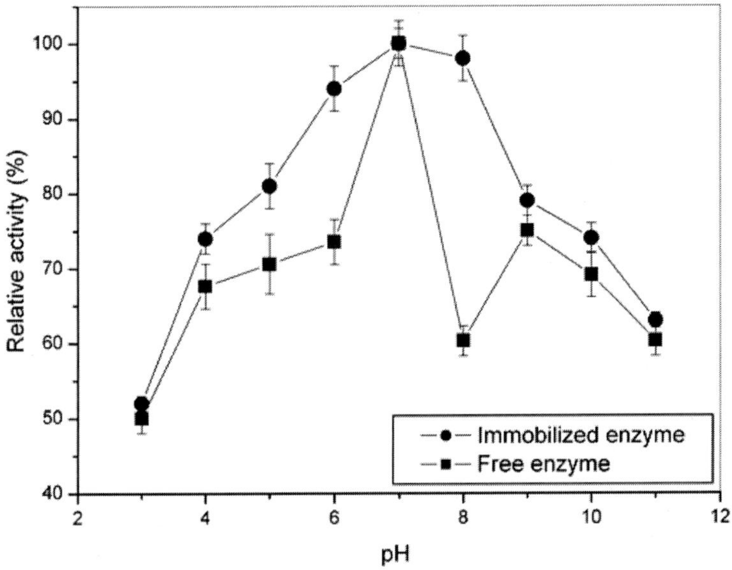

Figure 3. Effect of pH on the activity of the immobilized enzyme and free enzyme. The activity was assayed at various pH values using p-NNP as substrate. The following buffers were used: tartrate (pH 3.0 - 5.0), phosphate (pH 6.0 - 8.0), Tris-HCl (pH 9.0 - 11.0). The final results were presented in terms of relative enzyme activity as described above.

By comparative analysis between hexane and heptane, which have similar values of log P, it was possible to verify that hexane gave more stability to the relative enzyme activity values, even though the initial enzyme activity values with heptane as solvent were higher than with hexane. The initial enzyme activity with heptane was 98% while hexane showed 91% enzyme activity, as illustrated in Figure 4(b). This behavior can be explained by the fact that hydrophobic solvents do not lower the activity and can also promote enzymatic activation, since they do not remove water molecules from the enzyme's surface and can interact with the enzyme maintaining the hydrophobic lid open [44] . On the other hand, this argument cannot be applied to all situations, since, in some enzymes, the lid is not necessarily related to the interfacial activation [45] .

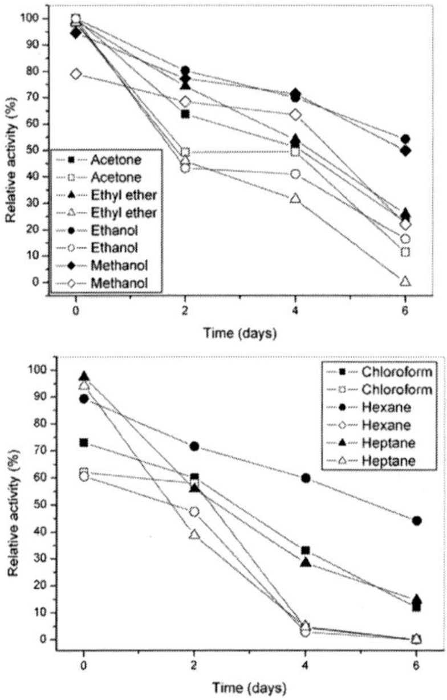

(a)

(b)

Figure 4. Effect of different solvents: (a) chloroform, hexane, heptane and (b) acetone, ethyl ether, ethanol, methanol on the activity of the immobilized enzyme and free enzyme. The solvents were tested in mixture of 50% (v/v in buffer phosphate pH 7.0) at 25°C. The enzyme activity was assayed after 0, 2, 4 and 6 days of incubation with solvents. The results were expressed as relative enzyme activity, using as reference the initial enzyme activity without addition of solvent, in phosphate buffer. The open symbols represent the free enzyme and the closed symbols represent the immobilized enzyme.

3.7. Effect of Various Substances on Lipase Stability

Salts bring about several effects on proteins properties, including activity, conformational stability and solubility. Such effects possibly appear due to ionic bonds to specific protein sites, changing its hydration level [46] . The results obtained for the effect of different ions on the stability of the enzyme obtained in this work are shown in Figure 5. In this work it was possible to verify the effect of enzyme activation when it is exposed to Fe^{2+}, Zn^{2+}, Ca^{2+} and Cu^{2+}. This effect was observed by other authors that found high values for enzyme activity after insertion of salts [42] [46] . These features can be associated with structural changes of the enzyme and to the fact that the salts increase the dispersion of enzyme molecules, facilitating mass transfer during the reaction. Other authors [42] observed increase in enzyme activity with the same salts used in this work. However, when there was an increase in the concentration of salts it was observed that there was a decrease in enzyme activity due to the fact that the salt covered the enzyme surface and hindered their contact with the substrate [46] . Furthermore, the EDTA showed an enzymatic inhibitory behavior indicating the importance of the presence of some of those ions for the enzyme activity. This inhibitory effect of the EDTA was also observed in the work by [47] on characterization of lipases from Burkholderia multivorans [48] .

3.8. Effect of Storage on Lipase Activity

It was observed that the dried immobilized enzyme was the one that exhibited the highest relative enzyme activity of 123% when compared to the same enzyme in its free form stored at 12°C, which showed activity of 77%. The results corroborate with the work of [41] that illustrates the study of stability in storage of lipase in its free form and immobilized in chitosan beads, and shows that the lipase in its free form lost all its activity after 2 days of storage at 30°C. On the other hand, when the same lipase was immobilized in chitosan beads and stored at the same temperature, the complete loss of its activity occurred 4 days after the storage and, when stored without water, the loss occurred only after 8 days of storage [42] . These data showed that the immobilized enzyme stored without any liquids was more stable than the immobilized enzyme stored with water, which indicates that, in aqueous media, the enzyme liberates its support and then denatures.

3.9. Lipolytic Activity with Different Substrates

From the data presented in Table 2 it was observed that higher enzyme activities were obtained using, firstly, soybean oil and, secondly, sunflower oil as substrates. The behavior observed can be attributed to the fact that the linoleic acid chain, abundant fatty acids in the soy and sunflower oil, was very adequate for the enzyme active site size and shape. Different substrates bind to the active site of the enzyme, releasing different amounts of energy. This energy is required to promote the necessary changes on the active site and make the most efficient catalyst. In general, substrates with small carbon chains have ability to release small amount of energy to modify the conformation of the lipase. However, substrates with long carbon chains release higher quantities of energy and promote conformational changes more desirable for the reaction [4] .

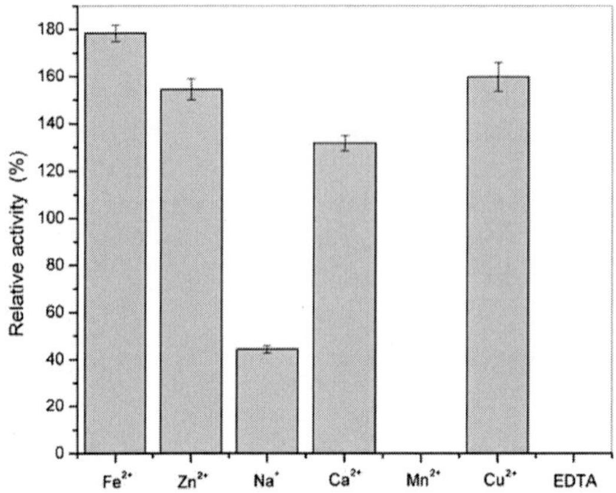

Figure 5. Effects of different substances on the activity of the immobilized enzyme which was incubated with phosphate buffer pH 7.0, containing 1 mM of different substances at 30°C for 30 min. Assays were performed in triplicate and mean values were reported.

Table 2. Effect of different substrates on the hydrolysis activity of the immobilized enzyme. Assays were performed in triplicate and mean values were reported.

Substrates	Hydrolysis activity (U/ml)
Soybean oil	361 ± 4.2
Castor oil	89 ± 2.8
Olive oil	129 ± 3.4
Sunflower oil	158 ± 2.7
Viscera oil	137 ± 1.9

One gramme (1 g) of substrates was incubated with 0.3 ml of Tween 80, 4 ml phosphate buffer pH 7.0 and 0.07 g of immobilized enzyme, at 30°C, agitation of 150 rpm for 30 min. Samples of 8 ml were taken and added in a Erlenmeyer flask containing 16 ml of acetone: ethanol 1:1. The activity was measured by titration of the free fatty acids with NaOH solution (50 mM) using 3 drops of phenolphthalein 1% as indicator. One unit (U) of enzyme activity was defined as the amount of enzyme which catalyzed the production of 1 μmol of fatty acid per minute under the experimental condition. Experiments were performed in triplicate and mean values are reported.

3.10. Reusability of Immobilized Enzyme

It was shown that the relative enzyme activity of the immobilized enzyme decayed over its use for hydrolysis reaction using p-NNP as substrate. This behavior was more expressive after the fourth cycle. After five cycles, the relative enzyme activity was 55%. This behavior indicates that the immobilization process resulted in an increased stability for the enzyme. The phenomenon observed can be explained by the fact that there is a possible release of enzyme molecules which eventually become physically bound to the support [42] .

3.11. Biodiesel Production

This experiment was performed to verify the potential application of the Trametes hirsuta lipase in the production of ethylic esters under specific conditions. The results were compared with a biocatalyst frequently used in several published studies: Novozym 435 [48] -[51] . Chicken viscera oil was the substrate used in this experiment. To perform comparative study widely used starting material, soybean oil was used in the reaction. Ethylic alcohol was the chosen solvent. The experimental conditions for biodiesel production were temperature of 30°C for 24 hours. This temperature was chosen after the characterization of the immobilized enzyme had shown that the enzyme showed higher activity at 30°C. Although many studies have used different experimental conditions for enzyme transterification reaction, generally, the reaction time did not exceed 24 h. In the enzyme transterification, the longer the reaction is carried out, the bigger the conversion rate will be, however a period of 24 h was used for this study. The results obtained revealed that the experiments using soybean oil with the commercial lipase and with the immobilized enzyme led to the higher values of ester conversion: 90% and 54%, respectively. The least values were obtained with systems consisting of chicken viscera oil and commercial lipase and chicken viscera oil and immobilized enzyme, which led to activities of 35% and 28%, respectively. The behavior observed might be related to the fact that the chicken viscera oil presents different characteristics from the soybean oil, such as the presence of phospholipids, salts, metallic ions and other substances that may affect the final yield. Nevertheless, it is important to highlight that lipase from Trametes hirsuta showed potential in the production of ethylic esters. The optimization of variables that affect directly the transterification reaction such as, water content, molar ratio alcohol-oil, temperature, enzyme concentration and the use of co-solvents will be the future steps in order to obtain better results for these systems.

4. CONCLUSION

It is possible to isolate a filamentous fungus from chicken slaughterhouse effluent, which is capable of producing copious amounts of lipase. This fungus was identified as Trametes hirsuta. Using submerged culture of the isolate, enzyme activity of 43.51 U/mL was achieved. The crude extract was immobilized in chitosan/clay beads. After 5 h incubation of the concentrated crude extract with the chitosan/clay beads activated with 3% (v/v) of glutaraldehyde, an immobilization yield of 80% was obtained. The crude extract enzyme characterization was performed both in the free and immobilized form. The results showed that the immobilization process allowed the increase of enzyme life span in different temperatures, pH(s), solvents, metallic ions and storage. The experiments performed to obtain biodiesel and revealed that reasonable values for ester conversion were obtained, thus indicating that the wild lipase and the viscera oil were highly promising substrates for biodiesel production.

ACKNOWLEDGEMENTS

The authors thank the academics Joel Pires Viana, Bruna Sampaio Slussarek and Nathália Correa Lima for their support in the experimental activities. CNPq is also gratified for financial support for this research, and CAPEs agency, for Américo Cruz pre doctoral fellowship.

REFERENCES

1. Marchetti, J.M., Miguel, V.U. and Errazu, A.F. (2007) Possible Methods for Biodiesel Production. Renewable and Sustainable Energy Reviews, 11, 1300-1311.
2. Illanes, A. (2008) Enzyme Biocatalysis: Principles and Applications. Springer, Netherlands.
3. Villeneuve, P., Muderhwa, M., Graille, J. and Haas, M. (2000) Customizing Lipases for Biocatalysis: A Survey of Chemical, Physical and Molecular Biological Approaches. Journal of Molecular Catalysis B: Enzymatic, 9, 113-148.
4. Tan, T., Lu, J., Nie, K., Deng, L. and Wang, F. (2010) Biodiesel Production with Immobilized Lipase: A Review. Biotechnology Advances, 28, 628-634.

5. Faber, K. (1997) Biotransformations in Organic Chemistry. Springer-Verlag, New York.
6. Kharrat, N., Ali, B.A., Marzouk, S., Gargouri, Y. and Chaabouni, M.K. (2011) Immobilization of Rhizopus oryzae Lipase on Silica Aerogels by Adsorption: Comparison with the Free Enzyme. Process Biochemistry, 46, 1083-1089.
7. Minovska, V., Winkelhausen, E. and Kuzmanova, S. (2005) Lipase Immobilized by Different Techniques on Various Support Materials Applied in Oil Hydrolysis. Journal of the Serbian Chemical Society, 70, 609-624.
8. Bayramoglu, G., Kaçar, Y., Denizli, A. and Arica, M.Y. (2002) Covalent Immobilization of Lipase onto Hydrophobic Group Incorporated Poly(2-Hydroxyethyl Methacrylate) Based Hydrophilic Membrane Matrix. Journal of Food Engineering, 52, 367-374.
9. Dave, R. and Madamwar, D. (2006) Esterification in Organic Solvents by Lipase Immobilized in Polymer of PVA-Alginate-Boric Acid. Process Biochemistry, 41, 951-955.
10. Yong, Y., Bai, Y., Li, Y., Lin, L., Cui, Y. and Xia, C. (2008) Characterization of Candida rugosa Lipase Immobilized onto Magnetic Microspheres with Hydrophilicity. Process Biochemistry, 43, 1179-1185.
11. Cruz, J.C., Pfromm, P.H. and Rezac, M.E. (2009) Immobilization of Candida antarctica Lipase B on Fumed Silica. Process Biochemistry, 44, 62-69.
12. Tzialla, A.A., Pavlidis, I.V., Felicissimo, M.P., Rudolf, P., Gournis, D. and Stamatis, H. (2010) Lipase Immobilization on Smectite Nanoclays: Characterization and Application to the Epoxidation of Alfa-Pinene. Bioresource Technology, 101, 1587-1594.
13. Ting, W.J., Tung, K.Y., Giridhar, R. and Wu, W.T. (2006) Application of Binary Immobilized Candida rugosa Lipase for Hydrolysis of Soybean Oil. Journal of Molecular Catalysis B: Enzymatic, 42, 32-38.
14. Foresti, M.L. and Ferreira, M.L. (2007) Chitosan-Immobilized Lipases for the Catalysis of Fatty Acid Esterifications. Enzyme and Microbial Technology, 40, 769-777.
15. 15. Tharanathan, R.N. and Kittur, F.S. (2003) Chitin—The Undisputed Biomolecule of Great Potencial. Critical Reviews in Food Science and Nutrition, 43, 61-87.

16. Chang, M. and Juang, R. (2004) Stability and Catalytic Kinetics of Acid Phosphatase Immobilized on Composite Beads of Chitosan and Activated Clay. Process Biochemistry, 39, 1087-1091.
17. Bormans, P. (2004) Ceramics Are More than Clay Alone. Cambridge International Science Publishing, Cambridge.
18. 18. Freire, D.M., Teles, E.M., Bon, E.P. and Sant'anna Jr., G.L. (1997) Lipase Production by Penicillium restrictum in a Bench-Scale Fermentor: Effect of Carbon and Nitrogen, Agitation and Aeration. Applied Biochemistry and Biotechnology, 63, 409-421.
19. Shelley, A.W., Deeth, H.C. and Macrae, I.C. (1987) Review of Methods of Enumeration, Detection and Isolation of Lipolytic Microorganisms with Special Reference to Dairy. Journal of Microbiological Methods, 6, 127-137.
20. Raeder, J. and Broda, P. (1985) Rapid Preparation of DNA from Filamentous Fungi. Letters in Applied Microbiology, 1, 17-20.
21. Thompson, J.D., Higgins, D.G., Gibson, T.J. and Clustal, W. (1994) Improving the Sensitivity of Progressive Multiple Alignment through Sequence Weighting, Positions-Specific Gap Penalties and Weight Matrix Choice. Nucleic Acids Research, 22, 4673-4680.
22. Tamura, K., Dudley, J., Nei, M. and Kumar, S. (2007) MEGA4: Molecular Evolutionary Genetics Analysis (MEGA) Software Version 4.0. Molecular Biology and Evolution, 24, 1549-1599.
23. Kimura, M. (1980) A Simple Model for Estimating Evolutionary Rates of Base Substitutions through Comparative Studies of Nucleotide Sequence. Journal of Molecular Evolution, 16, 111-120.
24. Saitou, N. and Nei, M. (1987) The Neighbor-Joining Method for Reconstructing Phylogenetic Trees. Molecular Biology and Evolution, 4, 406-425.
25. Zeraik, A.E., Souza, F.S., Fatibello-Filho, O. and Leite, O.D. (2008) Desenvolvimento de um Spot Test para Monitora- mento da Atividade da Peroxidase em um Procedimento de Purificação. Química Nova, 31, 731-734.
26. Chiou, S. and Wu, W. (2004) Immobilization of Candida rugosa Lipase on Chitosan with Activation of the Hydroxyl Groups. Biomaterials, 25, 197-204.
27. Bradford, M.M. (1976) A Rapid and Sensitive Method for the Quantification of Microgram Quantities of Protein Dye Binding. Analytical Biochemistry, 72, 248-254.

28. Hung, T., Giridhar, R., Chiou, S. and Wu, W. (2003) Binary Immobilization of Candida rugosa Lipase on Chitosan. Journal of Molecular Catalysis B: Enzymatic, 26, 69-78.
29. Abbas, H., Hiol, A., Deyris, V. and Comeau, L.C. (2002) Isolation and Characterization of an Extracellular Lipase from Mucor sp. Strain Isolated from Palm Fruit. Enzyme and Microbial Technology, 31, 968-975.
30. Kamini, N.R., Fujii, T., Kurosu, T. and Iefuji, H. (2000) Production, Purification and Characterization of an Extracellular Lipase from the Yeast, Cryptococcus sp. S-2. Process Biochemistry, 36, 317-324.
31. Pastore, G.M., Costa, V.S.R. and Koblitz, M.G.B. (2003) Production, Partial Purification and Biochemical Characterization of a Novell Rhizopus sp. Strain Lipase. Food Science and Technology, 23, 135-140.
32. Wang, D., Xu, Y. and Shan, T. (2008) Effects of Oils-Related Substrates on the Synthetic Activity of Membrane- Bound Lipase from Rhizopus chinensis and Optimization of the Lipase Fermentation Media. Biochemical Engineering Journal, 41, 30-37.
33. Lopes, C.R., D'oca, C.D.R.M., Duarte, R.C., Kurz, M.H.S., Primel, E.G., Clementin, R.M., Villarreyes, J.A.M. and D'oca, M.G.M. (2010) Síntese de novas amidas graxas a partir da aminólise de ésteres metílicos. Química Nova, 33, 1335-1341.
34. Tashtoush, G.M., Al-Widyan, M.I. and Al-Jarrah, M.M. (2004) Experimental Study on Evaluation and Optimization of Conversion of Waste Animal Fat into Biodiesel. Energy Conversion and Management, 45, 2697-2711.
35. Faccio, C., Rosa, C.D., Amroginski, C., Bender, J.P., Lipke, N., Oliveira, D., Oliveira, J.V., Luccio, M. and Dariva, C. (2005) Optimization of Alkaline Transesterification of Soybean Oil and Castor Oil for Biodiesel Production. Applied Biochemistry and Biotechnology, 122, 553-560.
36. Burkert, J.F.M., Maugeri, F. and Rodrigues, M.I. (2004) Optimization of Extracellular Lipase Production by Geotrichum sp Using Factorial Desing. Bioresource Technology, 91, 77-84.
37. Chaurasia, P.K., Yadav, R.S.S. and Yadava, S. (2014) Purification and Characterization of Yellow Laccase from Trametes hirsuta MTCC-1171 and Its Application in Synthesis of Aromatic Aldehydes. Process Biochemistry, 49, 1647- 1655.

38. Couto, S.R. (2007) Decolouration of Industrial Azo Dyes by Crude Laccase from Trametes hirsuta. Journal of Hazardous Materials, 148, 768-770.
39. Jiang, D., Long, S., Huang, J., Xiao, H. and Zhou, J. (2005) Immobilization of Pycnoporus sanguineus Laccase on Magnetic Chitosan Microspheres. Biochemical Engineering Journal, 25, 15-23.
40. Morrison, R. and Boyd, R. (1996) Química Organica. Fundação Calouste Gulbenkian, Lisboa.
41. Guisan, J.M. (2013) Immobilization of Enzymes and Cells. Third Edition, Methods in Molecular Biology, Vol. 1051, Humana Press, Totowa. http://link.springer.com/book/10.1007%2F978-1-62703-550-7
42. Desai, P.D., Dave, A.M. and Devi, S. (2006) Alcoholysis of Salicornia Oil Using Free and Covalently Bound Lipase onto Chitosan Beads. Food Chemistry, 95, 193-199.
43. Essamri, M., Deyris, V. and Comeau, L. (1998) Optimization of Lipase Production by Rhizopus oryzae and Study on Stability of Lipase Activity in Organic Solvents. Journal of Biotechnology, 60, 97-103.
44. Klibanov, A.M. (1997) Why Are Enzymes Less Active in Organic Solvents than in Water? Trends in Biotechnology, 15, 97-101.
45. Verger, R. (1997) Interfacial Activation of Lipases: Facts and Artifacts. Trends in Biotechnology, 15, 32-38.
46. Yu, H.W., Chen, H., Yang, Y.Y. and Ching, C.B. (2005) Effect of Salts on Activity, Stability and Enantioselectivity of Candida rugosa Lipase in Isooctane. Journal of Molecular Catalysis B: Enzymatic, 35, 28-32.
47. Demir, B.S. and Tukel, S.S. (2010) Purification and Characterization of Lipase from Spirulina platensis. Journal of Molecular Catalysis B: Enzymatic, 64, 123-128.
48. Dandavate, V., Jinjala, J., Keharia, H. and Madamwar, D. (2009) Production, Partial Purification and Characterization of Organic Solvent Tolerant Lipase from Burkholderia multivorans V2 and Its Application for Ester Synthesis. Bioresource Technology, 100, 3374-3381.
49. Mittelbach, M. (1990) Lipase Catalysed Alcoholysis of Sunflower Oil. Journal of the American Chemical Society, 67, 168-169.

50. Rodrigues, R.C., Volpato, G., Wada, K. and Ayub, M. (2008) Enzymatic Synthesis of Biodiesel from Transesterification Reactions of Vegetable Oils and Short Chain Alcohols. Journal of the American Oil Chemists' Society, 85, 925- 930.
51. Maceiras, R., Vega, M., Costa, C., Ramos, P. and Márquez, M.C. (2009) Effect of Methanol Content on Enzymatic Production of Biodiesel from Waste Frying Oil. Fuel, 88, 2130-2134.

4

CHAPTER

A STUDY ON ETHANOLYSIS AND METHANOLYSIS OF COCONUT OIL FOR ENZYMATICALLY CATALYZED PRODUCTION OF BIODIESEL

Livia M. O. Ribeiro, Albanise E. Silva, Margarete C. S. Silva, Renata M. R. G. Almeida*

Tecnology Center, Federal University of Alagoas, Avenida Lourival Melo Mota, Tabuleiro do Martins, Maceió, Brazil

ABSTRACT

Biodiesel production by enzymatic catalysis has been the subject of much research for developing processes that can potentially compete with other types of catalysis. The objective of this study was to investigate the variables that influenced the transesterification of coconut oil catalyzed by immobilized lipase for biodiesel production. A full 2^4 factorial design with the variables temperature (40°C - 60°C), enzyme concentration (3% - 7%), oil-ethanol ratio (1:6 - 1:10) and alcohol type (methanol-ethanol) was performed. The best conversion result (80.5%) was obtained using ethanol with a higher temperature, molar ratio and enzyme concentration. The obtained yields showed that the results attained with ethanol were more significant when compared with methanol.

Keywords: Biodiesel, Methanol, Ethanol, Lipases

1. INTRODUCTION

The demand for renewable fuels has recently been much intensified. In this scenario, biodiesel arises as an alternative to petroleum products, aiming to reduce pollutant emissions in the atmosphere.

Biodiesel is the name of a clean-burning fuel produced from domestic renewable resources. It does not contain petroleum, but can be added to it forming a blend [1] . It can be used in a compression-ignition engine (diesel) without need for modification. Biodiesel is simple to use, biodegradable, non-toxic and essentially free of sulfur and aromatic compounds [2] .

To produce biodiesel, vegetable oils are mixed with short-chain alcohols (ethanol or methanol) and then stimulated by a catalyst. This chemical process is known as transesterification, where glycerin is separated from the fat or oil. The process generates two products: esters (the chemical name for biodiesel) and glycerin (a valuable product in the soap industry). The oil is separated from the glycerin by filtration process [3] .

High-quality biodiesel must be produced following strict industry specifications, as the ASTM D6751 at international level. In USA, biodiesel is the only alternative fuel to obtain full approval in the Clean Air Act of 1990, which is authorized for sale and distribution by the USA Environmental Agency (EPA) [4] .

A very appropriate nomenclature has been used internationally to identify the concentration of biodiesel in the blend, defined as BX, wherein X refers to the percentage by volume of biodiesel. Therefore, B2, B5 and B20 refer to fuels at concentrations of 2%, 5% and 20% of added biodiesel, respectively [5] .

Although soybean oil is the most widely used in biodiesel production, other oilseed plants with a higher oil content, such as peanut, sunflower, corn, canola (rapeseed), castor oil and cotton have been reported in the literature as favorable for the production of biodiesel [6] - [8] . In this context, coconut oil has been identified as an important potential source of raw material for the local production of biodiesel in the northeast of Brazil [9] .

In the transesterification of vegetable oil, a triglyceride reacts with an alcohol in the presence of a catalyst, which may be an acid, a strong base or an enzyme, producing a mixture of alkyl esters of fatty acids and glycerol. The total process is a sequence of three consecutive and reversible reactions, where diglycerides and monoglycerides are formed as intermediates. For a complete stoichiometric transesterification, a 3:1

molar ratio of alcohol to triglyceride is required. Due to the reaction's reversible character, a transesterification agent (alcohol) is usually added in excess, thus contributing to increasing the ester yield, and to enabling its separation from the glycerol formed [10] .

Currently, the commercial production of biodiesel occurs via a chemical route, but the enzymatic process has attracted the interest of the scientific community. Although the biodiesel production through enzymatic transesterification has not yet been commercially developed, advances in this process have been reported in papers and patents. The common aspect of these studies is the optimization of reaction conditions (solvent, temperature, pH, type of microorganism that produces the enzyme, immobilization technologies, etc.), in order to establish the characteristics for industrial applications [11] .

However, once the enzymatic process is optimized, it may present some advantages over the chemical process (Table 1).

Methanol is used as the acyl group receptor on biodiesel production by transesterification of oils or fats using lipases, according to most recent studies found in the literature. Methanol is more reactive, and produces more volatile esters and it is cheaper on the international market, when compared with other short chain alcohols [12] .

However, some authors are considering other alcohols, such as ethanol, propanol, isopropanol and butanol for the production of biodiesel component esters, due to methanol's toxicity and by the fact that it is produced from petroleum. Ethanol is preferred because it is considered renewable [13] .

Table 1. Advantages and disadvantages of the chemical and enzymatic processes in biodiesel production.

Process	Advantages	Disadvantages
Chemical	Simplicity; High yield; Short reaction time.	Difficulty of separating the catalyst; Impossibility of reusing the catalyst; Difficulty of use hydrated ethanol; Obtaining products with lower purity.
Enzymatic	Ease of catalyst (support) separation; Obtaining products with higher purity; Possibility of using hydrated ethanol in the reaction.	Long reaction time; Enzymes cost; Low yield.

The objective of this work was to study the variables that influenced the transesterification of coconut oil in the production of biodiesel catalyzed by immobilized lipase. The transesterification was performed in sealed glass reactors stirred at 200 rpm and catalyzed by the commercial immobilized lipase Novozym 435 using ethanol and methanol. Assays were carried out in accordance with an experimental design where the variables were temperature (40°C - 60°C), enzyme concentration (3% - 7%), ratio oil-alcohol (1:6 - 1:10) and type of alcohol (ethanol and methanol).

2. MATERIALS AND METHODS

2.1. Materials

The samples of crudecoconutoil were provided by SOCOCO Coconut Food Industry, located in Maceió, Brazil. According to the supplier, the extraction of coconut oil was made by mechanical pressing of the dehydrated pulp.

Enzymes used in this study were provided by Novozymes, Latin America (Brazil). Other reagents used were of analytical grade.

2.2. Characterization of the Transesterification Products

To determine the biodiesel yield (%), the characterization of methyl or ethyl esters was carried out using gas chromatography (GC). Transesterification products were analyzed by gas chromatography using a VARIANCP- 3800 instrument equipped with a FID detector (Flame Ionization Detection) and a short capillary column (2.3 m). The temperature of the detector was 250°C and the injector was 240°C. The oven had temperature programmed from 150 to 260°C at a heating rate (ramp) of 10°C/min. Tricaprylin was used as the internal standard and hydrogen of high purity (99.95%) was used as the carrier gas.

The analyzed sample was prepared by mixing 0.15 mL of biodiesel previously purified with 1 mL of standard solution (tricaprylin plus hexane in desiccator). A 1 μL aliquot of the sample was then injected into the chromatograph, with a 10 mL glass syringe.

The yield calculation in esters was carried out based on the masses and areas under the peaks corresponding to the methyl or ethyl esters and to the internal standard, using Equation (1):

$$\text{Yield}(\%) = \frac{m_p A_b f}{m_b A_p} \times 100 \quad (1)$$

where:

m_p = internal standard weight (0.08 g);

A_b = sum of the peak are as referring to the esters in the sample (peak detected between 8 and 13 min);

f = response factor (0.78);

A_p = peak area referring to the internal standard (tricaprylin plus hexane-peak detected between 15 and 18 min);

m_b = sample weight (0.15 g).

The conversion analyses were performed in duplicate. The average conversion for each experiment was also calculated.

3. EXPERIMENTAL STAGE

3.1. Enzymatic Transesterification

The reactions, in laboratory scale, were carried out in 250 mL Erlenmeyer flasks closed with glass lids. The commercial lipase in pre-established concentrations was added to the oil-alcohol mixture. The flasks were then incubated in a rotating chamber at 200 rpm at a controlled temperature. The amount of coconut oil used was held at 25 g.

After the elapsed reaction time, the sample was filtered (in order to retain the enzymes for subsequent recovery and use) and then taken to a rotary evaporator at 90°C at 40 rpm, to evaporate the alcohol.

After this procedure, the resulting reaction mixture was placed in a separatory funnel, where the upper phase, concentrated in methyl or ethyl esters (biodiesel), was separated from the lower phase containing the remaining reaction products (mono and diglycerides, glycerol and impurities), as well as the unreacted substrates (alcohol and oil).

3.2. Experimental Design

The experimental design is a very useful tool for laboratory research. In addition to allowing research with several variables, it also enables an easy determination of the variation effects, since determining the influence of one or more variables over the others is a common problem.

A full factorial experimental design with two levels and four variables was performed to determine the experimental conditions that maximize the synthesis of biodiesel resulting from the reaction and to evaluate the influence of selected variables.

The variables studied at this stage were: oil-alcohol molar ratio, temperature, enzyme percentage and alcohol type. Based on preliminary studies, the agitation was kept constant, since it has been verified that no change results when increasing it from 200 to 500 rpm [14] . The reaction time was set at 24 hours. The limitations associated with each variable are shown in Table 2.

The experimental matrix for the 2^4 factorial design is presented in Table 3.

Through the obtained experimental data, an empirical model was developed with the variables coded by the letter x. This model was used to estimate the yield of coconut oil biodiesel by using coded values of the variables here studied. These coded values must belong to the interval $-1 \leq x \leq +1$, which was considered in this work.

Therefore, the following mathematical model, Equation (2), for the prediction of desired responses was adopted:

$$y = \beta_0 + \sum_{j=1}^{x} \beta_j x_j + \sum_{i<j}^{x} \beta_{ij} x_j x_i \qquad (2)$$

where:

y = percentage of the variable of interest (yield);

x = variable that shows the factors in coded scales. As in the present work only the low and high levels are being considered, x assumes −1 and +1 values;

β = parameters which values will be determined.

For practical purposes, the β_0 estimate is the general average of the responses, while the estimates of the other parameters were obtained from the regression table given by the STATISTICA7.0.

Table 2. Range of variables.

Variables	Extreme Negative (−1)	Extreme Positive (+1)
Molar ratio—oil:alcohol [R]	1:6	1:10
Temperature [T] (°C)	40	60
Enzyme percentage [E] (%p/p)	3	7
Alcohol type	Ethanol	Methanol

Table 3. Experimental design matrix.

Assay	Type	R (O:Alcohol)	T (°C)	E (%p/p)
1	Ethanol/−1	1:6/−1	40/−1	3/−1
2	Ethanol/−1	1:6/−1	40/−1	7/+1
3	Ethanol/−1	1:6/−1	60/+1	3/−1
4	Ethanol/−1	1:6/−1	60/+1	7/+1
5	Ethanol/−1	1:10/+1	40/−1	3/−1
6	Ethanol/−1	1:10/+1	40/−1	7/+1
7	Ethanol/−1	1:10/+1	60/+1	3/−1
8	Ethanol/−1	1:10/+1	60/+1	7/+1
9	Methanol/+1	1:6/−1	40/−1	3/−1
10	Methanol/+1	1:6/−1	40/−1	7/+1
11	Methanol/+1	1:6/−1	60/+1	3/−1
12	Methanol/+1	1:6/−1	60/+1	7/+1
13	Methanol/+1	1:10/+1	40/−1	3/−1
14	Methanol/+1	1:10/+1	40/−1	7/+1
15	Methanol/+1	1:10/+1	60/+1	3/−1
16	Methanol/+1	1:10/+1	60/+1	7/+1

4. RESULTS AND DISCUSSION

4.1. Transesterification Reactions

The experimental data obtained in the enzymatic transesterification of coconut oil using the commercial lipase Novozym 435 as a catalyst are shown in Table 4. Yield data referred to a 24-hour reaction. It is noteworthy that the experiments were done induplicate at each point.

Table 4 shows that the highest yield (80.5%) was obtained at the higher temperature (60°C), enzyme concentration (7%), molar ratio of oil:alcohol (1:10) and using ethanol.

Once the yield was determine, a statistical model was built for representing the experimental data, assessing the significance of the variables, and evaluating possible interactions between them.

The statistical model for calculating the yield of coconut biodiesel can be represented by Equation (3):

$$y = 54.97 + 2.57x_1 - 11.24x_2 + 4.35x_3 - 14.64x_4 - 3.82x_2x_3 - 15.14x_2x_4 - 2.66x_3x_4 \quad (3)$$

where:

The regression parameters 2.57, −11.24, 4.35, −14.64, −3.82, −15.14, and −2.66 were obtained from Table 5, and the parameter 54.97 is the mean of all responses obtained from the experiments:

y is the yield (response) estimate percentage;

x_1 is the coded variable of enzyme concentration which belongs to the interval $-1 \leq x \leq 1$;

x_2 is the coded variable of temperature which belongs to the interval $-1 \leq x \leq 1$;

x_3 is the coded variable of molar ratio which belongs to the interval $-1 \leq x \leq 1$;

x_4 is the coded variable of alcohol type which belongs to the interval $-1 \leq x \leq 1$.

The analysis of variance (ANOVA) (Table 6) shows that the model used in this study is statistically significant and suitable to represent the relation between response and the significant variables, because the coefficient of determination is satisfactory ($R^2 = 0.98$). This value indicates that 98% of the variation of the yield can be attributed to the independent variables and only 2% of the total variation cannot be explained by the model, suggesting that there was a good fit of the model to the experimental data.

Linearity and good fit of the model are shown in Figure 1 with a confidence interval of 95%.

To a significance level of 5%, the obtained results indicate that enzyme concentration, temperature, molar ratio, type of alcohol, as well as interactions between temperature and molar ratio, temperature and type of alcohol, and molar ratio and type of alcohol are significant. These significant values ($p < 0.05$) are best viewed through the Pareto chart in Figure 2:

Table 4. Yields obtained in the transesterification of coconut oil.

Assay	Type	R (O:Alcohol)	T (°C)	E (%p/p)	1st Run	2nd Run	Yield (%)
1	Ethanol	1:6	40	3	48	50.2	49.1
2	Ethanol	1:6	40	7	61	62.8	61.9
3	Ethanol	1:6	60	3	61.5	65.1	63.3
4	Ethanol	1:6	60	7	76.9	75.3	76.1
5	Ethanol	1:10	40	3	79.3	76.7	78
6	Ethanol	1:10	40	7	72.2	75.5	73.8
7	Ethanol	1:10	60	3	77.6	70.6	74.1
8	Ethanol	1:10	60	7	84	77.1	80.5
9	Methanol	1:6	40	3	58.4	55.9	57.1
10	Methanol	1:6	40	7	68.4	59.7	64
11	Methanol	1:6	60	3	19.2	19.6	19.4
12	Methanol	1:6	60	7	14.2	13.8	14
13	Methanol	1:10	40	3	68.2	67.2	67.7
14	Methanol	1:10	40	7	77.9	78.1	78
15	Methanol	1:10	60	3	10.3	10.6	10.4
16	Methanol	1:10	60	7	12.5	11.4	12

Table 5. Regression coefficients for the biodiesel yield.

Factors	Regression Coefficient	Standard Error	T	p	Confidence Interval −95%	Confidence Interval +95%
Mean	54.97	0.79	69.42	0.0000	53.32	56.62
(1) E%	2.57	0.79	3.25	0.004	0.93	4.22
(2) Temperature	−11.24	0.79	−14.2	0.0000	−12.89	−9.6
(3) R molar	4.35	0.79	5.49	0.0000	2.7	5.99
(4) Type	−14.64	0.79	−18.48	0.0000	−16.28	−12.99
1 × 2	−0.65	0.79	−0.83	0.4166	−2.3	0.99
1 × 3	−0.81	0.79	−1.07	0.3165	−2.46	0.83
1 × 4	−0.91	0.79	−1.15	0.2621	−2.56	0.73
2 × 3	−3.82	0.79	−4.82	0.0000	−5.46	−2.17
2 × 4	−15.14	0.79	−19.12	0.0000	−16.79	−13.5
3 × 4	−2.66	0.79	−3.36	0.0029	−4.31	−1.02

Table 6. Analyses of variance (ANOVA).

Variation Source	Sum of Squares	Mean Square	F	p
(1) E%	212.18	212.18	10.57	0.0038
(2) Temperature	4045.50	4045.50	201.62	0.0000
(3) R molar	605.52	605.52	30.18	0.0000
(4) Type	6856.21	6856.21	341.7	0.0000
2 by 3	466.65	466.65	23.26	0.0000
2 by 4	7338.66	7338.66	365.74	0.0000
3 by 4	226.85	226.85	11.31	0.0000
Error	421.36	20.065		
Total	20234.48			

$R^2 = 0.98$.

The values present in the Pareto chart represent the individual influence of each variable on the system response (the yield, in this case) and are interpreted as follows: The reaction yield decreases 30.29% on average when the temperature changes from the lower (40°C) to the higher level (60°C) and when ethanol (lower level) is exchanged for methanol (higher level);

The reaction yield decreases 29.27% on average when using methanol (higher level) instead of ethanol (lower level);

The yield of the reaction decreases 22.49% on average when the temperature changes from the lower (40°C) to the upper level (60°C);

The yield of the reaction increases 8.7% on average when the molar ratio changes from the lower (1:6) to the upper level (1:10);

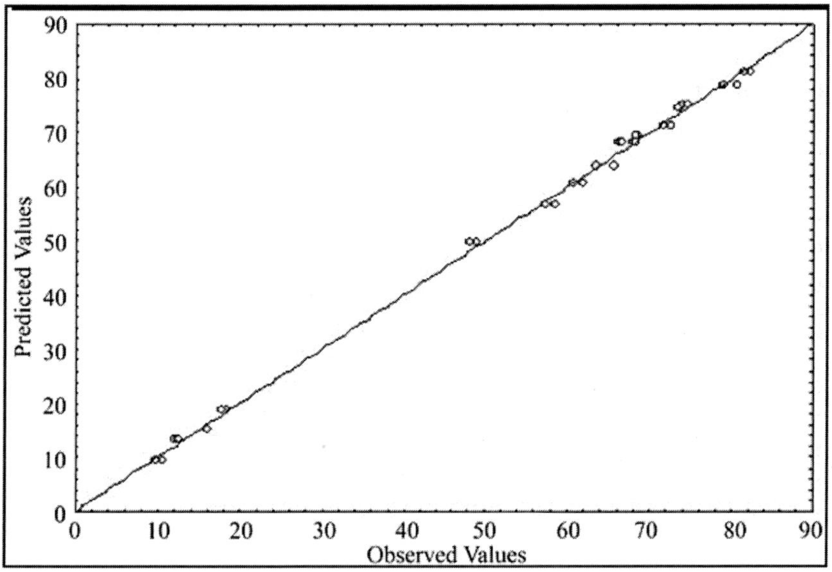

Figure 1. Confidence interval between the model and the experimental data at 95%.

The reaction yield decreases 7.64% on average when the temperature changes from the lower (40°C) to the top level (60°C) and the molar ratio changes from the lower (1:6) to the top level (1:10);

The reaction yield decreases 5.32% on average when the molar ratio changes from the lower (1:6) to the upper level (1:10) and when ethanol (lower level) is replaced by methanol (top level);

The reaction yield increases 5.15% on average, when the enzyme concentration changes from the lower (3%) to the upper level (7%).

The interactions between enzyme concentration and type of alcohol, enzyme concentration and molar ratio, and enzyme concentration and temperature had no significant effects, since $p > 0.05$. Although not significant, this result provides very interesting and attractive information. One of the problems involving enzymatic catalysis is the enzyme costs. A smaller amount of enzyme would make the process more feasible and appealing. The results obtained in this study show that the yield is not significantly influenced when 3% or 7% of enzyme are used. Hence, coconut oil biodiesel can be produced using enzyme concentrations below 7%, and still provide good yields.

4.2. Influence of Reaction Parameters

The higher the enzyme concentration, the greater the reaction yield [15] . In this study, the enzyme concentration had a positive effect, showing a higher yield when a greater amount of enzyme was used. However, when the assays 7 and 8 are compared it is notice that using 3% of enzyme, the yield did not decrease significantly; this variable can be optimized in future studies, keeping the other variables in the best conditions.

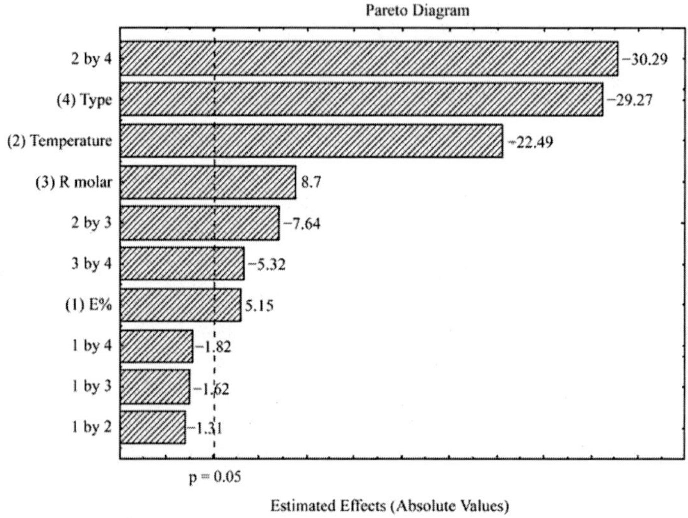

Figure 2. Pareto diagram with a significance level of 5%.

The molar ratio of substrate is a variable with great influence on the biodiesel synthesisreaction. Excess alcohol in the 3:1 stoichiometric ratio is used to ensure a high reaction rate and minimize diffusion limitations. However, excessive levels of alcohol can inhibit the enzyme activity, thus decreasing its catalytic activity during the transesterification reaction. Therefore, a high oil-alcohol ratio means a higher medium polarity produced by alcohol and water. This inactivates the enzyme and consequently the alkyl esters yield [16] .

As two types of alcohol were used in this study, these influences were evaluated separately.

When using ethanol, the molar ratio had a positive effect, demonstrating no enzyme inhibition for alcohol excess. This result corroborates [17] who observed similar effects when using castor oil.

Regarding the methanol, the molar ratio had negative effect, showing that the smaller molar ratio (1:6) yielded better results. This can be explained by the fact that there is a greater deactivation of enzymes by this alcohol [18] .

The reaction rate increases with temperature. In this work, at a temperature range between 40°C and 60°C, enzyme activity increased with temperature. However, when the temperature interacts with an alcohol, this behavior cannot be generalized. Temperature affects differently each type of alcohol. When using ethanol, the temperature had a positive effect, showing no enzyme inhibition at high temperatures.

When ethanol is replaced by highly volatile methanol, a significant yield drop was noticed. This can most likely occur due to the boiling point of methanol (64.70°C) to be very close to the maximum temperature used (60°C).

4.3. Influence of Alcohol Type

Methanol used in the transesterification reaction did not produce good yields when compared to ethanol. The results obtained for this alcohol corroborate those found in the literature. Methanol is a highly hydrophilic solvent, thus being able to solubilize and remove the essential water layer surrounding the enzymes, which can lead to a loss of catalytic activity of the lipase [16] - [19] . According to [19] , the poor results obtained with methanol may be related to the fact that the material used to immobilize enzymes (acrylic resin) can absorb polar compounds such as methanol. When the alcohol concentration in the mixture is high, it forms droplets that bind with the resin particles. The binding of alcohol molecules to the enzymes blocks the reaction (input) with triglycerides, resulting in low yields.

The use of ethanol in transesterification reactions to obtain coconut biodiesel had a positive effect. This result is very interesting and attractive, because ethanol has some advantages in relation to methanol, including: lower toxicity, renewability and availability in the region of this study.

Once discovered the best conditions for the production of biodiesel within the range studied in this work, it was possible to obtain the response surface for the yield using molar ratio and temperature as independent variables (Figure 3). The other two variables (enzyme concentration and type of alcohol) were fixed at their best results (7% and ethanol). Figure 3 shows that by using ethanol and 7% of enzyme, the yield increases when a molar ratio of 1:10 (+1) at 60°C (+1) is used.

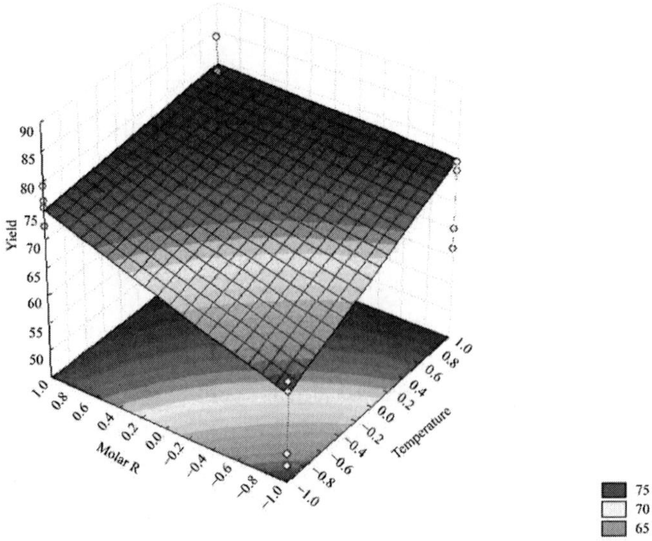

Figure 3. Response surface to yield in biodiesel as a function of the molar ratio and temperature.

The best yield obtained in this study was 80.5% with 60°C, using 7% of the enzyme, molar ratio oil-alcohol 1:10 and ethanol and this yield is too low when compared with alkali-catalysed process and the reaction rates are appreciably slower. Although, enzyme-based process has other advantages such as the capable of minimizing saponification problems due to a relatively high free fatty acid content [20] . Others papers that investigated biodiesel from coconut oil enzymatically catalyzed has obtained yield values between 35 and 90% [20] - [22] .

5. CONCLUSIONS

Biodiesel production from coconut oil was conducted by enzymatic catalysis in the presence of methanol or ethanol. The process conditions (temperature, molar ratio oil-alcohol, enzymeconcentration and type of alcohol) were varied in order to check their influence. To this end, a full 2^4 factorial design was performed.

This experimental design has proven efficient to study the influence of the process variables. The effects of temperature, type of alcohol, molar ratio, enzyme concentration and interactions between temperature and molar ratio, temperature and type of alcohol and the molar ratio and type of alcohol were very significant. The model was adjusted to the factorial design responses, showing the characteristics of a significant model under the statistical point of view, and the model has also predictive characteristics within the range studied for each variable.

The best yield in this study (80.5%) was obtained with 60°C, using 7% of the enzyme, molar ratio oil:alcohol

of 1:10 and ethanol. The results obtained when using methanol were not as significant when compared with ethanol. The results obtained using ethanol were very attractive, because ethanol was renewable and easily obtained in the region of this study.

REFERENCES

1. Zhang, Y., Dubé, M.A., Mclean, D.D. and Kates, M. (2003) Biodiesel Production from Waste Cooking Oil: Process Design and Technological Assessment. Bioresource Technology, 89, 1-16.
2. Worapun, I., Pianthong, K. and Thaiyasuit, P. (2012) Optimization of Biodiesel Production from Crude Palm Oil Using Ultrasonic Irradiation Assistance and Response Surface Methodology. Journal of Chemical Technology and Biotechnology, 87, 189-197.
3. Kumar, D., Kumar, G. and Singh, P.C.P. (2010) Fast, Easy Ethanolysis of Coconut Oil for Biodiesel Production Assistedby Ultrasonication. Ultrasonics Sonochemistry, 17, 555-559.
4. Ma, F. and Hanna, M.A. (1999) Biodiesel Production: A Review. Bioresource Technology, 70, 1-15.
5. Fukuda, H., Kondo, A. and Noda, H. (2001) Biodiesel Fuel Production by Transesterification of Oils. Journal of Bioscience and Bioengineering, 92, 405-416.
6. dos Santos, I.C.F., de Carvalho, S.H.V., Solleti, J.I., de La Salles, W.F., La Salles, K.T.S. and Meneghetti, S.M.P. (2008) Studies of Terminalia catappa L. Oil: Characterization and Biodiesel Production. Bioresource Technology, 99, 6545- 6549.
7. Silva, J.P.V., Serra, T.M., Gossmann, M., Wolf, C.R., Meneghetti, M.R. and Meneghetti, S.M.P. (2010) Moringa Oleifera Oil: Studies of Characterization and Biodiesel Production. Biomass Bioenergy, 34, 1527-1530.

8. La Salles, K.T.S., Meneghetti, S.M.P., La Salles, W.F., Meneghetti, M.R., dos Santos, I.C.F., da Silva, J.P.V., de Carvalho, S.H.V. and Soletti, J.I. (2010) Characterization of Syagrus coronata (Mart.) Becc. Oil and Properties of Methyl Esters for Use as Biodiesel. Industrial Crops and Products, 32, 518-521.
9. Ribeiro, L.M.O., Santos, B.C.S. and Almeida, R.M.R.G. (2012) Studies on Reaction Parameters Influence on Ethanolic Production of Coconut Oil Biodiesel Using Immobilized Lipase as a Catalyst. Biomass and Bioenergy, 47, 498-503.
10. Murugesan, A., Umarani, C., Chinnusamy, T.R., Krishnan, M., Subramanian, R. and Neduzchezhain, N. (2009) Production and Analysis of Bio-Diesel from Non-Edible Oils—A Review. Renewable and Sustainable Energy Reviews, 13, 825- 834.
11. Tan, T., Lu, J., Nie, K., Deng, L. and Wang, F. (2010) Biodiesel Production with Immobilized Lipase: A Review. Biotechnology Advances, 28, 628-634.
12. Talukder, M.M.R., Das, P., Fang, T.S. and Wu, J.C. (2011) Enhanced Enzymatic Transesterification of Palm Oil to Biodiesel. Biochemical Engineering Journal, 55, 119-122.
13. Akoh, C.C., Chang, S., Lee, G. and Shaw, J. (2007) Enzymatic Approach to Biodiesel Production. Journal of Agricultural and Food Chemistry, 55, 8995-9005.
14. Oliveira, D. and Alves, T.L.M. (2000) A Kinetic Study of Lipase Catalyzed Alcoholysis of Palm Kernel Oil. Applied Biochemistry and Biotechnology, 84-86, 59-68.
15. Oliveira, D., Oliveira, J.V., Faccio, C., Menoncin, S. and Amroginski, C. (2004) Influência das variáveis de processo na alcoólise enzimática de óleo de mamona. Ciência e Tecnologia de Alimentos, 24, 178-182.
16. Yu, D.H., Tian, L., Wu, H., Wang, S., Wang, Y., Ma, D.X. and Fang, X.X. (2010) Ultrasonic Irradiation with Vibration for Biodiesel Production from Soybean Oil by Novozym 435. Process Biochemistry, 45, 519-525.
17. Oliveira, D., Luccio, M., Faccio, C., Rosa, C.D., Bender, J.P., Lipke, N., Amroginski, C., Dariva, C. and Oliveira, J.D. (2005) Optimization of Alkaline Transesterification of Soybean Oil and Castor Oil for Biodiesel Production. Applied Biochemistry and Biotechnology, 122, 533-560.

18. Maleki, E., Aroua, M.K. and Sulaiman, N.M.N. (2013) Improved Yield of Solvent Free Enzymatic Methanolysis of Palm and Jatropha Oils Blended with Castor Oil. Applied Energy, 104, 905-909.
19. Chen, J. and Wu, W.T. (2003) Regeneration of Immobilized Cândida antarctica Lipase for Transesterification. Journal of Bioscience and Bioengineering, 95, 466-469.
20. Tupufia, S.C., Jeon, Y.J., Marquis, C., Adesina, A.A. and Rogers, P.L. (2013) Enzymatic Conversion of Coconut Oil for Biodiesel Production. Fuel Processing Technology, 106, 721-726.
21. Abigor, R.D., Uadia, P., Foglia, T.A., Haas, M.J., Jones, K.C., Okpef'a, E., Obibuzor, J.U. and Bafor, M.E. (2000) Lipase-Catalysed Production of Biodiesel Fuel from Some Nigerian Lauric Oils. Biochemical Society Transactions, 28, 979-981.
22. Sun, J., Yu, B., Curran, P. and Liu, S.Q. (2012) Lipase-Catalysed Transesterification of Coconut Oil with Fusel Alcoholsin a Solvent-Free System. Food Chemistry, 134, 89-94.

5
CHAPTER

BIODIESEL: PRODUCTION, CHARACTERIZATION, METALLIC CORROSION AND ANALYTICAL METHODS FOR CONTAMINANTS

Rodrigo A. A. Munoz[1], David M. Fernandes[1], Douglas Q. Santos[2], Tatielli G. G. Barbosa[1] and Raquel M. F. Sousa[1]

[1] Institute of Chemistry, Federal University of Uberlândia (UFU), Uberlândia-MG, Brazil

[2] Technical School of Health Sciences, Federal University of Uberlândia (UFU), Uberlândia-MG, Brazil

1. INTRODUCTION

In face of recent changes in the edaphoclimatic conditions (climate and soil) occurring worldwide, it has been necessary reflections on the need to exploit natural resources in a sustainable manner. Sustainability is a systemic concept, relating to the continuity of economic, social, cultural and environmental aspects of human society. It proposes to be a means of configuring the civilization and human activity so that the society, its members and its economies can fulfill its needs and express its greatest potential at present, while preserving biodiversity and natural ecosystems, planning and acting to achieve pro-efficiency in maintaining undefined these ideals [1].

According to the International Energy Outlook -2011 (IEO 2011), the energy consumption in the world was 505 quadrillion of British thermal units (Btu) in 2008, while in 2006 this consumption was 472 quadrillion Btu. According to 2011 bulletin, the outlook for energy consumption in 2020 will be 619 quadrillion Btu, and 770 quadrillion Btu in 2035 for the countries of the Organization for Economic Cooperation and Development (OECD) [2]. This gradual increase in worldwide energy consumption increases the search for renewable energy, once the existing conventional sources are exhaustible ones, such as for example the oil. In this sense, biofuels have appeared, such as ethanol (bioethanol) and biodiesel, which emerged very strongly due to large government incentives.

Within the energy issues, Brazil, for its favorable natural conditions, presents a huge potential for the production of biofuels, especially ethanol and biodiesel, a fact which makes it a strategic country in relation to the sustainability of such market. There is an effort to consolidate energy from renewable sources, as well as the use of byproducts from these industries, with the creation of programs and incentives by the federal government, such as Proálcool [3].

This text reports a brief historic background of biodiesel in Brazil, methods for biodiesel production indicating the main raw materials, and the regulated physical-chemical properties for the quality control of biodiesel discussing the consequences of cases of non-conformity and their regulated methods by European, American and Brazilian norms. A special topic is dedicated to metallic corrosion which is closely related to storage stability of biodiesel, and finally a comprehensive review of analytical methods developed for monitoring contaminants (glycerol and trace metal) in biodiesel is presented.

1.1. Historic Background of Biodiesel in Brazil

The first use of vegetable oil in diesel engine was tested at the request of the French government with the intention of stimulating energy self-sufficiency in its colonies in Africa, minimizing the costs relating to imports of coal and liquid fuels. The oil selected for the tests was from peanut, whose culture was abundant in tropical countries. The diesel engine produced by the French company Otto, powered by peanut oil, was presented at the Paris Exhibition in 1900. Other experiments conducted by Rudolph Diesel were held in St. Petersburg with locomotives powered by castor oil and animal oils. In both cases the results were very good and the engines showed good performance.[3] In chapter "Liquid Fuel" from Diesel's book "Die

Entstehung des Dieselmotors" (The Emergence of Diesel Engines)[4], it mentions:

For completeness, it is important that, back in 1900, vegetable oils were already being used successfully in diesel engines. During the Paris Exhibition of 1900, the French company Otto demonstrated the operation of a small diesel engine with peanut oil. This experiment was so successful that only some of the people present realized the circumstances in which it was conducted. The engine, which had been built to consume oil, was operated with vegetable oil without any modification. It was also observed that the consumption of vegetable oil resulted in a use of heat literally identical to the oil.

Vegetable oils were also used as emergency fuel, among other applications, during the Second World War. For example, Brazil has banned the export of cottonseed oil because this product could be used to replace imports of diesel oil. Reductions in imports of liquid fuels were also reported in Argentina, which required greater commercial exploitation of vegetable oils. China produced "diesel", lubricating oils, "gasoline" and "kerosene", the last two by cracking processes, from tung oil and other raw material oilseeds. However, the requirements of war forced installation of cracking units of unusual technology base. Quickly, research activities with new oil sources were expanded, but with the subsequent decline in the price of crude oil barrel (post World War II), reaching more affordable price, these researches were abandoned, as happened in India [5].

The use of vegetable oil as an alternative renewable fuel to compete with diesel oil was proposed in the early 1980. The more advanced study with the sunflower oil happened in South Africa because of diesel oil embargo, and the first International Conference on Plants and Vegetable Oils was held in Fargo, North Dakota, in August 1982 [6].

In Brazil, since the 1930s, efforts have been made to incorporate renewable fuels in the energy matrix, which are mainly accomplished by government authorities, universities and research institutes. In addition to the PRO-ALCOHOL (created in 1975, this program was designed to ensure the supply of ethanol from sugarcane in the process of replacing gasoline, as government initiative to tackle the successive increases in oil prices), it was established the Production Plan of Vegetable Oils for Energy Purposes (PRO-OIL), which from the 1980s was named as the National Program of Vegetable Oils for Energy Purposes. This program was designed to generate a significant surplus of vegetable oils, able to make their production costs competitive with mineral oil. It was envisaged by legislation a mixture of 30% of vegetable oil in diesel oil, with prospects for full replacement in the long term. At this time, it was proposed as a technological alternative the transesterification or alcoholysis of vegetable oils, highlighting the studies conducted at Federal University of Ceará, using different sources of vegetable oils such as soybean, babassu, peanut, cottonseed and sunflower, among others. Unfortunately, this program was abandoned by the government in 1986, when the oil price fell again in the international market along with the high cost of production and the crushing of oilseeds, which were decisive factors for the slowdown of the program. However, even after the end of PRO-OIL, there was a considerable progress in researches on the production and use of biodiesel in Brazil, which were conducted in different universities and research centers, particularly the registration of the first Brazilian patent deposited by Chemical Engineer Expedito José de Sá Parente [7]. In accordance with Parente he did not developed a new method: *The transesterification process has been known for many years. What I have patented was the production of esters for use as fuel in diesel cycle engines, which is entirely different from what Rudolf Diesel did. Modern engines could not run for a long time using a vegetable oil under the conditions tested by Diesel.* Table 1 shows the evolution of fuels in Brazil since the 1970s.

In Brazil, especially from the year 2005, when there was the beginning of the National Biodiesel Program, these surveys were intensified due to growing importance in using this material, and three years later, in January became mandatory the addition of 2% (v/v) of biodiesel to commercial diesel, whose mixture was called B2 (B for blend). In March 2008, the National Energy Research Council (CNPE) determined the mandatory addition of 3% (v/v) from July. Since January 1, 2010, the diesel oil sold in Brazil contains 5% biodiesel. This rule was established by Resolution number 6/2009 of the National Energy Policy Council (CNPE), published in the Official Gazette (DOU) in October 26, 2009, which increased from 4% to 5% the mandatory blend percentage of biodiesel to diesel oil. Until mid-2011, two more increases were conducted over, and at that time the addition was 5% (v/v) of biodiesel to diesel, and such addition was put forward to year 2011, previously proposed for the year 2013, due mainly to the increase in diesel consumption and the consequent increase of the fleet of cars using this fuel.

This same law, published on January 13, 2005, introduced the biodiesel in the Brazilian energy matrix and expanded the administrative competence of the National Agency of Petroleum and Natural Gas (ANP), which became, since then, the National Agency of Petroleum, Natural Gas and Biofuels. Since the publication of the abovementioned law, ANP took over the assignment to regulate and supervise the activities related to production, quality control, distribution, sale and marketing of biodiesel and diesel-biodiesel blend [8]. The continued rise in the percentage of biodiesel added to diesel demonstrates the success of the National Program for Production and Use of Biodiesel and the experience accumulated by Brazil in production and large-scale use of biofuels, especially the biodiesel.

Table 1. Evolution of fuels in Brazil since the 1970s.

Year	Event
1973	First Oil Shock
1974	Creation of the Pro-Alcohol
1977	Addition of 4.5% Ethanol to Gasoline
1979	Addition of 15% Ethanol to Gasoline
1980	Second Oil Shock
1983	Alcohol cars account for over 90% of total sales
1985	Percentage of Alcohol added to gasoline reaches 22%
1989	Oil prices fall and gasoline equates with alcohol
1992	Rio 92: Signing of the milestone about climate change
90's	Alcohol comes to represent 20-25% of Gasoline
2005	Founded PNPB
2007	Third Oil Shock
January 2008	Beginning of mandatory B2
March 2008	CNPE determines the mandatory use of B3 from July 2008
April 2008	Alcohol Consumption equates to Gasoline
July 2009	Validity of B4
January 2010	Validity of B5

1.2. Aspects of the biodiesel market in Brazil

According to the Monthly Bulletin of Renewable Fuels of the Ministry of Mines and Energy (MME), based on deliveries of auctions promoted by ANP, it shows that the estimated production in May 2012 was 173,000 m^3. The total of the year up to this month, the cumulative production was 958,000 m^3. The installed production capacity, in May 2012, stood at 6.092 million m^3/year (507,000 m^3/month) of which 88% is related to companies holding the social seal [9]. The Social Fuel Seal is a component of identification created from the Decree No. 5297 of December 6, 2004, awarded by MDA to biodiesel producer who meets the criteria described in Instruction No. 01 of February 19, 2009. The Seal gives to the possessor the character of promoter for social inclusion of family farmers classified in the National Program of Family Agriculture (PRONAF).

2. RAW MATERIALS FOR BIODIESEL PRODUCTION

In general, biodiesel can be produced from any source having oil, either of animal or vegetable source; however, to ensure the quality of the final product, some factors must be observed, as, perennial crops, oil content and preferably with no potential for the food industry, as occurs primarily for soybean in Brazil. Furthermore, it should be noted the productivity per unit area, the agronomic balance and other aspects related to the life cycle of the plant (seasonal). These raw materials, along with the production processes, depend on the region concerned. The economic, environmental and social diversities have given distinct regional motivations for the production and consumption. Due to the favorable conditions of climate and soil, Brazil presents numerous possibilities for use as raw material for the biodiesel industry. It can be highlighted the use of soybean, castor, palm, cottonseed, sunflower, macauba, rapeseed, jatropha, animal fat (tallow) and residual oils, among which, the latter presents itself as an excellent alternative, because it is a material whose reuse at industrial level was insignificant until its application in the biodiesel industry. Oil contents of some of the oil seeds used in Brazil are shown in Figure 1.

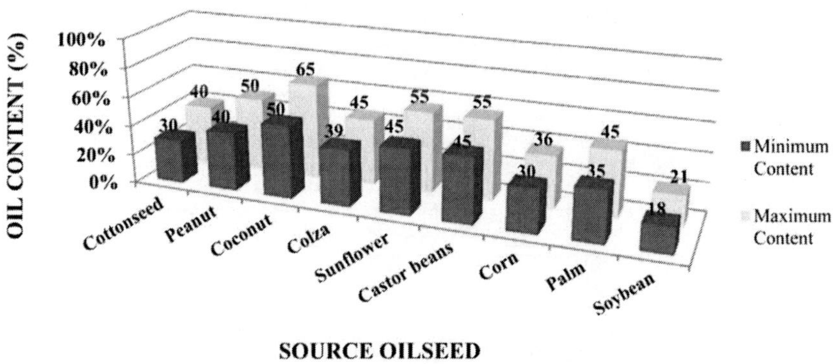

Figure 1. Oil content of some of the seeds used in Brazil.

In Brazil the most relevant is the use of soybean for biodiesel production, followed by beef tallow and cottonseed oil as described by Monthly Bulletin of Biodiesel (published by ANP - reference month May 2012) as shown in Figure 2.

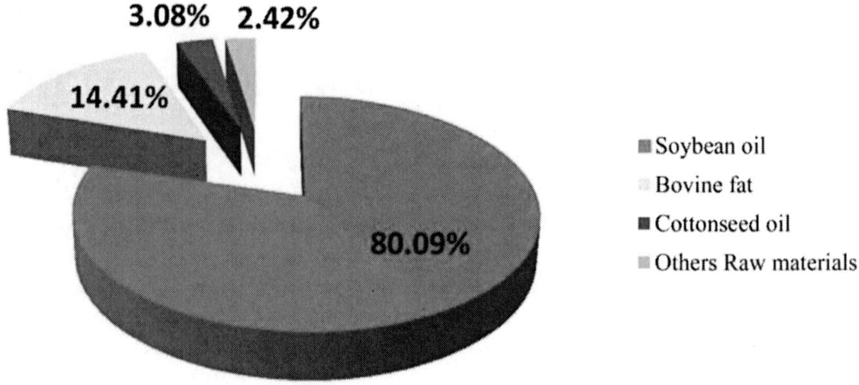

Figure 2. Oil sources used for biodiesel production in Brazil.

In Brazil, other species of oilseeds are cultivated and they are distributed by region as shown in Table 2.

Table 2. Distribution of oilseeds in Brazilian regions.Adapted from ref. [7].

Region	Raw materials that are available
Southeast	Soybean, castor, cottonseed, sunflower, jatropha and macauba
South	Soybean, corn, rapeseed, sunflower, jatropha and cottonseed
Centralwest	Soybean, castor, cottonseed, sunflower, jatropha, macauba, palm and animal fat
North	Palm, babassu and soybean
Northeast	Babassu, soybean, castor, palm, jatropha, cottonseed and coconut

Despite the many advantages in the use of biodiesel, there are two serious problems from the viewpoint of production of vegetable oils which can retard or impede the use of derivatives of vegetable oils as fuel. The first problem is the amount of oil produced. Some of the key raw materials used in Brazil and the corresponding harvest months and oil yields (t/ha) are listed in Table 3. The second problem is the quality of the extracted oil. Oils with high concentrations of polyunsaturated fatty acids are undesirable for biodiesel production, for decreasing their stability to oxidation [12]. Generally, the more unsaturated acids, such as linoleic and linolenic acids, respectively, with two and three unsaturations, are more susceptible to oxidation. In addition, these oils induce a higher carbon deposit than oils with high content of monounsaturated or saturated fatty acids as occurs to the palm oil.

Table 3. Name, Scientific name, harvests months and yields of raw materials for biodiesel production in Brazil. Adapted from refs. [10] and [11]*.

Name	Scientific name	Harvest months/ years	Yield (ton/ha)
Peanut	*Arachis hypogea L.*	3	0.6-0.8
Cottonseed	*Gossipum Hirsutum L.*	3	0.1-0.2
Babassu	*Orbiguya phalenta Mart.*	12	0.1-0.3
Rapeseed	*Brassica napus L.*	3	0.5-0.9
Coconut	*Cocos mucifera L.*	12	1.3-1.9
Palm	*Elaesis guineenses L.*	12	3-6
Sunflower	*Helianthus annus L.*	3	0.5-1.9
Castor beans	*Ricinus communis L.*	3	0.5-0.9
Jatropha	*Jatropha Curcas*	12	0.5-2.5*
Soybean	*Glycine Max (L.) Merril*	3	0.2-0.4

2.1. Cottonseed

The cotton is currently produced by more than 60 countries. China, India, U.S.A., Pakistan and Brazil emerge as major producers. Together, they account for 80% of world production of around 23.3 million tons. Only these five countries have more than 22 million hectares of planted areas, out of 35 million on average that are planted in different points of the planet [13]. The cottonseed oil shows 13-44% oleic acid (18:1) and 33-59% linoleic acid (18:2) [14], and gossypol, the main byproduct of the cotton industry, is a polyphenolic compound, yellow color, toxic, non-interfering in the biodiesel production process [15].

2.2. Palm

Over 70% of the world's oils come from 4 plant species: soybean, palm, sunflower and rapeseed, and currently, the palm oil (known in Brazil as "dendê") is the most traded oil in the world. It is used in the food industry in margarine, ice cream, biscuits, pies, and others. It is also used in the industry of hygiene and cleanliness in the composition of soaps, detergents and cosmetics. In the chemical industry, it is part of the lubricant composition and can also be used as biofuel [16].

In Brazil, the rise of the use of this oilseed is so great that the Federal Government launched in 2010, the Program for Sustainable Production of Oil Palm, which aims to regulate the expansion of oil production and offer tools to ensure a production in sustainable environmental and social bases [16].

This oilseed has the largest oil content among the raw materials for the biodiesel production reaching the yield of 6000 kg/hectare/year. This yield is about 15 times larger than the main raw material in Brazil, the soybean.

The palm oil presents approximately 35-47% palmitic acid (16:0) and 36-47% oleic acid (18:1)[14], and the main problem in using it as fuel is its high viscosity (≈ 20 Cst), about 10 times greater than oil diesel (≈ 2 Cts) (ASTM D445).

2.3. Rapeseed

The rapeseed (*Brassica napus*) is a winter oilseed of Brassica genus, from Cruciferae family and it has 39-45% oil with excellent quality for fatty acid composition. It is the main oil used for biodiesel production in Europe, and is usually known as "colza". It started in Canada in 1974 and with plant breeding developed a seed without erucic acid, a substance which limited its use for human consumption [14]. It represents, in percentage, 15.32% of world production of vegetable oils, considering the 2011/2012 crop, behind only the production of palm oil (33.14%) and soybean (27.28%) [17]. The rapeseed oil contains mostly oleic acid (18:1), 62% followed by 32% linoleic acid (18:2) [18].

2.4. Jatropha

The jatropha (*Jatropha curcas*) is a drought-resistant plant, multipurpose, having a large amount of oil compared to other inedible plants [19]. The highest incidence and the center of diversity of this genus are the Central America and the Caribbean Islands. The genus has over 210 species growing in different regions of the globe [20-21]. This oleaginous has a different composition when it comes to saturated fatty acids [22-23], due to the presence of 16.4% palmitic acid (16:0) against 2.3% soybean and 7% corn [24]. The jatropha contains mostly 40.3% linoleic acid (18:1), 37% linolenic acid (18:2) which indicates a good quality of biodiesel to be produced from this source because the lower the amount of raw material unsaturation, the better is the stability to oxidation [25].

2.5. Soybean

In 2010/2011 harvest, Brazil produced about 75 million tons of soybeans, 65% of this production was exported, equivalent to 48.7 million tons (29.1 million tons of grain, 13.7 million tons of bran and 1.6 million tons of oil). China is the main destination of Brazilian exports of grain (66%) and oil (60%) and the European Union is responsible for the majority of imported bran (70%) [26]. In 2011/12 harvest, the soybean had a stake in percentage terms in the world production of vegetable oils of 27.28% and is the second after only to oil palm, with 33.14% [17].

In Brazil, in May 2012, the soybean had a participation of 80.09% in the production of biodiesel. Taking into account that the biodiesel production in Brazil in that same month was 188,367 x 10^3 m^3 (ANP) then about 150,863 x 10^3 m^3 of the biodiesel produced was derived from soybean oil. This fact is due to the growth of agricultural productivity of soybean and also the technical-industrial feasibility in this oilseed cultivation.

The soybean has about 19.0-30.0% of oleic acid (18:1) and 44.0-62.0% of linoleic acid (18:2) [14].

One aspect widely discussed in relation to this oilseed is the competition between the production for the biofuel market and food industry. In 2011/12 harvest, Brazil exported approximately 44.9% of the soybeans produced [10], due mainly to the increased buying power of emerging countries such as China and India and a small part (about 8%) was for national biodiesel production. China in this season was the largest buyer of Brazilian soybeans (66%) and processed refined oil (60%).

Regarding to soybean oil, Brazil in 2010/11 harvest produced about 7,434 x 10^3 tons, and 5,495 x 10^3 tons (74.85%) were used for domestic market and 1,758 x 10^3 tons (23.95%) for export [26]. In Brazil, this situation has fluctuated over the months, due to the offseason of soybeans and also because many other oilseeds arose that can be intended for biodiesel production, such as jatropha, cottonseed, moringa, among other.

2.6. Waste Cooking Oil

This raw material represents currently a potential use for biodiesel production worldwide since it is a residue from cooking process of industries, restaurants, bars, among other outlets. The high cost of biodiesel in the national and international market is mainly due to the soaring cost of industrialized vegetable oils and virgin oils [27,28].

The biodiesel made from these raw materials is most often cheaper than from sources such as soybean, rapeseed and sunflower, but the quality of the oil used must be monitored to obtain a high quality biodiesel. The free fatty acid (FFA) content and water content (moisture content) are the main parameters to be analyzed. For a reaction catalyzed by alkali, a FFA content of less than 3% is required, since the use of oil with a high FFA content produces the hydrolysis of triglycerides at high temperatures during the frying process [29]. The residual oil has as striking characteristic a high acidity index and it is previously necessary a transesterification reaction, an acid catalysis using a strong acid (generally H_2SO_4) to promote the esterification of fatty acids and thereby reduction of FFA to less than 1% [27,28]. The residual cooking oil has mostly in its composition, oleic acid (18:1) and linoleic acid (18:2) [30].

The use of recycled cooking oil for Biodiesel production has been relevant in recent times, and the flight KL705, originated from Amsterdam with destination to Rio de Janeiro, was carried out in part with this raw material, fueling a Boeing 777-200 [31].

2.7. Moringa Oleifera

The *Moringa oleifera* belongs to the family Moringaceae, consisting of only one genus (Moringa) and fourteen known species, native to northern India. This species is found in natural conditions in India, Africa, Asia, Arabia, South America and the Pacific and Caribbean Islands [32]. Moringa is a multipurpose plant from the leaves to the seeds, showing different properties. The Moringa leaves are the source of a diet rich in protein for both humans and animals. Moreover, oil extraction from its seed enables the utilization of this raw material to produce biodiesel having a remarkable oxidation resistance, with shelf time between 4 and 5 years [33]. The seeds of *Moringa Oleifera* contain between 33 and 41% m/m oil [33]. This value is considered a good natural emollient for cosmetic, based on tactile properties, a lower natural occurrence of color and odor besides a high concentration of oleic acid (>73%) [34]. Recently studies showing the potential of Moringa oil extracted from seeds of India and Pakistan for the biodiesel production were published [35,36]. The Moringa has in its composition about 7% of palmitic acid (16:1) and 78% of oleic acid (18:1) [32]. Variations in oil content within countries and species are attributed to possible changes in environmental and geological conditions of the regions [37].

3. METHODS FOR BIODIESEL PRODUCTION

Vegetable oils and animal fats contain, in addition to triacylglycerols, free fatty acids, phospholipids, sterols, water and other impurities. These compounds give special properties to these raw materials which prevent their use directly as a fuel, as, for example, the blockage point and the high viscosity [38]. These problems can be overcome with adjustments to the compression engines; however, such possibility makes the process very expensive and often impractical. The adaptation of the fuel becomes more interesting and it can be used directly to the existing fleet, without adjusting existing technologies by those that cause chemical modifications (cracking, esterification and transesterification). In the last decade some reviews on the different methods for biodiesel production were reported in the literature [39-47].

The transesterification (also called alcoholysis) is generally a term used to describe the important class of organic reactions in which a triacylglycerol reacts with alcohol in catalytic environment and becomes ester and glycerol [39]. The overall process is a sequence of three consecutive and reversible reactions, in which the mono-and diacylglycerols are formed as intermediates of reaction. In the transesterification reaction three moles of alcohol are needed for each mol of triacylglycerol. In practice, it is used an excess of alcohol in order to increase the ester productivity (shifting the reaction towards the product side) and to allow the separation of the obtained glycerol (the impure fraction containing glycerol is called glycerin).

Glycerin makes the oil more dense and viscous, therefore, during the transesterification process, the glycerin is removed resulting in a product of lower viscosity. In the transesterification, the reaction of biodiesel production creates alkyl esters and glycerol, and the glycerol layer is denser than the ester one and it deposits at the bottom of the reactor [39]. The process is based on the stoichiometric reaction of the alkyl glycerol with alcohol, which in most cases has short organic chain in the presence of a catalyst [39-42]. Alcohols such as methanol, ethanol, propanol or butanol can be used in the transesterification reaction and the (produced) monoesters are known as methyl, ethyl, propyl and butyl esters, respectively. The raw materials used as triacylglycerol source for biodiesel production were discussed previously in the text (Section 2).

The technology for the biodiesel production prevailing in the world is methyl transesterification, where, vegetable oils or animal fat are mixed with methanol that, along with a catalyst, produce the biodiesel. This option occurs mainly because of high ethanol cost and operational facilities. The advantages of biodiesel production using methanol include: ester phase separation (biodiesel) of glycerol that occurs instantaneously; the alcohol recovery that is completed and can be returned to the process; the obtaining of glycerol which is feedstock to the chemical industry; the synthesis is more attractive under the industrial viewpoint, because it is faster and cheaper than the others.

In Brazil, the enterprises that are in operation adopt the methylic route; however, due to ethanol production in Brazil is established beyond favorable environmental factors, there are enterprises that adopt the ethylic route (at industrial level). The ethyl route is recognized as more ecologically friendly, because ethanol is a renewable source. Production of ethyl ester is slightly more complex and requires more steps and the use of specific centrifugal pump sand optimized for a good separation of glycerol from esters. The more alcohol is added to the oil, the faster is the conversion to ester, however, an excess can stabilize the emulsion and complicate the separation of glycerol.

The catalysis used for biodiesel production can be chemical or biological, homogeneous or heterogeneous, acidic or basic. The process for producing biodiesel by basic catalysis is faster than the process of acid catalysis, the biodiesel produced presents less corrosivity, [43,44] and the most used catalysis are the potassium hydroxides, sodium hydroxide, potassium methoxide and methoxide sodium. In Brazil, KOH is more expensive than NaOH, however, there is less soap formation using KOH [45]. The base catalysis is the preferred procedure when there are oils with low water content and low acidity index. On the other hand, if oils and fats have a high content of free fatty acids, it is recommended a pre-treatment or the acid catalysis, followed by an alkaline transesterification [46].

In general, the transesterification of oils or fats can be affected by several factors such as [48]: presence of free fatty acids, moisture [50], type of alcohol used, the molar ratio of alcohol/oil, concentration and type of catalyst, time, temperature of reaction [49] and the intensity of shaking [41].

The content of free fatty acids and moisture are important parameters to determine the feasibility of the basic process of transesterification of vegetable oils and animal fats, requiring low levels of free fatty acids and moisture in the raw material for higher conversion efficiency. Studies have shown [43] that the transesterification of beef tallow catalyzed by NaOH in the presence of free fatty acids (FFA) and water has its yield compromised. Therefore, the raw material that has a high content of free fatty acids can be purified by saponification or using acid catalysis for previous esterification of these acids [51].

The production of ethyl esters via basic catalysis becomes more difficult compared with the production of methyl esters, due to the undesired formation of stable emulsion during the reaction. Methanol and ethanol are immiscible in triglyceride at room temperature, and the reaction media are commonly kept under mechanical stirring to increase mass transfer. In methanolysis, these emulsions easily form two layers: a bottom one rich in glycerol and a top one rich in esters; in ethanolysis the phases are more stable complicating the separation and purification of esters [39,50].

The molar ratio of alcohol/oil is one of the most important variables affecting the yield of esters in the transesterification reaction. The transesterification is a reaction in equilibrium, which requires an excess of alcohol to drive the reaction in the formation of esters. For a maximum conversion of esters, the molar ratio should be greater than or equal to 6:1 [39], but a very high molar ratio of alcohol/oil interferes with the separation of glycerol since its solubility increase takes place. When glycerin remains in solution, it favors the reaction equilibrium for the reactants and then decreasing the yield of esters [52]. Some researchers [42,53] have studied the molar ratios between 3:1 and 15:1 with ethanol, and they have observed arise in the yield of esters with molar ratio increase up to a value 12:1. The best results were between 9:1 and 12:1 [48].

The catalyst concentration is another factor extremely important for the alkaline transesterification [53] since the excessive addition of catalyst can promote an acidity reduction, but, on the other hand, it leads to the formation of soap, hampering the separation of glycerol from esters and consequently decreasing the yield of the reaction.

The catalysts most used during transesterification are the alkoxides [54,55], hydroxides [56] and carbonates of sodium or potassium. The alkoxides of alkali cations such as potassium methoxide (CH_3ONa) are the most reactive catalysts, since they exhibit high yields (> 98%) in a short reaction time (30 min), even at low molar concentrations (0.5 mol L^{-1}). The greater efficiency of the catalyst CH_3ONa relative to NaOH is described by Freedman and colleagues [48,54] which had a similar conversion of oil at concentrations of 1% NaOH and 0.5% CH_3ONa, while in the work developed by Ma and colleagues [45], NaOH showed better performance in ester than CH_3ONain the transesterification of beef tallow. Vincent and colleagues [57] reported good yields obtained with methoxide catalyst, but higher conversion rate was obtained with NaOH and lower with CH_3OK at 65° C, in methanol/oil ratio of 6:1 and catalyst concentration of 1%. Hydroxides of alkaline cations (KOH and NaOH) are more accessible in price than the respective alkoxides, but are less reactive.

Stirring of the reaction medium is an important factor in the transesterification process. Stirring should be intense to transfer amounts of mass of triglycerides from oil phase to the interface with methanol, as the reaction mixture is heterogeneous, consisting of two phases. In this case, the greater the stirring the higher is the mass transesterification [58]. Ma and colleagues [45] added NaOH and MeOH to the tallow beef melted in a reactor and found that after a certain reaction time without stirring anything occurred, suggesting the need of stirring for the reaction be initiated.

Some oils and fats, which may be used as raw materials for the production of biodiesel, have high levels of free fatty acids. The presence of free fatty acids impairs the synthesis of biodiesel via homogeneous basic catalysis [59]. In this sense, heterogeneous acid catalysts, which simultaneously promote reactions of alcoholysis of triglycerides and esterification of free fatty acids, present themselves as promising substitutes of homogeneous basic catalysts [60]. Moreover, such catalysts have the advantages inherent to heterogeneous catalysis, such as significantly reducing the number of purification steps of products as well as the possibility of reuse and enable the production of biofuel by a continuous process with fixed bed reactors [60].

For the production of biodiesel there are other technological routes besides the transesterification, as for example, the esterification catalyzed by an acid, preferably the sulphonic or sulfuric acids. The obtained yield is very high (99%), but the reaction is slow, requiring high temperatures (above 100°C) over 3 hours to reach the mentioned output [45,46,61,62]. Furthermore, it is necessary to use a large excess of alcohol to ensure a high yield reaction. The acid catalysis is suitable for oils with high content of free fatty acids and moisture. In this case the process is the esterification of free acids and not the transesterification of triacylglycerol.

The enzymatic catalysis (biological) allows the simple recovery of glycerol, the transesterification of triglycerols with high content of fatty acids, the total esterification of free fatty acids, and the use of mild conditions in the process, with yields of at least 90%, making it a commercially viable alternative. In this type of catalyst there are no side reactions that result in by products, which reduce the expenditure of further purification. Some enzymes require cofactors, metal ions or organic compounds (coenzymes). These co-factors will influence the activity of biological catalyst [47]. The advantages of the process are: lack of aqueous alkaline waste, lower production of other contaminants, greater selectivity and good yields. The main drawbacks of this methodology are the high cost of pure enzymes, the high cost of extraction and purification process of macromolecules and their instability in solution, which represent an obstacle to the recovery of the biocatalyst after its use [47]. On the other hand, the immobilization of enzymes allows their reuse, reducing the process cost. In the case of biocatalysis in nonaqueous media, immobilization also results in improvement in enzyme activity. Thus, many processes of transesterification using immobilized lipases have been developed [51,63,64].

Another possible and effective reaction is the supercritical transesterification using methanol, which allows a conversion of 60-90% in only 1 minute and more than 95% in 4 minutes. The better reaction conditions are: temperature of 350°C, pressure of 30 MPa and the volumetric ratio between methanol and oil of 42:1 for 240 seconds. The supercritical treatment of lipids with suitable solvent such as methanol depends on the relationship between temperature, pressure and thermo physical properties such as dielectric constant, viscosity, specific mass (density) and polarity [65,66]. The process is attractive since it overcomes problems such as oil/fat waste that is rich in free fatty acids and also the problem of the presence of water that often favors the formation of soap. However, side reactions involving the unsaturated esters occur when the reaction temperature exceeds 300°C, resulting in loss of material. There is also a critical residence time value in high temperature above which the efficiency decreases [51,67].

The H-BIO process was developed to introduce the processing of renewable raw materials in the scheme of petroleum refining and allow the use of existing facilities. The vegetable or animal oil is blended with fractions of petroleum diesel to be converted to units of Hydrotreating (HDT), which are employed in refineries, especially for reducing the sulfur content and improving the diesel oil quality, adjusting the fuel characteristics to the legal specifications. It was already conducted tests with up to 30% vegetable oil in the HDT load, mixed with diesel fractions, generating a product that has the same characteristics as petroleum diesel, but the use of such a high proportion of vegetable oil, in existing industrial HDT units meet operational constraints and limitations of some equipment that were not rated for such task in its original design. A patent of this method was registered (INPI PI0900789-0 A2) and a summary of this method is available at the webpage of Petrobras (Brazilian Oil Company) [68].

4. PHYSICAL-CHEMICAL PROPERTIES OF BIODIESEL

Because of the importance of biodiesel and regulations for its use in the country, the concern of the ANP (National Agency of Petroleum and Natural

Gas and Biofuels) is to ensure a quality fuel in any situation, through the establishment of quality standards for biodiesel. The federal law 11.097/2005 establishes in the country the introduction of minimum percentages of mixture of biodiesel to diesel, and also the monitoring of such insertion. The Brazilian specification of biodiesel is similar to European and American ones, with some flexibility to meet the characteristics of domestic raw materials. This specification is issued by ANP decree. This agency has developed standards in recent years with respect to biodiesel, among them there are the resolutions 15, 41 and 42. Resolution No. 42 of ANP [69] provides a specification for biodiesel (B100) according to the provisions contained in the Technical Regulation No. 4/2004, part that composes such Resolution. B100 can be added to diesel oil in proportions defined by volume, sold by various economic agents authorized throughout the national territory. Any vegetable oil can be used as fuel for diesel engines, but some oils have better performance in terms of their thermodynamic properties [70].

Visual observation of biodiesel formed which should be presented clear and free of impurities, either in suspension, precipitated material or any other. This parameter is adopted only by ANP and it is a simple but important test which is performed in a tube without graduation because this parameter is a qualitative indication of quality.

Tables 4 and 5 present the specifications for biodiesel in accordance with Brazilian, European and American standards and the corresponding analytical methods, respectively.

The specific mass (density) is connected with the molecular structure, i.e. the longer the carbon chain of the ester alkyl, the greater is the density which is, however, reduced by the presence of unsaturations. It is an important parameter for the vehicle injection system. Biodiesel has a specific mass greater than the diesel. This parameter is variable according to the raw material, alcohol excess, among others; very high values may indicate contamination with soap and/or vegetable oil, and alcohol excess causes a decrease in density. The European standard presents as specification limit the values of 860-900 kg m^{-3}, according to the manual method EN ISO 3675 using glass hydrometers and the automatic method EN ISO 12 185 using digital densimeters, and the later with better repeatability. ASTM D6751 norm does not consider the specific mass as a measure of quality of biodiesels. ANP provides the specification range with values between 850 to 900 kg m^{-3}, which adopts the European standard methods, in addition to ASTM D1298 (manual) and ASTM D4052 (automatic), corresponding to NBR 7148 and NBR 14065, respectively. The Brazilian specification differs from European only for this parameter. However, as the reference temperature in Brazil is 20ºC and assuming that the lower the temperature of the test, the higher is the density, it can be concluded that the European specification is more restrictive.

The kinematic viscosity (measurement of internal resistance of liquid flow) is an important parameter for the vehicle injection system and fuel pumping system. It depends on the efficiency of the process (reduction of viscosity of raw material). The viscosity reaches high levels with polymerization processes and/or thermal or oxidative degradation. The EN 14214 standard provides an acceptable range from 3.5 to 5.0 mm^2/s (EN ISO 3104 method), while the ASTM D6751 standard allows a broader range from 1.9 to 6.0 mm^2/s (ASTM D445 method). The Brazilian standard adopts, besides the methods already mentioned, also the ABNT NBR 10441 method, with the allowed viscosity range from 3.0 to 6.0 mm^2/s. The bottleneck for this parameter is the upper limit of the specification which is the value of 6.0 mm^2/s for ANP and ASTM, while the standard EN 14214 has a limit 5.0 mm^2/s, which may restrict the use of some raw materials as for example the biodiesel from castor beans.

The flash point corresponds to the lowest temperature at which the product generates steam enough to ignite when a flame is applied under controlled conditions. This analysis measures the power of self-igniting the fuel and is essential for safety in stocking, handling, transport and storage of fuel. Biodiesel has a flash point much higher than diesel, and low flash point is commonly linked to the presence of alcohol residue in the process. ANP states that the flash point presents at least 100°C and adopts the EN ISO 3679 method, the same of European standard, ASTM D93, the same used by ASTM D6751, and also recommends ABTN NBR 14598.

Water and sediment in biodiesel are usually higher than in diesel, as biodiesel is hygroscopic. The water may generate an unwanted reaction, producing free fatty acids, growth of microorganisms, corrosion and malfunctions of the engine [71]. ASTM D6751 standard adopted this parameter as a quality control for biodiesel using ASTM D2709 method, while ANP and European standard recommends coulometric method (Karl Fischer) EN ISO 12937, and the Brazilian standard also recommends ASTM D6304 method. Comparing the methods, it seems that the coulometric method has greater sensitivity, higher repeatability and lower response time compared to the volumetric method (ASTM D2709).

The total contamination is mainly originated from the raw material, from soaps formed during the process and from unsaponifiables such as wax, hydrocarbons, carotenoids, vitamins and cholesterol (animal origin and used oils). The unsaponifiables present higher boiling point and create waste and soaps in engines and which result in sulphated ashes and finally in abrasion. ASTM standard has not adopted this parameter that is indicated by the European and Brazilian standards, which recommend the same method of analysis described by EN 12662, with maximum specification limit of mg kg^{-1}.

Table 4. Comparison of limits for the quality control of biodiesel in accordance with Brazilian, European and American standards.

Property	Unit	Limits		
		ANP 07/2008	**EN 14214**	**ASTM D6751**
Aspect	---	Limpid and without impurities	---	---
Density	kg/m^3	850-900 (20°C)	860-900 (15°C)	---
Kinematic viscosity (40°C)	mm^2/s	3.0-6.0	3.5-5.0	1.9-6.0
Water and sediment, max.	%vol.	---	---	0.050
Flash point, min.	°C	100	101	130
Distillation, 90% recovered vol., max.	°C	---	---	360
Carbon residue, max.	% mass	0.050 (100% of the sample)	0.3 (10% Distillation residual)	0.05 (100% of the sample)
Sulfated ash, max.	% mass	0.020	0.02	0.020
Sulfur content, max.	mg/kg	10	10	15
Copper strip corrosion, 3h at 50°C, max.	Rating	1	1	3
Cetane number	---	Note	51 (min.)	47 (min.)
Cold soak filterability	°C	By region	By region	---
Pour point	°C	---	By region	---
Cloud point	°C	---	---	Note
Sodium and potassium, max.	mg/kg	5	5	5
Calcium and magnesium, max.	mg/kg	5	5	---
Phosphorus content, max.	mg/kg	10	4.0	10
Total contamination, max.	mg/kg	24	24	---
Ester content, min.	% mass	96.5%	96.5%	---
Acid value, max.	mg KOH/g	0.50	0.5	0.5
Free glycerol, max.	% mass	0.02	0.02	0.020
Total glycerol, max.	% mass	0.25	0.25	0.240
Monoglycerides	% mass	0.80 (max.)	0.80 (max.)	---
Diglycerides	% mass	0.20 (max.)	0.20 (max.)	---
Triglycerides	% mass	0.20 (max.)	0.20 (max.)	---
Methanol or Ethanol, max.	% mass	0.20	0.20	0.20
Iodine value	g I_2/100 g	Note	120 (max.)	---
Oxidation stability at 110 °C, min.	H	6	6	3
Water content, max.	mg/kg	380	500	---
Linolenic acid	% mass	---	12 (max.)	---
Polyunsaturated methyl esters (with more than four double bonds)	% mass	---	1 (max.)	---

The ester content is the main property as it indicates the degree of purity of the biodiesel produced and the efficiency of the production process used. It depends on the amount of unsaponifiables in the raw material and process variables (time, temperature, agitation speed, molar ratio, catalyst concentration, type of catalyst, water, free fatty acids, and alcohol). The low content of esters indicates a low yield of the transesterification reaction, i.e., most of the triacylglycerols has not reacted, which can cause difficulties in combustion and carbonization of engine cylinders. This parameter is required by the ANP and EN 14214 standards, whose minimum percentage is 96.5 % wt. using the EN ISO 14103 method. Brazilian standard also recommends ABNT NBR 15342 method for biodiesel from animal origin or for blends in which there is the presence of biodiesel from castor beans.

The carbon residue indicates the tendency of a fuel to form carbon deposits in engines. These residues are deposited in the nozzles and other parts of the engine, reducing their useful life. They correspond to the amount of triacylglycerols, soaps, leftover of catalyst and unsaponifiables, which are present in the final biodiesel. American and Brazilian standards adopt the same ASTM D4530 method, also having the same specification limit, no more than 0.050 % wt. Now the European standard indicates EN ISO 10370 method in which the maximum allowed is 0.3 % wt.

Sulphated ashes cause saturation of filters and wear on various parts of the engine and may be present in the form of abrasive solids, soluble metal soaps and catalyst residues. The maximum content of sulphated ashes in biodiesel is 0.020 % wt. set by the EN 14214 standard (EN ISO 3987 method), ANP (EN ISO 3987, ABNT NBR 6294 and ASTM D874 methods) and by ASTM D6751 (ASTM D874 method).

The sulfur content derived from the raw material generates toxic emissions affecting the performance of vehicle emission control system. Brazilian standard adopts methods of analysis described by ASTM D5453 (molecular fluorescence), EN ISO 20 846 (also by molecular fluorescence) and EN ISO 20 884 (dispersive X-ray fluorescence), with the maximum acceptable limit of 50 mg kg^{-1}. As to ASTM D6751, the maximum limit is 15 mg kg^{-1} (ASTM D5453 method). The most restrictive is the European standard which recommends the methods given by EN ISO 20846 and EN ISO 20884, with the maximum value of 10 mg kg^{-1}.

Group I (Na + K) and group II (Ca + Mg) metal ions cause the formation of deposits of insoluble soaps, as well as catalyze polymerization reactions. They derive from catalysts employed in the production of biodiesel, as KOH, NaOH and / or CH_3ONa or CH_3OK. Calcium and magnesium may also be present as impurities in the NaOH or KOH used.

Table 5. Established methods for the quality control of biodiesel in accordance with Brazilian, European and American standards.

Property	Methods				
	ANP 07/2008			EN 14214	ASTM D6751
	ABNT NBR	ASTM D	EN/ISO		
Aspect	---	---	---	---	---
Density	7148 14065	1298 4052	EN ISO 3675 EN ISO 12185	EN ISO 3675 EN ISO 12185	---
Kinematic viscosity (40°C)	10441	445	EN ISO 3104	EN ISO 3104	D445
Water and sediment	---	---	---	---	D2709
Flash point	14598	93	EN ISO 3679	EN ISO 2719 EN ISO 3679	D93
Distillation, 90% recovered vol.	---	---	---	---	D1160
Carbon residue	---	4530	---	EN ISO 10370	D4530
Sulfated ash	6294	874	EN ISO 3987	EN ISO 3987	D874
Sulfur content	15867	5453	EN ISO 20846 EN ISO 20884	EN ISO 20846 EN ISO 20884	D5453
Copper strip corrosion, 3h at 50°C	14359	130	EN ISO 2160	EN ISO 2160	D130
Cetane number	---	613 6890	EN ISO 5165	EN ISO 5165	D613
Cold soak filterability	14747	6371	EN 116	---	D7501
Pour point	---	---	---	---	---
Cloud point	---	---	---	---	D2500
Sodium and potassium	15553 15554 15555 15556	---	EN 14108 EN 14109 EN 14538	EN 14108 EN 14109 EN 14538	EN 14538
Calcium and magnesium	15553 15556	---	EN 14538	EN 14538	---
Phosphorus content	15553	4951	EN 14107	EN 14107	D4951
Total contamination	---	---	EN 12662	EN 12662	---
Ester content	15342	---	EN 14103	EN 14103	---
Acid value	14448	664	EN 14104	EN 14104	D664
Free glycerol	15341	6584	EN 14105 EN 14106	EN 14105 EN 14106	D6584
Total glycerol	15344	6584	EN 14105	EN 14105	D6584
Monoglycerides	15342 15344 15908	6584	EN 14105	EN 14105	---
Diglycerides	15342 15344 15908	6584	EN 14105	EN 14105	---
Triglycerides	15342 15344 15908	6584	EN 14105	EN 14105	---
Methanol or Ethanol	15343	---	EN 14110	EN 14110	EN 14110
Iodine value	---	---	EN 14111	EN 14111	---
Oxidation stability at 110 °C	---	---	EN 14112	EN 15751 EN 14112	EN 15751
Water content	---	6304	EN ISO 12937	EN ISO 12937	---
Linolenic acid	---	---	---	EN 14103	---

Phosphorus can damage catalytic converters used in emission control systems of the engine. They comes mainly from the raw material and, eventually, from residues of the phosphoric acid used in the neutralization. Phosphorus is determined in biodiesel via optical emission spectroscopy with inductively coupled plasma. Both the European standard (maximum 4.0 mg kg^{-1}) as the Brazilian (maximum 10.0 mg kg^{-1}) recommends the analytical method EN ISO 14107. The Brazilian standard also recommends the method ABNT NBR 15553 and ASTM D4951. The latter is also indicated in the American standard (maximum 10.0 mg kg^{-1}).

Corrosivity to copper (copper strip corrosion) due to sulfur compounds, as well as due to free fatty acids, can lead to problems of corrosion in storage tanks and some engine parts. As acids are included in this parameter, it keeps a relationship with the acid index. The maximum acceptable values are degree 1, for standard EN 14214 and ANP, and degree 3 for ASTM D6751 (the degree number corresponds to a visual comparison with the copper strip established by ASTM D4951).

The acid number can increase or accelerate corrosion of the engine. It also measures the presence of free fatty acids and other acids and is related to the quality of the process. In injection systems that work at higher temperatures, faster biodiesel degradation may occur increasing the level of acidity and causing problems in filters. All standards have adopted as the specification limit for that parameter the value of 0.5 mg de KOH/g. The ANP standard recommended the methods ABNT NBR 14448, ASTM D664 and EN ISO 14104, and these last two methods are also adopted by European standard and American standard, respectively.

The cetane number measures the ignition quality of fuel. A low cetane number indicates a poorer ignition, which can form deposits and wear on pistons as well as provide greater fuel consumption. It depends upon the feedstock besides the oxygenate content in biodiesel. The number of cetane is measured with the aid of a special motor and the cetane index is calculated. The cetane index is a useful tool to estimate the number of cetanes according to ASTM standard [51].

The content of total glycerol and free glycerol as well as monoglyceride, diglyceride and triglyceride, reflects the quality of biodiesel. A high content of these can cause problems ranging from the formation of crystals, crusts inside the fuel storage tank, contributing to the formation or deposit of waste on pistons, injectors, valves, rings (of segments), filters, up to clogging the nozzles, decreasing the engine life. These are intermediate products of the process that ended up not reacting. The free glycerin (a byproduct) depends on the efficiency of separation of esters/glycerin.

The presence of residual alcohol in the biodiesel may cause corrosion on items of aluminum and zinc, and also can influence the flash point, reduce the cetane number and decrease the lubricity of the engine [51]. The alcohol content is determined by the chromatographic method EN ISO 14110 indicated by the standards EN 14214, ASTM D6751 and ANP. The Brazilian standard also recommends the method ABNT NBR 15343.

The iodine value is related to viscosity and cetane number, and indicates the quantitative degree of unsaturation of esters that form the biodiesel. The method suggested by the Brazilian and European standards is the same EN ISO 14111, though the latter is the only one to provide a limit to the parameter of 120 g I_2/100 g.

The oxidation stability determines the degradation of biodiesel, and mixtures thereof. It is related to the time required to degrade the biodiesel under controlled heating and in the presence of oxygen. The method adopted by the ANP standard and also EN 14214 is given in EN 14112, where the specification limit for both is 6 h at least (induction time by Rancimat method). For ASTM D6751 standard this limit is 3 hours and the recommended method is EN 15751.

In addition to the mentioned properties, other parameters may also be important and relate directly to the raw material used in the production of biodiesel: the soap content, the boiling range of esters, the peroxide index (that express the oil oxidation degree), the filterability (that express the difficulty to filter the oil before injecting into the engine) and the gum content, which expresses the amount of gums formed by polymerization of unsaturated oil components during the combustion [51].

In Brazil, the National Program for Production and Use of Biodiesel in one of its aspects provides the Program for Engine Test and Experiment, which is coordinated by the Ministry of Science and Technology (MCT). This program establishes test performances on vehicles and stationary engines and the gradual use increase of the blend biodiesel/diesel and evaluating its consequent technical feasibility. Currently in Brazil, B5 blend is used and it is extremely important to carry out tests to ensure commercial guarantees to vehicles moved with this blend and the quality of the blend commercialized. Currently in Brazil, B5 blend is used and test performing is extremely important to ensure commercial guarantees to vehicles moved with this blend and the quality of the blend commercialized.

At low temperatures, biodiesel tends to partially solidify or increase its viscosity (fluidity loss) inhibiting or even stopping the flow of fuel with the consequent clogging of filters, damaging the starting system of the engine. In tropical countries like Brazil, where most states have high temperatures throughout the year, the effects of these engines operating at low temperatures are minimized, and such problems are mainly present in countries in North America and Europe. Some properties analyzed in such conditions are: cloud point (ASTM D2500), which is the temperature at which, in a process of fuel cooling, it is observed the formation of first crystals, cold filter plugging point (EN 116), which is the temperature where the fuel ceases to be filtered when a cooling occurs and pour point (EN 3016) which is the temperature that the fuel loses its fluidity at established test conditions.

Thus, a major problem is the biodiesels from raw materials that have a high content of saturated compounds in its chain, as occurs for the tallow and the palm oil [72]. Biodiesels from raw materials of animal source when compared to those of vegetable origin have higher pour point, cold filter plugging point and cloud point.

Despite having standards for performing tests, there are no fixed and established values for these parameters. In Europe, they are measured for each country in terms of its climate. In U.S.A., they are dependent on the seasons and climate and in Brazil they depend on the states and months of the year, as shown in Table 6.

Despite the problems that can arise with the use of this fuel in cold seasons and countries with low average temperature, the biodiesel has many advantages over petroleum diesel as, it is a renewable fuel with lower emissions of particulate matters, polycyclic aromatic hydrocarbons, and sulfur compounds; storage and handling of this material is safer, since it has a flash point much higher than petroleum diesel, and also a higher flammability point which ensures a greater safety in loading/unloading and handling of this material by the drivers and operators [72].

Table 6. Maximum limits of cold filter plugging point for the Brazilian Federation states according to the months of the year in accordance with the ANP resolution number 14 (May 2012). For the other federation states (not included in the table), the value remains in 19ºC.

Brazilian Federation Units	Maximum limit, ºC											
	JAN	FEB	MAR	APR	MAY	JUN	JUL	AUG	SEP	OCT	NOV	DEC
SP - MG – MS	14	14	14	12	8	8	8	8	8	12	14	14
GO - DF - MT - ES – RJ	14	14	14	14	10	10	10	10	10	14	14	14
PR - SC - RS	14	14	14	10	5	5	5	5	5	10	14	14

5. METALLIC CORROSION IN BIODIESEL

Although biodiesel has many properties that assist in good yield as a fuel as a relatively high flash point and good lubricating properties compared to diesel, some of these properties facilitate its self-oxidation, and oxidation of the metallic materials which they are in contact. The metallic corrosion becomes extremely important since many of the engine parts are composed of variety metals such as aluminum, copper, stainless steel and alloys. The percent of aluminum in engine parts includes piston (100%), cylinder heads (70%), and engine blocks (19%). Pumps and injectors are composed of copper and its alloys. Parts composed of stainless steel include fuel filter, valve bodies, nozzle and pump ring [73-75]. The metallic corrosion may occur due to the following factors:

a) biodiesel is an ester so makes hydrogen bonds with water; then it becomes much more hygroscopic compared to diesel which is composed by hydrocarbons. Water acts on the corrosion of metallic materials, or it causes the hydrolysis of biodiesel, resulting in fatty acids and glycerol which increase metallic corrosion, or it promotes microbial growth and thereby microbial corrosion [73,75-77].

b) the presence of impurities like water, methanol, free glycerol, free fatty acid, catalyst residues (Na and K) due to incomplete conversion or inadequate purification can also result in metallic corrosion [73,75].

c) due to its good lubricity, biodiesel dissolves more metallic parts than diesel, and these trace metals in solution enhance biodiesel degradation and promote metallic corrosion [73,75].

d) metals into biodiesel like brass, copper and aluminum act as catalysts for biodiesel oxidation. Therefore, the acid number of biodiesel increases proportionally with the corrosion rate for different metals [78-81].

Generally, the corrosion tests are performed by immersion tests with metallic coupons in biodiesel [79,80,82,83]. After immersion, the weight of the coupons was measured and the corrosion was analyzed by measurement of corrosion rate, according Equation 1:

$$Corrosion\ rate = \frac{W\ x\ 534}{D\ x\ T\ x\ A} \qquad (1)$$

Where:

Corrosion rate = mpy (stands for mils (0.001 inch) per year);

W = weight loss (mg);

D = density (g cm^{-3});

A = exposed surface area (square inch);

T = exposure time (h).

Kaul et al. [80] studied the corrosivity of biodiesel from different oil sources by static immersion tests using metallic (aluminum alloy) piston for 300 days at 15 to 40 ° C. In this study the authors observed that the corrosion rate varied with the chemical composition of each oleaginous. The corrosion tests were performed with *Jatropha curcas*, *Karanja*, *Madhuca indica* and *Salvadora* biodiesels. Corrosion rates values were 0.0117, 0.0058, 0.0058, and 0.1236 mpy, respectively. The *Salvadora oleoides* biodiesel presented the highest rate due to its higher content of total sulfur (1200 ppm), while other biodiesels presented lower concentrations of total sulfur (around 1 ppm), except *Madhuca indica* biodiesel (164.8 ppm). *Jatropha curcas* biodiesel was slightly more corrosive than *Karanja* and *Madhuca indica* biodiesels because of its elevated concentration of C18:2 (19–41%) fatty acid, which is more prone to oxidation due to presence of two double bonds.

The metal corrosivity also depends on the nature of the metal exposed to the biofuel. Haseeb et al. [82] performed static immersion tests with coupons of copper and leaded bronze (87% Cu, 6% Sn, 6% Pb) in palm biodiesel at room temperature (25-30 °C) for 840 h; the corrosion rates for copper and bronze were 0.042 and 0.018 mpy, respectively. Additionally, at 60 °C in 2640 h, the corrosion rates of both metals were relatively higher, 0.053 mpy for copper and 0.023 mpy for bronze. The corrosion resistance of bronze was believed to be related to the presence of alloying elements such as tin (Sn) in the alloy. These results clearly show that copper is more prone to corrosion by biodiesel.

Similar behavior was observed by Fazal et al. [83], which performed static immersion tests and evaluated the corrosion rates of copper, brass (Cu: 58.5 %; Zn:41.5), aluminum, cast iron (C:3 %; Si:1.84 %; Mn: 0.82 %; P:0.098%; S:0.089; Fe: balance) at room temperature (25-27 °C) and for 120 days in palm biodiesel. The authors verified that copper presented higher corrosion rates (0.39278 mpy) followed by brass (0.209898 mpy), which contained zinc its composition that probably reduced its corrosion. Aluminum presented higher corrosion rates (0.173055 mpy) than cast iron (0.112232 mpy). This is accordance with Geller et al. [84], which reported that copper alloys are more prone to corrosion in biodiesel than ferrous alloys.

Recently, Hu et al. [85] compared the corrosion rates of several metals in rapeseed biodiesel and proposed corrosion mechanisms through static immersion tests at 43 °C for 60 days. The obtained corrosion rates of copper, carbon steel, aluminum, and stainless steel were 0.02334, 0.01819, 0.00324, and 0.00087 mmy, respectively (which correspond to 0.9336, 0.7276, 0.1296, and 0.0348 mpy, respectively). This study indicates that copper and carbon steel presented higher corrosion rates than aluminum and stainless steel.

Using scanning electron microscope with energy dispersive X-ray analysis and X-ray photoelectron spectroscopy to analyze the effects of biodiesel on the corrosion of different metallic materials, Hu et al. [85] reported that the corrosion process of metal surfaces in biodiesel was mainly attributed to the chemical corrosion and the products after corrosion were primarily fatty acid salts or metal oxides, depending on the studied metal. Elements of copper and iron are catalysts for the decomposition of biodiesel because they enabled various chemical reactions to easily take place. According to Hu et al. [85], metals were oxidized by oxygen and active oxygen atom dissolved in biodiesel, resulting in the formation of metal oxides (CuO, Cu_2O, Fe_2O_3, etc.). Copper and carbon steel were easily oxidized but aluminum and stainless steel were protected by films of metal oxide and then their corrosion rates were lower. The protective metal oxide layer prevented the metal surface from contact with the oxygen and atom oxygen as well as from contact with the oil sample.

Fazal et al. [86] has also found that the oxygen concentration increases with increasing temperatures, which may explain the high corrosion rates at higher temperatures. However, water and fatty acids are also responsible for metallic corrosion and need to be considered [80].

Similarly to Hu et al. [85] which carried out static immersion tests at 80 °C for 600 h in rapeseed biodiesel, Norouzi et al. [78] verified corrosion rates of 0.9 mpy for copper and 0.35 mpy for aluminum and confirmed that increase in temperature enhanced corrosion rates.

Another limiting factor that should be considered is the stirring. Fazal et al. [79] obtained corrosion rates of 0.586, 0.202, and 0.015 mpy for copper, aluminum, stainless steel, respectively, in immersion tests at 80 °C for 50 days under 250 rpm stirring in palm biodiesel. The corrosion rates presented superior values than those obtained in static immersion tests. Table 7 presents corrosion rates of different metallic materials in various biodiesels.

Analyzing the data in Table 7, palm biodiesel is less corrosive than rapeseed biodiesel. This different behavior can be correlated by the chemical composition of each biodiesel; rapeseed biodiesel presents 68.821 % oleic acid (C18:1) and 19.5927 % linoleic acid (C18:2) [78], whilst palm biodiesel presents 41.8 % oleic acid (C18:1) and 9.10 % linoleic acid (C18:2) [87].

Table 7. Corrosion rate of some metals in different biodiesels.

Operation	Biodiesel	METALS (mpy)							Ref.
		Aluminum	Copper	Bronze	Carbon steel	Stainless steel	Brass	Cast iron	
300 days 15-40 °C static	*Jatropha curcas*	0.0117	-	-	-	-	-	-	[80]
	Karanja,	0.0058	-	-	-	-	-	-	[80]
	Madhuca	0.0058	-	-	-	-	-	-	[80]
	Salvadora	0.1236	-	-	-	-	-	-	[80]
60 days 43 °C static	Rapeesed	0.1296	0.9336	-	0.7276	0.0348	-	-	[85]
25 days 80 °C static	Rapeesed	~ 0.35	~ 0.9	-	-	-	-	-	[78]
50 days 80 °C 250 rpm	Palm	0.202	0.586	-	-	0.015	-	-	[78]
120 days 25-27 °C static	Palm	0.173055	0.39278	-	-	-	0.209898	0.112232	[83]
35 days 23-30 °C static	Palm	-	0.042	0.018	-	-	-	-	[82]
		-	0.053	0.023	-	-	-	-	[82]

The content of metal released to biodiesel during corrosion can be quantified. In the static immersion tests in rapeseed biodiesel performed by Hu et al. [85], 41.088 mg L^{-1} Cu (copper coupon), 3.544 mg L^{-1} Fe (carbon steel coupon) and 2.756 mg L^{-1} Fe and 9.02 mg L^{-1} Cr (stainless steel coupon) were obtained. On the other hand, Haseeb et al. [80] quantified lower amounts of metals in palm biodiesel (copper: 5 ppm Cu, bronze: 5 ppm Cu, 4 ppm Pb and 10 ppm Zn) because this biodiesel is less corrosive. According to McCormick et al. [88] the high concentration of metals in biodiesel generates higher oxidation of the biofuel. Biodiesel containing more than 6 ppm of metal exhibits very short OSI (oil stability index) induction time [88]. The negative effect of the presence of metal contaminants on the biodiesel oxidation stability was also reported by other authors who performed experiments adding organometallic standards or powdered metals into different biodiesels [25,89-94].

Fazal et al. [79] and Hu et al. [85] have reported that biodiesel is more corrosive than diesel oil (based on the corrosion rates in different metallic materials). Figure 3 displays data from both reports [79,85] on the corrosion rate of copper, aluminum, stainless steel, and carbon steel in biodiesel and diesel. See that the first work only evaluated the corrosion of copper, aluminum, and stainless steel [79], while the second work also evaluated all the previous materials and in addition carbon steel [85]. The higher corrosion rates were also confirmed by the metal release verified in diesel and biodiesel, especially copper and iron [85]. Figure 3 also indicates that biodiesel presents more corrosive behavior than diesel oil and copper is not compatible with biodiesel.

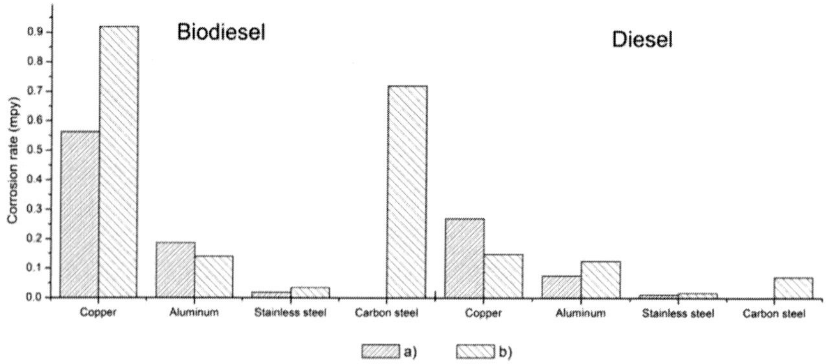

Figure 3. Corrosion rate of metals in biodiesel and diesel oil: a) adapted from Ref. [79]; b) adapted from ref. [85].

An alternative way to reduce the corrosive behavior of biodiesel is to use it as blends with diesel oil. According to Norouzi et al. [78], the greater the amount of diesel in blends, the lower corrosion rate as well as the lower total acid number (TAN). Figure 4 shows the variation of (a) corrosion rate of aluminum and copper coupons and of (b) TAN number of biodiesels exposed to these metallic materials through static immersion tests.

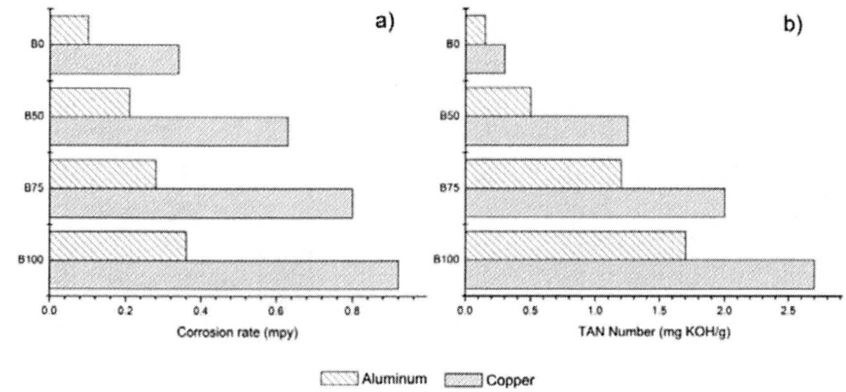

Figure 4.a) Corrosion rate of the Al and Cu and b) TAN of the rapeseed biodiesel-diesel blends in contact with both metals (adapted from ref. [78]).

There is a clear relationship between corrosion and TAN of biodiesel; the higher the corrosion rates the higher the TAN numbers. These results show that copper was more corrosive than aluminum and that as long the biodiesel-diesel blend was richer in biodiesel, higher corrosion rates were verified and consequently higher TAN numbers.

Some studies reported that the corrosiveness of biodiesel can be reduced by using corrosion inhibitors or antioxidants. Corrosion inhibitors act by the formation of adsorbed monolayer films at the metal-solution interface. The common corrosion inhibitors in oil are imidazolines, primary amines, diamines, amino-amines, oxyalkylated amines, naphthaneic acid, phosphate esters, dodecyl benzene sulfonic acids, etc [95]. Amine inhibitors can reduce the dissolution of metal oxide layers into biodiesel by forming stable oxide layers on the metal surface [95]. Fazal et al. [95] reported the inhibition effect on the corrosion of grey cast iron (C: 3%, Si: 1.84%, Mn: 0.82%, P: 0.098%, S: 0.089%, Fe: balance) immersed (static) in biodiesel containing 100 ppm of ethylenediamine (EDA), n-butylamine (nBA), tert-butylamine (TBA) at room temperature for 1200 h. The corrosion rates in each case are shown in Figure 5. The results indicate that EDA is the most powerful corrosion inhibitor under the experimental conditions. However, the analysis of fuel properties revealed that the biodiesel containing EDA was more degraded.

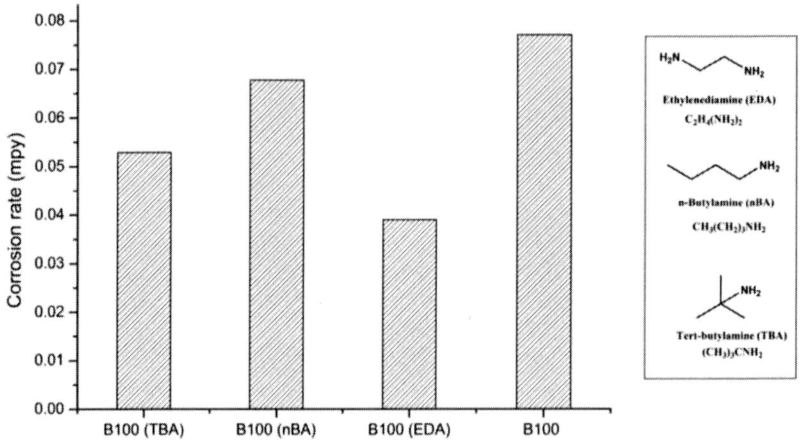

Figure 5. Corrosion rate of cast iron (duplicate and respective standard deviation) in palm biodiesel with and without corrosion inhibitors (EDA, nBA and TBA) and their respective chemical structures (adapted from ref. [95]).

Antioxidants can be classified into two groups: chain breakers and hydroperoxide decomposers. The chain breakers have two most common types of antioxidants: phenolic and amine-types. These antioxidants present a highly labile hydrogen that is able to abstract a peroxy radical and thus stops oxidation reactions in the ester chain [96]. The chain breakers antioxidants can be re-classified into natural antioxidants (tocopherols present in vegetables oils) and synthetic antioxidants such as butyl-hydroxytoluene (BHT), tert-butylhydroquinone (TBHQ), butyl-hydroxyanisol (BHA), pyrogallol (PY), and propyl gallate (PG) [96].

Liang et al. [97] have reported that synthethic antioxidants presented superior antioxidant activity in palm biodiesel than natural antioxidants. Other authors investigated BHT, BHA, TBHQ, and natural antioxidants as potential antioxidants in soybean oil biodiesel and TBHQ presented a superior antioxidant activity [98,99]. Jain and Sharma compared the efficiency of eight synthetic antioxidants in different biodiesels and they concluded that only three antioxidants significantly increased the stability of biodiesel in the order of TBHQ > PY > PG [96]. The addition of antioxidants can also overcome the low oxidation stability of biodiesels promoted by metal contamination (experiments performed by adding organometallic standards) in order to re-establish the required 6-hour induction time [25,90-94]. This statement is especially essential considering the presence of metals in biodiesel due to corrosion of containers and engine components.

Almeida et al. [100] evaluated the effect of the synthethic antioxidant TBHQ on the corrosive character of biodiesel against copper coupon through static immersion tests. Due to the strong catalytic effect of copper towards biodiesel oxidation, the oxidation stability of biodiesel decreased tremendiously after 24 h of exposure even in the presence of TBHQ. The copper content in biodiesel continuously increased with the exposure time and the metal concentration was much higher in the non-stabilized biodiesel. Then, the presence of TBHQ decreased the corrosion rate of the copper coupon and the authors claimed that the antioxidant may have acted as a corrosion inhibitor through the formation af a protective layer on the metallic surface. Performing mass-spectrometry (MS) and MS-MS analysis of deteriorated biodiesel containing TBHQ (after the metal corrosion for 24 h and 168 h), the authors have also identified the formation of new molecules of high molecular weight formed by the association between oxidized antioxidants molecules and free radicals of long-chain molecules (fatty acid derivatives). These results give light to possible side-reactions between phenolic antioxidants and esters molecules of biodiesel under oxidative conditions.

6. ANALYTICAL METHODS FOR CONTAMINANTS IN BIODIESEL

6.1. Free and Total Glycerol

The determination of the content of total glycerol and free glycerol in biodiesel is required by the different regulatory agencies and upper limits are established as listed in Table 4. Glycerol is the major by-product in the biodiesel production and its removal is necessary. A high content of total glycerol (sum glycerol, mono-, di- and triacylglycerols) and free glycerol can cause problems ranging from the formation of deposits in injectors, sediments inside the fuel storage tank, reducing the engine life. The recommended method for the determination of total and free glycerol in biodiesel is gas chromatography (GC) (Table 5). Both ASTM and EN methods require a derivatization step (silylation reaction) using N-methyl-N-trimethylsilyltrifluoroacetamide (MSTFA) with a few differences in the standard solution concentrations and column temperature conditions. Table 8 and 9 present a comprehensive list of analytical methods developed for the determination of free glycerol and total glycerol, respectively. Generally, all analytical methods listed in Tables 8 and 9 required a sample

Table 8.Methods for free glycerol determination in biodiesel.

Method	Derivatization reagents	Analyte extraction procedures	Detection limit	Ref.
Enzymatic assay	No	2 g sample + 6.0 mL of 0.1 mol L^{-1} HCl	Not cited	[101]
GC-FID/ MS	BSTFA	No	10^{-4}-10^{-5} % wt	[102]
HPSEC	No	No	Not cited	[103]
GC-FID	MSTFA/ pyridine	No	No cited	[104]
GC-FID/ MS	BSTFA	No	10^{-4} % wt	[105]
HPLC-PAD	No	4 g sample + 45 g H_2O + 50 mL hexane stirred for 30 min at 40°C, 2 h standing	1 μg g-1	[106]
GPC-RI	No	PTFE membrane filtration	Not cited	[107]
Greenhill enzymatic assay	No	Not cited	5 ppm	[108]
UV-Vis	$NaIO_4$, acetylacetone	1 g sample + 4 mL hexane + 2 mL distilled water + 2 mL ethanol; 5 min vortex-stirring, 15 min centrifugation at 2000 rpm	Not cited	[109]
HPSEC	No	No	Not cited	[110]
HPLC-RI	No	4-20 g sample + 4.5 mL distilled water; 30 min stirring, 2 h standing	$4x10^{-4}$ % wt	[111]
GC-FID	MSTFA/ pyridine	No	Not cited	[112-114]
CE-UV-Vis-DAD	$NaIO_4$	200 mg sample + 800 mg water + 200 μL chloroform; 10 min vortex-stirring, 15 min centrifugation at 2000 rpm	4.3 mg L^{-1}	[115]
SFC-MS-UV-ELSD	No	Methanol dissolution	Not cited	[116]
IC-PAD	No	5 g sample + 45 g distilled water; 5 min shaking, 5 min standing	$7x10^{-5}$ % wt	[117]
UV-Vis-DAD	$NaIO_4$, acetylacetone	Aqueous extraction, 30 min heating	0.011 % wt	[118]
Cyclic voltammetry	No	2 g sample + 6.0 mL water; 5 min vortex, 10 min centrifugation; C18 filtration	2.3 mg L^{-1}	[119]
FIA-UV-Vis	KIO_4, acetylacetone	1 g sample + 4 mL distilled water; 30 min shaking, 5 min centrifugation at 3000 rpm	$4x10^{-4}$ % wt	[120]

Method	Derivatization reagents	Analyte extraction procedures	Detection limit	Ref.
Enzymatic assay with amperometry	No	400 μL sample + 800 μL distilled water + 800 μL ethanol + 1600 μL heptane; 2 min vortex-stirring and centrifugation	$1x10^{-5}$ % wt	[121]
Enzymatic assay with amperometry	No	400 μL sample + 800 μL distilled water + 800 μL ethanol + 1600 μL heptane; 2 min vortex-stirring and centrifugation	0.013 % wt	[122]
Titration	$NaIO_4$, $C_2H_4(OH)_2$	Aqueous Extraction	$9x10^{-4}$ % wt	[123]
GC-MS	BSTFA, TMCS, MSTFA	No	0.04 μg mL^{-1}	[124]
HPAE-PAD	No	Aqueous extraction	0.5 μg kg-1	[125]
UV-Vis	$NaIO_4$, acetylacetone	Aqueous extraction	0.5 mg L^{-1}	[126]
UV-Vis	No	Solid phase extraction	0.004 % wt	[127]
FIA-Amperometry	No	250 mg sample + 5.0 mL water; 5 min vortex, 10 min centrifugation	5 mg kg^{-1}	[128]
Enzymatic assay with colorimetric detection	No	400 μL sample + 800 μL ethanol + 800 μL distilled water + 1600 μL heptane; 1 min vortex-mixing, 2 min centrifugation	$7.1x10^{-6}$ % wt	[129]
FIA-PAD	No	1.0 g sample + 4.0 mL water; 5 min vortex-stirring and centrifugation	44.2 μg L^{-1}	[130]
Cyclic voltammetry	No	Not cited	33 μmol L^{-1}	[131]
GC-FID-MS	MSTFA	No	0.053 % wt	[132]
GC-FID	MSTFA	No	0.02-0.09 % wt	[133]

CE: Capillary Electrophoresis; DAD: Diode Array Detector; ELSD: Evaporative Light Scattering Detector; FIA: Flow Injection Analysis; FID: Flame Ionization Detector; GC: Gas Chromatography; GPC: Gel Permeation Chromatography; HPAE: High Performance Anion Exchange Chromatography; HPLC: High Performance Liquid Chromatography; HPSEC: High Performance Size Exclusion Chromatography; IC: Ion Chromatography; IR: Infrared Spectrophotometry; PAD: Pulsed Amperometric Detection; RI: Differential Refractive Index Detector; SEC: Size Exclusion Chromatography; SFC: Supercritical Fluid Chromatography; UV-Vis: Ultraviolet Spectrophotometry; BSTFA: N, O-bis (trimethylsilyl)-trifluoroacetamide; MSTFA: N-methyl-N-(trimethylsilyl)-trifluoroacetamide; PTFE: Polytetrafluoroethylene; TMCS: Trimethylchlorosilane.

Table 9. Methods for total glycerol determination in biodiesel.

Method	Derivatization reagents	Sample preparation	Detection limit	Ref.
Enzymatic assay	No	Saponification; Solid phase extraction with C8	Not cited	[101]
Greenhill enzymatic assay	No	Saponification	75 ppm	[108]
GC-FID	MSTFA	No	Not cited	[114]
IC-PAD	No	Saponification; 5 g sample + 45 g distilled water; 5 min shaking, 5 min standing	$7x10^{-4}$ % wt	[117]
UV-Vis-DAD	$NaIO_4$, acetylacetone	Saponification; Aqueous extraction, 30 min heating	0.064 % wt	[118]
Enzymatic assay with amperometry	No	Transesterification; 0.1 g sample + 3.9 mL distilled water + 5.0 mL heptane; 2 min vortex-stirring and centrifugation	$1x10^{-4}$ % wt	[121]
Titration	CH_3NaO, $NaIO_4$, $C_2H_4(OH)_2$	Aqueous Extraction	0.0046 % wt	[123]
HPLC-UV-Vis	9,9-dimetoxifluorene	No	0.05 % wt	[134]
HPSEC-PAD	No	Saponification	0.5 μg kg^{-1}	[125]
UV-Vis	$NaIO_4$, acetylacetone	Saponification; Aqueous extraction	1.4 mg L^{-1}	[126]
Enzymatic assay with colorimetric detection		Transesterification; 0.5 g sample + 2.0 mL distilled water + 2.0 mL heptane; 1 min vortex-mixing, 2 min centrifugation	$7.1x10^{-4}$ % wt	[129]
GC-FID-MS	MSTFA	No	2.458 % w/w	[132]

preparation step. GC methods often required a derivatization step (modification of a functional group of the analyte such as the silylation reaction) which improves the chromatographic performance. Derivatization reagents (when required) are cited and in a separate column other sample preparation steps (analyte extraction procedures) are detailed in accordance with information contained in the literature.

Several GC methods coupled with flame ionization (FID) or mass spectrometric (MS) detectors are listed in Tables 8 and 9 including works published since the early Nineties. The most recent contribution reported a new GC method with reduced sample preparation and analysis time (25 min elution) that can be applied for a wide range of oilseed-derived biodiesels including biodiesel from tallow, babassu, and palm kernel, which contain shorter chain fatty acids and then cannot be accurately analyzed by the EN 14105 method [133]. The EN14105 (and also the ASTM 6584 method) is based on the GC-FID method developed by Plank and Lorbeer [104], who reported the simultaneous determination of glycerol, mono-, di-, and triacylglycerides in C18 methylic esters produced from vegetable oils by the silylation of the free hydroxyl groups using MSTFA. Analytical methods employing different modes of liquid chromatography coupled with varied detectors were also reported but in less extension. The first HPLC method employed pulsed-amperometric detection and thus it was not necessary the derivatization step [106]. Additionally, this HPLC method allows the determination of residual alcohol (methanol or ethanol) [106]. A high performance size exclusion chromatography method was developed for the simultaneous determination of the total amounts of mono-, di-, and triacylglycerides, fatty acid methyl esters, free glycerol and methanol [110]. The method is simple, robust, relatively fast, and required minimal sample treatment (no derivatization steps); however, the SEC columns are quite expensive. Preliminary results using a supercritical fluid chromatography coupled with three different detectors (MS, UV, and ELSD) were reported [116]. Separation of fatty acid methyl esters, free fatty acids, and glycerol was obtained in less than 5 min [116].

The oxidation reaction of glycerol with periodate is the basis of the first spectrophotometric method developed for glycerol determination in biodiesel [109]. The oxidation of glycerol results in the formation of formaldehyde, which is reacted with acetylacetone leading to the formation of 3,5-diacetyl-1,4-dihydrolutidine that can be measured at 410 nm [109]. Other spectrophotometric methods using a similar approach were reported in the literature [118,120,126,127]. The great advantage of the spectrophometric methods is their low-cost, rapidness, accuracy and moderate sensitivity. A capillary electrophoresis method was developed based on the UV detection of iodate generated by the oxidation of glycerol by periodate [115]. Similarly, glycerol in biodiesel was determined by alkaline titration of formic acid generated by the oxidation of glycerol by periodate, which can be considered a very simple method easily assessed by local producers [123].

Electrochemical approaches (cyclic voltammetric, amperometric and pulsed-amperometric detectors) were also developed and coupled with chromatographic techniques, flow-injection methods, and enzymatic reactions. Electrochemical methods present high sensitivity, selectivity, can be easily miniaturized and require portable commercially-available instrumentation. Additionally, glycerol can be electrochemically detected without the use of derivatization reagents. Amperometric detection was often applied due to the easiness of its association with flow techniques such as chromatography or flow-injection methods [106,117,125,128,130]. The selection of amperometry or pulsed-amperometry is more related to the electrochemical oxidation of glycerol at different working electrode materials. Enzymatic assays employed enzymes which specifically convert glycerol to dihydroxyacetone phosphate generating H_2O_2 and consuming oxygen. Then, H_2O generation or oxygen consumption can be easily monitored by electrochemical techniques (similarly to portable glucose-sensors) as well as using colorimetric assays (the commercial-available Greenhill assay provides the formation of a quinoneimine dye that shows maximum absorbance at 540 nm) [108]. The development of such kits for fast and low-cost monitoring of glycerol in biodiesel is an alternative to GC methods which employ bulky instrumentation and organic solvents during the derivatization step. However, enzymes require special condition of storage and limited shelf-life time. To overcome such a drawback, the development of (electro)chemical sensors based on the direct detection of glycerol is promising.

Analytical methods developed for the determination of total glycerol in biodiesel (Table 9) typically reported the conversion of mono-, di-, and triglycerides into glycerol by saponification or transesterification reactions except GC or HPLC methods which separate and directly quantify each component present in biodiesel. Other analytical methods developed for the determination of free glycerol (listed in Table 8) can also be readily adapted to perform the determination of total glycerol.

6.2. Trace Metals

The presence of metals in biodiesel can be arisen from catalyst residues (Na, K, Ca, Mg) and due to corrosion of storage tanks and automotive engine parts (e.g. Al, Cu, Cr, Fe, Mn, Pb, Zn, etc.). Metal ions cause the formation of deposits of insoluble soaps, as well as catalyze polymerization reactions of biodiesel degradation. In this way the different regulatory agencies establishes upper limits as listed in Table 4 for group I (Na and K) and group II (Ca and Mg) cations. However, no limits are established for transition metals which are strong catalysts for biodiesel oxidation even at trace concentrations as previous works have reported [89-94]. Therefore, analytical methods capable of monitoring trace metals in biodiesel are required and this information can be correlated with biodiesel oxidation stability. Table 10 presents a list of analytical methods developed for the determination of metals in biodiesel including sample preparation steps when required and detection limits.

Table 10. Methods for trace metal determination in biodiesel.

Method	Analyte	Sample preparation	Detection limit	Ref.
Inductively coupled plasma optical emission spectrometry	Ca, K, Mg, Na	Sample dilution in kerosene	0.4 – 0.9 mg kg^{-1}	[135]
Inductively coupled plasma mass spectrometry	31 elements	Sample dilution in kerosene	0.0109 – 22.7 µg kg^{-1}	[136]
Inductively coupled plasma optical emission spectrometry (axial viewing)	Ca, Mg, K	Sample dilution in ethanol	0.005 – 0.1 µg g^{-1}	[137]
Inductively coupled plasma optical emission spectrometry (axial viewing)	Ca, Cu, Fe, Mg, Mn, Na	Sample emulsion with Triton X-100 and water	0.007 – 0.165 µg g^{-1}	[138]
Flame atomic absorption spectrometry	Na, K	Sample microemulsion with n-pentanol and Triton X-100	0.1 and 0.06 µg g^{-1}	[139]
Flame atomic emission spectrometry	Na, K	Sample microemulsion with n-pentanol and aqueous acid solution	0.1 µg g^{-1}	[140]
Graphite-furnace atomic spectrometry	As	Sample microemulsion with n-pentanol and aqueous acid solution	0.3 mg kg^{-1}	[141]
Graphite-furnace atomic spectrometry	Cu, Pb, Ni, Cd	Focused microwave wet digestion	Not cited	[142]
Cold-vapor atomic fluorescence spectrometry	Hg	Sample microemulsion with n-pentanol and Triton X-100	0.2 µg kg^{-1}	[143]
Inductively coupled plasma optical emission spectrometry	28 elements	Microwave acid digestion	0.1 – 136.5 µg g^{-1}	[144]
Inductively coupled plasma optical emission spectrometry and mass spectrometry	19 elements	Microwave acid digestion	0.001 – 0.4 µg g^{-1}	[145]
Graphite-furnace atomic spectrometry	Ni, Cd	Sample microemulsion with Triton X-100 and acid aqueous solution	0.9 and 0.1 µg L^{-1}	[146]
Flame atomic emission spectrometry	Cu, Ni, Zn	Pre-concentration by adsorption on chitosan microspheres	Not cited	[147]
Flame atomic emission spectrometry	Na, K	Sample dilution in ethanol	2.16 and 2.00 mg kg^{-1}	[148]
Inductively coupled plasma mass spectrometry	32 elements	Microwave acid digestion	10^{-6} mg kg^{-1}	[149]
Potentiometry	K	Not required	0.01 ppm	[150]
Potentiometry	K	Liquid-liquid extraction	1.9×10^{-3} mol L^{-1}	[151]
Ion chromatography	Na, K, Ca, Mg	Liquid-liquid extraction, heating, sonication	0.11 – 0.42 mg kg^{-1}	[152]
Ion chromatography	Na, K	Liquid-liquid extraction	Not cited	[153]
Capillary electrophoresis	Ca, K, Mg, Na	Liquid-liquid extraction	0.07 – 0.14 mg L^{-1}	[154]
Square-wave stripping voltammetry	Sn	Dry-ashing decomposition	0.14 µg L^{-1}	[155]
Potentiometric stripping analysis	Cu	Sample dilution in hydroethanolic electrolyte	200 ng g^{-1}	[81]

Atomic absorption spectrometry (AAS) and inductively coupled plasma optical emission spectrometry (ICP OES) are the recommended technique by Brazilian and European standards (Table 5). The first analytical methods reported for the determination not only of Na, K, Ca and Mg as well as other metals were the spectrometric methods such as the recommended techniques AAS and ICP OES. AAS coupled to graphite furnace (electrothermal atomization) provides higher sensitivity and for this reason it has been applied for metal determination in biodiesels, although relatively higher cost and longer analysis time are verified. The main advantage of ICP OES is the lower chemical and spectral interferences and multi-element determination; dozens of elements can be analyzed at once after simple sample dilution, including phosphorus and sulfur which are also regulated by European, American and Brazilian norms (Table 4 and 5). However, the overall analysis costs can be so elevated that this method is not available to all analytical laboratories of quality control. ICP MS provides higher sensitivity than ICP OES if sub-ppb levels are required. Most of the spectrometric methods require very simple preparation steps such as sample dilution or sample microemulsion with surfactants, which is an elegant strategy to avoid sample digestions using of concentrated acids. Pre-concentration steps can be applied for AAS with flame atomization which is a low-cost spectrometer but does not provide low detection limits generally required for trace metal determinations. Then pre-concentration steps supplies the low sensitivity of flame atomic absorption (or emission) spectrometers. More details on the use of spectrometric methods to determine metals and metalloids in automotive fuels (including biodiesel) can be found in a recent review [156].

Separation techniques such as ion chromatography and capillary electrophoresis were applied for simultaneous determination of cations in biodiesel after a simple liquid-liquid extraction [152-154]. The capillary electrophoretic (CE) method presented faster separation of ions, employed very low sample volumes and capillaries presents much lower cost than IC columns. Moreover, the CE method can also be used for glycerol determination after its chemical conversion by periodate as a previous method described [115].

Electroanalytical methods for metal determination in biofuels were recently reviewed [157,158]. Potentiometry is a well-known technique which allow sensitive detection of K in aqueous solutions and were also applied for biodiesel analysis after a liquid-liquid extraction [151]. An ion-selective electrode sensor associated with cellophane semi-permeable membrane was applied for K determination in biodiesel without a sample preparation step [150]. The elimination of the sample preparation step is a tendency in modern analytical methods since this step may provide sample contamination, analyte losses, and high analysis time, whose characteristics are avoided when developing an analytical method for routine analyses. Electroanalytical potentiometric stripping analysis was applied for Cu determination in biodiesel after its dilution in hydroethanolic electrolyte [81]. This is the first report on the use of electroanalysis for metal determination in biodiesel. Electroanalytical methods provide real advantages for routine analysis such as high sensitivity and selectivity employing a low-cost portable instrumentation. Similar technology applied for gluco-sensors (using disposable sensors) can be extended to on-site analysis of biodiesel, aiming not only the determination of trace metals but also other species in the biofuel [157,158].

6.3. Other Trace Contaminants

Other trace contaminants can be found in biodiesel such as water (moisture), residual alcohol (typically methanol and ethanol), sterols, phosphorus and sulfur. Table 4 shows the upper limits established for water and residual alcohol and Table 5 lists the recommended methods for the analysis of each parameter. The GC method developed Mittelbach et al [105] can also be applied for the determination of alcohol residues in biodiesel as well as the HPLC-PAD method [106]. A flow analysis method coupled to a membrane extraction was developed for methanol determination in biodiesel [159]. Methanol was detected by spectrophotometry (at 240 nm) after reaction with alcohol oxidase in aqueous solution [159].

Sterols are minor components found in animal fats and vegetable oils and that occur in biodiesel due to their solubility in the biofuel. Sterol glycosides can accelerate precipitate formation in biodiesel even at room temperature and block fuel filters [160-162]. HPLC methods [160,161], mass spectrometry [161], and a GC MS method [162] have been exploited for the determination of sterol glycosides. More information about the use of chromatographic techniques for biodiesel and biodiesel blends (not only including the analysis of sterols) can be found in a recent review [163].

The contents of phosphorus and sulfur in biodiesel are additional parameters regulated by European, American and Brazilian norms. Some ICP OES methods reported for metal determinations in biodiesel (Table 10) were also applied for the determination of phosphorus and sulfur [136,144,145,149]. A graphite-furnace AAS method was developed for the direct determination of phosphorus in biodiesel using a solid sampling accessory [164]. Simpler methodologies using spectrophotometry were developed [165,166]. In the first work biodiesel samples were mineralized (dry ashing) and their residue containing phosphate was reacted with 1-amino-2-naphthol-4-sulfonic acid to form a blue molybdenum complex [165]. In the second work biodiesel samples were digested using an acid mixture and the obtained solution containing phosphate was mixed with ammonium molybdate and potassium and antimony tartrate ion to form phosphomolybdic acid (yellow) [166]. An electroanalytical method was developed for phosphorus determination in biodiesel using a phosphomolybdic modified electrode [167]. X-ray fluorescence [168], and improvements on ICP OES [169] and ICP MS [170] methods for sulfur determination in biodiesel have been reported.

Analytical methods for monitoring of the transesterification reaction and for the determination of fatty mono-alkyl esters in biodiesel-diesel blends or in pure biodiesel as well as other parameters such as biodiesel oxidation and thermal stability are well-addressed in the review by Monteiro et al. [171] and is not discussed in this text.

7. CONCLUSION AND PERSPECTIVE

Much effort has been dedicated to the development of new methods for biodiesel production and improvement of the traditional ones. A large variety of raw materials have been investigated for biodiesel production in the world, and Brazil has favourable environment and climate conditions in this scenario for the development of biodiesels from new oilseeds although the large Brazilian soybean biodiesel production. The production of biodiesels of new oilseeds has economic impact since small local producers can contribute for biodiesel production and the special characteristics of these biodiesels may improve the physical-chemical properties (e.g. acidity, oxidation stability, viscosity, etc.) of other biodiesels by using blends of biodiesels. Microalgae for biodiesel production is a promising sustainable source which has received incresing interest from researchers worldwide. Metallic corrosion is a real problem in storage stability of biodiesels and inside diesel engines and thus requires constant investigation. The monitoring of trace metals in the biofuel may provide information on how trace metals really affect biodiesel oxidation stability. Moreover, the presence of antioxidants and corrosion inhibitors in biodiesel plays key role on the metallic corrosion and deeper investigations would indicate the real need of additives in the biofuel and their required concentration. Modern analytical methods developed for monitoring contaminants in biodiesel have been reported and the creation of portable devices for the quality control of biofuels is a promising tendency. In the future, any local producer would have access to fast and reliable technologies to certificate the quality of its biodiesel during the production process. Additionaly, the real-time and on-site analysis of biofuels is desirable at local gas stations which receive large volumes of biodiesel and need to quickly check the real quality of the biofuel.

ACKNOWLEDGMENTS

The authors are grateful to FAPEMIG (APQ-00236-12 and APQ-02276-09), CNPq (476667/2008-9 and 305227/2010-6) and CAPES for the financial support.

REFERENCES

1. Network of Sustainability. What is Sustainability? Available in http://www.sustentabilidade.org.br. Accessed on September 10, 2012.
2. Energy Information Administration (EIA). International Energy Outlook 2011 – "World Energy Demand and Economy Outlook". Available in (http://www.oecd.org). Accessed on September 10, 2012.
3. SEBRAE – Spelling Book on Biodiesel. Available in www.biblioteca.biodiesel.sebrae.com.br Accessed on September 10, 2012.
4. Diesel R. Die Entstehung des Dieselmotors. Berlin: Verlag von Julius Springer; 1913
5. Knothe G, Gerpen JV, Krahl J, Ramos LP. The Biodiesel Handbook, Berlin: Ed. Edgard Blücher; 2006
6. Fangrui M, Hanna MA. Biodiesel: A review. Bioresource Technology 1999; 70 1-15.
7. Expedito Parente – Avalaible in http://www.inovacao.unicamp.br /report/entre-expedito.php Accessed on September 10, 2012.
8. ANP – National Agency of Petroleum, Natural Gas and Biofuels. The ANP and the Biofuels. Available in http://www.anp.gov.br. Accessed on September 10, 2012.
9. Secretary of Petroleum, Natural Gas and Renewable Fuels and Department of Renewable Fuels. Monthly Bulletin of Renewable Fuels, Issue No. 53, June 2012.
10. CONAB - National Supply Company. Available in http://www.conab.gov.br. Accessed on September, 10, 2012.

11. Jatropha Project EPAMIG /FINEP. Final Report relating 1st period finalized in March 31, 1985. Available in http://www.epamig.br /index.php?option=com_docman&task=doc_download&gid=14&Itemid=109 Accessed on September 10, 2012.
12. Borsato D, Maia ECR, Dall'Antonia LH, da Silva, HC, Pereira JL. Kinetics of oxidation of biodiesel from soybean oil mixed with TBHQ: determination of storage time. Química Nova 2012; 35 733-737.
13. ABRAPA – Brazilian Association of Cotton Producers. Management Report, Biennium 2008-2010, 2010, 1, 12.
14. ANVISA – National Agency of Sanitary Surveillance. Resolution No. 482, September 23, 1999.
15. Jorge FAS. Study of adsorption of gossypol existing in cottonseed oil. Dissertation of Master's degree, Fortaleza, 2006.
16. EMBRAPA - Brazilian Agricultural Research Company. Agroenergy review No 2. EMBRAPA Agroenergy May, 2011.
17. USDA – United States Department of Agriculture. Canola Production and Processing. Available in http://www.ers.usda.gov. Accessed on July 30, 2012.
18. Georgogianni KG, Katsoulidis AK, Pomonis PJ, Manos G, Kontominas MG. Transesterification of rapeseed oil for the production of biodiesel using homogeneous and heterogeneous catalysis. Fuel Processing Technology 2009; 90 671-676.
19. Endalew AK, Kiros Y, Zanzi R. Heterogeneous catalysis for biodiesel production from *Jatropha curcas* oil (JCO). Energy 2011; 36 2693–2700.
20. Heller J. Physic Nut: jatropha curcas L. International Plant Generic Resources Institute, 1996.
21. Bártoli JAA. Physic Nut (*jatropha curcas*) Cultivation in Honduras Handbook, Agricultural Communication Center of the Honduran Foundation for Agricultural Research, 2008.
22. Oliveira JS, Leite PM, Souza LB, Mello VM, Silva EC, Rubim JC Meneghetti SMP, Suarez PAZ. Characteristics and composition of Jatropha gossypol and *jatropha curcas* L. oils and application for biodiesel production. Biomass and Bioenergy 2009;33 449-453.
23. Kumar A, Sharma S. An evaluation of multipurpose oil seed crop for industrial uses (*jatropha curcas* L.): A review, Industrial Crops and Products 2008;28 1-10.

24. Koris A, Vatai G. Dry degumming of vegetable oils by membrane filtration, Desalination 2002;(148) 149-153.
25. Sarin, A.; Arora, R.; Singh, N.P.; Sarin, R.; Malhota, R.K.; Sharma, M. Synergistic effect of metal deactivator and antioxidant on oxidation stability of metal contaminated Jatropha biodiesel. Energy 2010;35 2333-2337.
26. ABIOVE - Brazilian Association of Oil Industries Vegetables. Monthly Statistic. Reference Month – July, 2012.
27. Kulkarni MG, Dalai AK. Waste Cooking Oil An Economical Source for Biodiesel: A Review. Industrial & Engineering Chemistry Research 2006;45 2901-2913.
28. Zheng S, Kates M, Dubé MA, McLean DD. Acid-catalyzed production of biodiesel from waste frying oil. Biomass and Bioenergy 2006;30 267-272.
29. Refaat AA, Attia NK, Sibak HA, El Sheltawy ST, El Diwani GI. Production optimization and quality assessment of biodiesel from waste vegetable oil. International Journal of Environmental Science and Technology 2008;5 75-82.
30. Canakci M. The potencial of restaurant waste lipids as biodiesel feedstocks. Bioresource Tecnhology 2007;98 183-190.
31. Tang H, Salley SO, Simon Ng KY. Fuel properties and precipitate formation at low temperature in soy-, cottonseed-, and poultry fat-based biodiesel blends. Fuel 2008;87 3006-3017.
32. Da Silva JPV, Serra TM, Gossmann M, Wolf CR, Meneghetti MR, Meneghetti SMP. Moringa Oleira oil: Studies of characterization and biodesel production. Biomass and Bioenergy 2010;34 1527-1530
33. Rashid, U.; Anwar, F.; Moser, B.R.; Knothe, G. Moringa oleifera oil: A possible source of biodiesel. Bioresource Technology 2008;99 8175-8179
34. Ayerza, R. Seed yield components, oil content, and fatty acid composition of two cultivars of Moringa (Moringa oleifera Lam.) growing in the Arid Chaco of Argentina. Industrial Crops and Products 2011;33 389-394.
35. Azam MM, Waris A, Nahar NM. Prospects and potential of fatty acid methyl esters of some non-traditional seed oils for use as biodiesel in India. Biomass and Bioenergy 2005;29 293-302.

36. Lalas S, Tsaknis J. Characterization of Moringa Oleifera Seed oil Variety "Priyakulam". Journal of Food Composition and Analysis 2002;15 65-77.
37. Ibrahim SS, Ismail, M, Samuel G, Kamal E, El Azhari T. Benseed: A potential oil source. Agriculture Research and Review. 1974;52 47-50.
38. Kegl B. Effects of biodiesel on emissions of a bus diesel engine. Bioresource Technology 2008;99 863-873.
39. Meher LC, Vidya Sagar D, Naik SN. Technical aspects of biodiesel production by transesterification--a review. Renewable and Sustainable Energy Reviews 2006;10 248-268.
40. Srivastava A, Prasad R. Triglycerides-based diesel fuels. Renewable and Sustainable Energy Reviews 2000;4 111-133.
41. Refaat AA, El Sheltawy ST, Sadek KU. Optimum reaction time, performance and exhaust emissions of biodiesel produced by microwave irradiation. International Journal of Environmental Science and Technology 2008;5 315-322.
42. Encinar JM, González JF, Rodríguez-Reinares A. Ethanolysis of used frying oil. Biodiesel preparation and characterization. Fuel Processing Technology 2007;88 513-522.
43. Ma F, Hanna MA. Biodiesel production: a review. Bioresource Technology 1999;70 1-15.
44. Gerpen JV. Biodiesel processing and production. Fuel Processing Technology 2005;86 1097-1107.
45. Fukuda H, Kondo A, Noda H. Biodiesel fuel production by transesterification of oils. Journal of Bioscience and Bioengineering 2001;92 405-416.
46. Marchetti JM, Miguel VU, Errazu AF. Possible methods for biodiesel production. Renewable and Sustainable Energy Reviews 2007;11 1300-1311.
47. Singh SP, Singh D. Biodiesel production through the use of different sources and characterization of oils and their esters as the substitute of diesel: A review. Renewable and Sustainable Energy Reviews 2010;14 200-216.
48. Sinha S, Agarwal AK, Garg S. Biodiesel development from rice bran oil: Transesterification process optimization and fuel characterization. Energy Conversion and Management 2008;49 1248-1257.

49. Crabbe E, Nolasco-Hipolito C, Kobayashi G, Sonomoto K, Ishizaki A. Biodiesel production from crude palm oil and evaluation of butanol extraction and fuel properties. Process Biochemistry 2001;37 65-71.
50. Faccini CS, Da Cunha ME, Moras MAS, Krause Lc, Manique MC, Rodrigues MRA, Benvenutti EV, Caramao EB. Journal of the Brazilian Chemical Society 2011; 22 558-563.
51. Mittelbach M. Diesel fuel derived from vegetable oils, VI: Specifications and quality control of biodiesel. Bioresource Technology 1996;56 7-11.
52. Tomasevic AV, Siler-Marinkovic SS. Methanolysis of used frying oil. Fuel Processing Technology 2003;81 1-6.
53. Lima AL, Lima AP, Portela FM, Santos DQ, Neto WB, Hernández-Terrones MG, Fabris JD. Parameters of transesterification reaction of corn oil with ethanol for biodiesel production. Eclética Química 2010;35 101-106.
54. Freedman B, Pryde EH, Mounts TL. Journal of the American Oil Chemists' Society 1984;61 1638-1643.
55. Schwab AW, Bagby MO, Freedman B. Preparation and properties of diesel fuels from vegetable oils. Fuel 1987;66 1372-1378.
56. Aksoy HA, Becerik I, Karaosmanoglu F, Yatmaz HC, Civelekoglu H. Utilization prospects of Turkish raisin seed oil as an alternative engine fuel. Fuel 1990;69 600-603.
57. Vicente G, Martínez M, Aracil J. Integrated biodiesel production: a comparison of different homogeneous catalysts systems. Bioresource Technology 2004;92 297-305.
58. Stamenkovic OS, Lazic ML, Todorovic ZB, Veljkovic VB, Skala DU. The effect of agitation intensity on alkali-catalyzed methanolysis of sunflower oil. Bioresource Technology 2007;98 2688-2699.
59. Arzamendi G, Campo I, Arguiñarena E, Sánchez M, Montes M, Gandía LM. Synthesis of biodiesel with heterogeneous NaOH/alumina catalysts: Comparison with homogeneous NaOH. Chemical Engineering Journal 2007;134 123-130.
60. Xie W, Huang X, Li H. Soybean oil methyl esters preparation using NaX zeolites loaded with KOH as a heterogeneous catalyst. Bioresource Technology 2007;98 936-939.

61. Al-Widyan MI, Al-Shyoukh AO. Experimental evaluation of the transesterification of waste palm oil into biodiesel. Bioresource Technology 2002;85 253-256.
62. Zheng S, Kates M, Dubé MA, McLean DD. Acid-catalyzed production of biodiesel from waste frying oil. Biomass and Bioenergy 2006;30 267-272.
63. Shieh CJ, Liao HF, Lee CC. Optimization of lipase-catalyzed biodiesel by response surface methodology. Bioresource Technology 2003;88 103-106.
64. Iso M, Chen B, Eguchi M, Kudo T, Shrestha S. Production of biodiesel fuel from triglycerides and alcohol using immobilized lipase. Journal of Molecular Catalysis B: Enzymatic 2001;16 53-58.
65. Saka S, Kusdiana D. Biodiesel fuel from rapeseed oil as prepared in supercritical methanol. Fuel 2001;80 225-231.
66. Demirbas AH, Demirbas I. Importance of rural bioenergy for developing countries. Energy Conversion and Management 2007;48 2386-2398.
67. He H, Wang T, Zhu S. Continuous production of biodiesel fuel from vegetable oil using supercritical methanol process. Fuel 2007;86 442-447.
68. PETROBRAS – Brazilian Oil Company. Available in http://www.petrobras.com.br/pt/quem-somos/perfil/atividades/producao-biocombustiveis/downloads/pdf/Cartilha-Biocombustiveis-PORTUGUES.pdf. Accessed on September 10, 2012.
69. ANP, Resolução n0 42 de novembro de 2004, in, 2010.
70. Sharma YC, Singh B, Upadhyay SN. Advancements in development and characterization of biodiesel: A review. Fuel 2008;87 2355-2373.
71. Benjumea P, Agudelo J, Agudelo A. Basic properties of palm oil biodiesel-diesel blends. Fuel 2008;87 2069-2075.
72. Sharon H, Karuppasamy K, Soban Kumar, Sundaresan D R A. A test on DI diesel engine fueled with methyl esters of used palm oil. Renewable Energy 2012;47 160-166.
73. Singh B, Korstad J, Sharma YC. A critical review on corrosion of compression ignition (CI) engine parts by biodiesel and biodiesel blends and its inhibition. Renewable and Sustainable Energy Reviews 2012;16 3401-3408.

74. Díaz-Ballote L, López-Sansores JF, Maldonado-López L, Garfias-Mesias LF. Corrosion behavior of aluminum exposed to a biodiesel. Electrochemistry Communications 2009;11 41-44.
75. Haseeb ASMA, Fazal MA, Jahirul MI, Masjuki HH. Compatibility of automotive materials in biodiesel: A review. Fuel 2011;90 922-931.
76. Fazal MA, Haseeb ASMA, Masjuki HH. Biodiesel feasibility study: An evaluation of material compatibility; Performance; emission and engine durability. Renewable and Sustainable Energy Reviews 2011;15 1314-1324.
77. Ambrozin ARP, Kuri SE, Monteiro MR. Metallic corrosion related to mineral fuels and biofuels utilization. Quimica Nova 2009;32 1910-1916.
78. Norouzi S, Eslami F, Wyszynski ML, Tsolakis A. Corrosion effects of RME in blends with ULSD on aluminium and copper. Fuel Processing Technology 2012; 104 204-210.
79. Fazal MA, Haseeb ASMA, Masjuki HH. Comparative corrosive characteristics of petroleum diesel and palm biodiesel for automotive materials. Fuel Processing Technology 2010;91 1308-1315.
80. Kaul S, Saxena RC, Kumar A, Negi MS, Bhatnagar AK, Goyal HB, et al. Corrosion behavior of biodiesel from seed oils of Indian origin on diesel engine parts. Fuel Processing Technology 2007;88 303-307.
81. Almeida ES, Monteiro MANA, Montes RHO, Mosquetta R, Coelho NMM, Richter EM, Munoz RAA. Direct Determination of Copper in Biodiesel Using Stripping Analysis. Electroanalysis 2010; 22 1846-1850.
82. Haseeb ASMA, Masjuki HH, Ann LJ, Fazal MA. Corrosion characteristics of copper and leaded bronze in palm biodiesel. Fuel Processing Technology 2010;91 329-334.
83. Fazal MA, Haseeb ASMA, Masjuki HH. Degradation of automotive materials in palm biodiesel. Energy 2012;40 76-83.
84. Geller DP, Adams TT, Goodrum JW, Pendergrass J. Storage stability of poultry fat and diesel fuel mixtures: Specific gravity and viscosity. Fuel 2008;87 92-102.
85. Hu E, Xu Y, Hu X, Pan L, Jiang S. Corrosion behaviors of metals in biodiesel from rapeseed oil and methanol. Renewable Energy 2012;37 371-378.

86. Fazal MA, Haseeb ASMA, Masjuki HH. Effect of temperature on the corrosion behavior of mild steel upon exposure to palm biodiesel. Energy 2011;36 3328-3334.
87. Salamanca M, Mondragón F, Agudelo JR, Benjumea P, Santamaría A. Variations in the chemical composition and morphology of soot induced by the unsaturation degree of biodiesel and a biodiesel blend. Combustion and Flame 2012;159 1100-1108.
88. McCormick RL, Ratcliff M, Moens L, Lawrence R. Several factors affecting the stability of biodiesel in standard accelerated tests. Fuel Processing Technology 2007;88 651-657.
89. Knothe G, Dunn RO. Dependence of oil stability index of fatty compounds on their structure and concentration and presence of metals. Journal of the American Oil Chemists' Society 2003;80 1021-1026.
90. Sarin A, Arora R, Singh NP, Sharma M, Malhotra RK. Influence of metal contaminants on oxidation stability of Jatropha biodiesel. Energy 2009;34 1271-1275.
91. Sarin A, Arora R, Singh NP, Sarin R, Malhotra RK. Oxidation Stability of Palm Methyl Ester: Effect of Metal Contaminants and Antioxidants. Energy & Fuels 2010;24 2652-2656.
92. Sarin A, Arora R, Singh NP, Sarin R, Sharma M, Malhotra RK. Effect of Metal Contaminants and Antioxidants on the Oxidation Stability of the Methyl Ester of Pongamia. Journal of the American Oil Chemists' Society 2010;87 567-572.
93. Santos NA, Damasceno SS, De Araujo PHM, Marques VC, Rosenhaim R, Fernandes VJ, Queiroz N, Santos IMG, Maia AS, Souza AG, Energy & Fuels 2011; 25 4190-4194.
94. Jain S, Sharma MP. Correlation development for effect of metal contaminants on the oxidation stability of Jatropha curcas biodiesel. Fuel 2011;90 2045-2050.
95. Fazal MA, Haseeb ASMA, Masjuki HH. Effect of different corrosion inhibitors on the corrosion of cast iron in palm biodiesel. Fuel Processing Technology 2011;92 2154–2159.
96. Jain S, Sharma MP. Stability of biodiesel and its blends: A review. Renewable and Sustainable Energy Reviews 2010;14 667-678.
97. Liang YC, May CY, Foon CS, Ngan MA, Hock CC, Basiron Y. The effect of natural and synthetic antioxidants on the oxidative stability of palm diesel. Fuel 2006;85 867-870.

98. Domingos AK, Saad EB, Vechiatto WWD, Wilhelm HM, Ramos LP. The influence of BHA, BHT and TBHQ on the oxidation stability of soybean oil ethyl esters (biodiesel). Journal of the Brazilian Chemical Society 2007;18 416–423.
99. Tang H, De Guzman RC, Simon-Ng K, Salley SO. Effect of Antioxidants on the Storage Stability of Soybean-Oil-Based Biodiesel. Energy & Fuels 2010;24 2028–2033.
100. Almeida ES, Portela FM, Sousa RMF, Daniel D, Terrones MGH, Richter EM, Munoz RAA. Behaviour of the antioxidant tert-butylhydroquinone on the storage stability and corrosive character of biodiesel. Fuel 2011;90 3480-3484.
101. Bailer J, de Heuber K. Determination of saponifiable glycerol in bio-diesel. Fresenius Journal of Analytical Chemistry 1991;340 186-186.
102. Mittelbach M. Diesel fuel derived from vegetable oils, V (1): Gas chromatographic determination of free glycerol in transesterified vegetable oils. Chromatographia. 1993;37 623-626.
103. Fillieres R, Mlayah BB, Delmas M. Ethanolysis of rapeseed oil: quantitation of ethyl esters, mono-, di-, and triglycerides and glycerol by high performance size-exclusion chromatography. Journal of the American Oil Chemists' Society. 1995;72(4):427-432.
104. Plank C, Lorbeer E. Simultaneous determination of glycerol, and mono-, di- and triglycerides in vegetable oil methyl esters by capillary gas chromatography. Journal of Chromatography A. 1995;697 461-468.
105. Mittelbach M, Roth G, Bergmann A. Simultaneous gas chromatographic determination of methanol and free glycerol in biodiesel. Chromatographia. 1996;42 431-434.
106. Lozano P, Chirat N, Graille J, Pioch D. Measurement of free glycerol in biofuels. Fresenius Journal of Analytical Chemistry 1996; 354 319-322.
107. Darnoko D, Cheryan M, Perkins EG. Analysis of vegetable oil transesterification products by gel permiation chromatography. J Liq Chrom Rel Technol. 2000;23 2327-2335.
108. Greenhill SJ, inventor Stepan Company, assignee. Method for determination of free and combined glycerin in biodiesel. 2004.
109. Bondioli P, Della Bella L. An alternative spectrophotometric method for the determination of free glycerol in biodiesel.

European Journal of Lipid Science and Technology. 2005;107 153-157.

110. Gandia LM, Arzamendi G, Arguinarena E, Campo I. Monitoring of biodiesel production: Simultaneous analysis of the transesterification products using size-exclusion chromatography. Chemical Engineering Journal. 2006;122 31-40.

111. Hajek M, Skopal F, Machek J. Determination of free glycerol in biodiesel. European Journal of Lipid Science and Technology. 2006;108 666-669.

112. Munari F, Cavagnino D, Cadoppi A. Determination of free and total glycerin in pure biodiesel (B100) by GC in compliance with EN 14105. Thermo Fisher Scientific. 2007(10215):8.

113. Nguyen N, Kelly K, Countryman S. Reliable analysis of glycerin in biodiesel using a high temperature non-metal GC column. LC/GC North America. 2007;26(2) 60.

114. McCurry JD, Wang C-X. Improving the analysis of biodiesel using capillary flow technology: Agilent Technologies; 2007 Contract No.: Document Number|.

115. Goncalves-Filho LC, Micke GA. Development and validation of a fast method for determination of free glycerol in biodiesel by capillary electrophoresis. Journal of Chromatography A. 2007;1154 477-480.

116. Cole J, Lefler J, Rui C. Fast separation of FFA, FAME and glycerol for biodiesel analysis by supercritical fluid chromatography. LC-GC Europe. 2008; 40-41.

117. Gandhi J, Wille A, Steinbach A. Ion Chromatographic Determination of Free and Total Glycerol in Biodiesel and Biodiesel Blends. Lc Gc Europe. 2009: 8-14.

118. de Castro MDL, Pinzi S, Capote FP, Jimenez JR, Dorado MP. Flow injection analysis-based methodology for automatic on-line monitoring and quality control for biodiesel production. Bioresource Technology 2009;100 421-427.

119. Lourenco LM, Stradiotto NR. Determination of free glycerol in biodiesel at a platinum oxide surface using potential cycling technique. Talanta. 2009;79 92-96.

120. Rocha FRP, Silva SG. A flow injection procedure based on solenoid micro-pumps for spectrophotometric determination of free glycerol in biodiesel. Talanta. 2010;83 559-564.

121. D'Elia E, Luetkmeyer T, dos Santos RM, da Silva AB, Amado RS, Vieira ED. Analysis of Free and Total Glycerol in Biodiesel Using an Electrochemical Assay Based on a Two-Enzyme Oxygen-Electrode System. Electroanalysis. 2010;22 995-999.
122. D'Elia E, Pêgas MM, Amado RS, de Castro EV. Analysis of free glycerol in biodiesel using an electrochemical assay based on a two-enzyme platinum microelectrode system. Journal of Applied Electrochemistry. 2010;40 2061-2063.
123. Pisarello ML, Dalla Costa BO, Veizaga NS, Querini CA. Volumetric method for free and total glycerin determination in biodiesel. Industrial & Engineering Chemistry Research 2010;49 8935-8941.
124. Hollebone BP, Yang ZY, Wang ZD, Yang C, Landriault M. Determination of polar impurities in biodiesels using solid-phase extraction and gas chromatography-mass spectrometry. Journal of Separation Science 2011;34 409-421.
125. Christison TT, De Borba BM, Rohrer JS. Determination of Free and Total Glycerol in Biodiesel. Lc Gc North America. 2011:39.
126. Morales-Rubio A, Silva SG, de La Guardia M, Rocha FRP. Sequential spectrofluorimetric determination of free and total glycerol in biodiesel in a multicommuted flow system. Analytical and Bioanalytical Chemistry.2011;401 365-371.
127. Halaweish F, Mercer EJ. Determination of Free Glycerol in Biodiesel via Solid-Phase Extraction and Spectrophotometric Analysis. Journal of the American Oil Chemists' Society 2011;88 655-659.
128. Paixão TRLC, Maruta AH. Flow injection analysis of free glycerol in biodiesel using a cooper electrode as an amperometric detector. Fuel 2012;91 187-191.
129. Valdez HdC, Amado RS, Souza FCd, D'Elia E, Vieira EdC. Determinação de glicerol livre e total em amostras de biodiesel por método enzimático com detecção colorimétrica. Química Nova 2012;35 601-607.
130. Barbosa TGG, Richter EM, Muñoz RAA. Flow-injection pulsed-amperometric determination of free glycerol in biodiesel at a gold electrode. Electroanalysis 2012 May 2012;24 1160-1163.
131. Tehrani RMA, Ghani SA. Electrocatalysis of free glycerol at a nanonickel modified graphite electrode and its determination in biodiesel. Electrochimica Acta 2012;70 153-157.

132. Dias AN, Cerqueira MBR, de Moura RR, Kurz MHS, Clementin RM, D'Oca MGM, et al. Optimization of a method for the simultaneous determination of glycerides, free and total glycerol in biodiesel ethyl esters from castor oil using gas chromatography. Fuel 2012;94 178-183.
133. Prados CP, Rezende DR, Batista LR, Alves MIR, Antoniosi Filho NR. Simultaneous gas chromatographic analysis of total esters, mono-, di- and triacylglycerides and free and total glycerol in methyl or ethyl biodiesel. Fuel 2012;96 476-481.
134. Chadha A, Reddy SR, Titu D. A Novel Method for Monitoring the Transesterification Reaction of Oil in Biodiesel Production by Estimation of Glycerol. Journal of the American Oil Chemists' Society 2010 ;87 747-754.
135. Edlund M, Visser H, Heitland P. Analysis of biodiesel by argon–oxygen mixed-gas inductively coupled plasma optical emission spectrometry. Journal of Analytical Atomic Spectrometry 2002; 17 232-235.
136. Woods GD, Fryer FI. Direct elemental analysis of biodiesel by inductively coupled plasma–mass spectrometry. Analytical and Bioanalytical Chemistry 2007; 389 753-761.
137. Santos EJ, Herrmann AB, Chaves ES, Vecchiatto WWD, Schoemberger AC, Frescura VLA, Curtius AJ. Simultaneous determination of Ca, P, Mg, K and Na in biodiesel by axial view inductively coupled plasma optical emission spectrometry with internal standardization after multivariate optimization. Journal of Analytical Atomic Spectrometry 2007; 22 1300-1303.
138. Sousa RM, Leocadio LG, Silveira CLP. ICP OES Simultaneous Determination of Ca, Cu, Fe, Mg, Mn, Na, and P in Biodiesel by Axial and Radial Inductively Coupled Plasma-Optical Emission Spectrometry. Analytical Letters 2008; 41 1615-1622.
139. Jesus A, Silva MM, Vale MGR. The use of microemulsion for determination of sodium and potassium in biodiesel by flame atomic absorption spectrometry. Talanta 2008; 74 1378-1384.
140. Chaves ES, Saint Pierre TD, Santos EJ, Tormen L, Bascunan VLAF, Curtius AJ. Determination of Na and K in Biodiesel by Flame Atomic Emission Spectrometry and Microemulsion Sample Preparation. Journal of the Brazilian Chemical Society 2008; 19(5) 856-861.

141. Vieira MA, Oliveira LCC, Gonçalves RA, Souza V, Campos RC. Determination of As in Vegetable Oil and Biodiesel by Graphite Furnace Atomic Absorption SpectrometryEnergy Fuels 2009; 23 5942-5946.

142. Lobo FA, Gouveia D, Oliveira AP, Pereira-Filho E, Fraceto LF, Dias-Filho NL, Rosa AH. Comparison of the univariate and multivariate methods in the optimization of experimental conditions for determining Cu, Pb, Ni and Cd in biodiesel by GFAAS. Fuel 2009; 88 1907-1914.

143. Aranda PR, Pacheco PH, Olsina RA, Martinez LD, Gil RA. Total and inorganic mercury determination in biodiesel by emulsion sample introduction and FI-CV-AFS after multivariate optimization. Journal of Analytical Atomic Spectrometry 2009; 24 1441-1445.

144. Iqbal J, Carney WA, LaCaze S, Theegala CS. Metals Determination in Biodiesel (B100) by ICP-OES with Microwave Assisted Acid Digestion. The Open Analytical Chemistry Journal 2010, 4 18-26.

145. Chaves ES, Santos EJ, Araujo RGO, Oliveira JV, Frescura VLA, Curtius AJ. Metals and phosphorus determination in vegetable seeds used in the production of biodiesel by ICP OES and ICP-MS. Microchemical Journal 2010; 96 71-76.

146. Lobo FA, Gouveia D, Oliveira AP, Romao LPC, Fraceto LF, Dias-Filho NL, Rosa AH. Development of a method to determine Ni and Cd in biodiesel by graphite furnace atomic absorption spectrometry. Fuel 2011; 90 142-146.

147. Prado AGS, Pescara IC, Evangelista SM, Holanda MS, Andrade RD, Suarez PAZ, Zara LF. Adsorption and preconcentration of divalent metal ions in fossil fuels and biofuels: Gasoline, diesel, biodiesel, diesel-like and ethanol by using chitosan microspheres and thermodynamic approach. Talanta 2011; 84 759-765.

148. Barros AI, Oliveira AP, Magalhães MRL, Villa RD. Determination of sodium and potassium in biodiesel by flame atomic emission spectrometry, with dissolution in ethanol as a single sample preparation step. Fuel 2012; 93 381-384.

149. Pillay AE, Elkadi M, Fok SC, Stephen S, Manuel J, Khan MZ, Unnithan S. A comparison of trace metal profiles of neem biodiesel and commercial biofuels using high performance ICP-MS. Fuel 2012; 97 385-389.

150. Rapta P, Paligova J, Rotheneder H, Cvengros J. Determination of potassium in fatty acid methyl esters applying an ion-selective potassium electrode. Chemical Papers 2007; 61, 337-341.
151. Castilho MS, Stradiotto NR. Determination of potassium ions in biodiesel using a nickel(II) hexacyanoferrate-modified electrode. Talanta 2008; 74 1630-1634.
152. Steinbach A, Willie A, Subramanian NH. Biofuel analysis by ion chromatography. LG GC Northe America 2008, S 32-35.
153. Caland LB, Silveira ELC, Tubino M. Determination of sodium, potassium, calcium and magnesium cations in biodiesel by ion chromatography. Analytica Chimica Acta 2012; 718 116-120.
154. Nogueira T, Lago CL, Microchemical Journal 2011, 99, 267.
155. Frena M, Campestrini I, Braga OC, Spinelli A. In situ bismuth-film electrode for square-wave anodic stripping voltammetric determination of tin in biodiesel. Electrochimica Acta 2011; 56, 4678-4684.
156. Korn MGA, Dos Santos DSSS, Welz B, Vale MGR, Teixeira AP, Lima DC, Ferreira SLC. Atomic spectrometric methods for the determination of metals and metalloids in automotive fuels – A review. Talanta 2007; 73 1-11.
157. Santos AL, Takeuchi RM, Fenga PG, Stradiotto NR. Electrochemical Methods in Analysis of Biofuels. Rijeka: In Tech; 2011; 451-494.
158. Santos AL, Takeuchi RM, Munoz RAA, Angnes L, Stradiotto NR. Electrochemical Determination of Inorganic Contaminants in Automotive Fuels. Electroanalysis 2012; 24 1681-1691.
159. Araujo ARTS, Saraiva LMFS, Lima JLFC, Korn MGA. Flow methodology for methanol determination in biodiesel exploiting membrane-based extraction. Analytica Chimca Acta 2008; 613 177-183.
160. Wang H, Tang H, Salley S, Simon Ng KY. Analysis of Sterol Glycosides in Biodiesel and Biodiesel Precipitates. Journal of the American Oil Chemists' Society 2010; 87 215-221.
161. Moreau RA, Scott KM, Haas MJ. The Identification and Quantification of Steryl Glucosides in Precipitates from Commercial Biodiesel. Journal of the American Oil Chemists' Society 2008; 85 761-770.

162. Bondioli P, Cortesi N, Mariani C. Identification and quantification of steryl glucosides in biodiesel. European Journal of Lipid Science Technology 2008; 110 120-126.
163. Pauls RE. A Review of Chromatographic Characterization Techniques for Biodiesel and Biodiesel Blends. Journal of Chromatographic Science 2011; 49 384-396.
164. Lyra FH, Carneiro MTWD, Brandao GP, Pessoa HM, De Castro EVR. Direct determination of phosphorus in biodiesel samples by graphite furnace atomic absorption spectrometry using a solid sampling accessory. Journal of Analytical Atomic Spectrometry 2009; 24 1262-1266.
165. Silveira ELC, De Caland LB, Tubino M. Molecular absorption spectrophotometric method for the determination of phosphorus in biodiesel. Fuel 2011; 90 3485-3488.
166. Lira LFB, Dos Santos DCMB, Guida MAB, Stragevitch L, Korn MDA, Pimentel MF, Paim APS. Fuel 2011; 90 3254-3258.
167. Zezza TRC, Castilho MD, Stradiotto NR. Determination of phosphorus in biodiesel using 1:12 phosphomolybdic modified electrode by cyclic voltammetryFuel 2012; 95 15-18.
168. Barker LR, Kelly R, Guthrie WF. Determination of Sulfur in Biodiesel and Petroleum Diesel by X-ray Fluorescence (XRF) Using the Gravimetric Standard Addition Method-II. Energy & Fuels 2008; 22 2488-2490.
169. Young CG, Amais RS, Schiavo D, Garcia EE, Nobrega JA. Determination of sulfur in biodiesel microemulsions using the summation of the intensities of multiple emission lines. Talanta 2011; 84 995-999.
170. Amais RS, Donati GL, Nobrega JA. Interference Standard Applied to Sulfur Determination in Biodiesel Microemulsions by ICP-QMS. Journal of the Brazilian Chemical Society 2012; 12 797-803.
171. Monteiro MR, Ambrozin ARP, Liao LM, Ferreira AG. Critical review on analytical methods for biodiesel characterization. Talanta 2008; 77 593-605.

6

CHAPTER

THERMOECONOMIC ANALYSIS OF BIODIESEL PRODUCTION FROM USED COOKING OILS

Emilio Font de Mora *, César Torres and Antonio Valero

Centre of Research for Energy Resources and Consumption – CIRCE, University of Zaragoza, Mariano Esquillor 15, 50018 Zaragoza, Spain

ABSTRACT

Biodiesel from used cooking oil (UCO) is one of the most sustainable solutions to replace conventional fossil fuels in the transport sector. It can achieve greenhouse gas savings up to 88% and at the same time reducing the disposal of a polluting waste. In addition, it does not provoke potential negative impacts that conventional biofuels may eventually cause linked to the use of arable land. For this reason, most policy frameworks favor its consumption. This is the case of the EU policy that double-counters the use of residue and waste use to achieve the renewable energy target in the transport sector. According to different sources, biodiesel produced from UCO could replace around 1.5%–1.8% of the EU-27 diesel consumption. This paper presents an in-depth thermoeconomic analysis of the UCO biodiesel life cycle to understand its cost formation process. It calculates the ExROI value (exergy return on investment) and renewability factor, and it demonstrates that thermoeconomics is a useful tool to assess life cycles of renewable energy systems. It also shows that UCO life cycle biodiesel production is more sustainable than biodiesel produced from vegetable oils.

Keywords: biodiesel; UCO; exergy; exergy cost; ExROI; renewability

1. INTRODUCTION

As its name indicates, used cooking oils (UCO) are vegetable oils that have been used in cooking, normally for frying purposes. Once the consumer considers the oil to have been sufficiently used, the oil is replaced and becomes a waste. UCO produced in the EU are considered non-hazardous wastes according to the Consolidated European Waste Catalogue. They are classified as municipal wastes (household waste and similar commercial, industrial and institutional wastes), including separately-collected fractions, under the code 20 01 25 (edible oils and fats).

The wrong disposal of UCO can have important negative effects on the environment: the increased amount of nutrients in the drains augments the number of rats and other plagues. If the waste is not eliminated in waste water treatment plants, it can produce a negative effect on the capacity of the auto-regeneration of rivers, fauna and flora and can increase the population of jelly fish on coastlines [1]. Other negative effects are the increase in the cost of sewage treatment and clogging problems in drains with the consequent repair costs [1].

With the current legislation in place in the European Union, the use of UCO for biodiesel production has gained importance. According to Article 21 of the Renewable Energy Directive [2], the contribution made by biofuels to the 2020 targets produced from wastes, residues, non-food cellulosic material and ligno-cellulosic material is considered to be twice that made by other biofuels. Biofuels produced from these resources are called advanced or second generation biofuels. Biodiesel from UCO is one of them. The importance of UCO-derived biodiesel will increase even more when a limit to the production of crop-based biofuels will be introduced in the European legislation, as recently agreed by the European Parliament and the Member States.

This regulation has had unexpected counter effects [3,4,5]: as international imports of UCO have started to take place, the price of UCO has scaled up to even exceed the price of edible vegetable oils, such as palm oil, and finally, it has opened the door to fraud. As there is no real definition of used cooking oil and no possible way nowadays to determine when a vegetable oil is used or not, some actors are claiming to have UCO when, in reality, the vegetable oil has only been used once, and they have used high quantities of oil to fry a small portion of food. In light of the revision of the Renewable Energy Directive, there are voices suggesting eliminating UCO from the list of advanced biofuels benefiting from double counting [4].

Be it considered or not an advanced biofuel, biodiesel produced from UCO will still play a role in achieving the renewable energy in transport 2020 target in the EU, in view of its beneficial lower GHG emissions. In the Annex V of the Renewable Energy Directive, while the typical GHG emission reduction for biodiesel produced from unused vegetable oils ranges from 40 to 62%, biodiesel from UCO shows reductions of 88%, which still makes it attractive for mandated operators [2].

2. UCO-BASED BIODIESEL POTENTIAL

Estimating the potential of UCO collection that could be used in biodiesel production is a challenge. In the EU, the consumption of vegetable oils in 2011 was about 19 million tons for rapeseed, soybean, sunflower and palm oils [6]. The worldwide vegetable oil production was about 145 million tons for the same oils plus olive oil [7]. However, from these data, knowing the quantities of UCO that could be effectively recovered is not easy. This is especially the case in the household sector.

There are three types of producers of used cooking oils: households, the HORECA sector (i.e., hotel, restaurants and caterers) and the food industry. According to the Waste Framework Directive [8], EU Member States shall take measures to encourage the separate collection of bio-waste and treatment in a way that fulfills a high level of environmental protection. In line with these requests, normally, the HORECA sector and food industry have a high degree of implementation of selective collection systems by means of authorized collectors. These sectors are easily controllable from the waste authority viewpoint.

In households, the collection of UCO is not as manageable as in the other sectors, and countries do not have established collection plans in place. The existing collection systems are generally encouraged by regional or local governments or stem from private initiatives. In this sector, the consideration of used oil is left to the discretion of the owner, and normally, the used oil is thrown away through the drain or disposed at the bin in a used container (empty plastic bottle or brick).

Some initiatives and institutions have made attempts to calculate how much biodiesel could be produced from UCO following two methodologies. One of these methodologies follows a top-down approach. Knowing the vegetable oil consumption of an area, a certain percentage of losses is assumed (e.g., around 85%) in terms of the oil ingested by the consumer and potential shortfalls in oil collection and recycling.

The other methodology follows a bottom-up approach. By means of practical experiences, it obtains the quantity of vegetable oil consumed per person/year. Knowing the population of a defined area with the same cultural conditions, one can calculate the quantity of vegetable oil consumed in the entire region. Assuming a certain loss in UCO collection and recycling, one obtains a final number.

The weakness in both methodologies is related to the assumptions made in terms of UCO losses in collection and recycling. The bottom-up approach also adds uncertainty to the data obtained per person/year and the extrapolation of these data to an entire region. As an example, the quantities obtained in the Intelligent Energy–Europe (IEE) project OILECO range from 4 to 18 kg/person/year [9] in EU countries (Italy, Spain, Hungary, Slovakia and Bulgaria).

Being this the situation, some projects have estimated the potential biodiesel production in the EU. The IEE project RecOil states that biodiesel produced from UCO could replace 1.5% of the EU-27 diesel consumption [10]. The final report of the IEE project BioDieNet [11] states that the total amount of recoverable UCO available in the EU-27 in 2009 was about 3.5 million tons (3950 million liters). As the EU-27 diesel consumption was 214 billion liters in 2007, around 1.8% of the EU-27 diesel consumption could have potentially been replaced with biodiesel produced from UCO.

3. UCO-BASED BIODIESEL LIFE CYCLE

The life cycle (see Figure 1) starts by the collection from households, HORECA and industry and is followed by recycling, where the oil is filtered and decanted in order to separate solid particles and water. Once the oil is refined, it is sent to the transesterification plant.

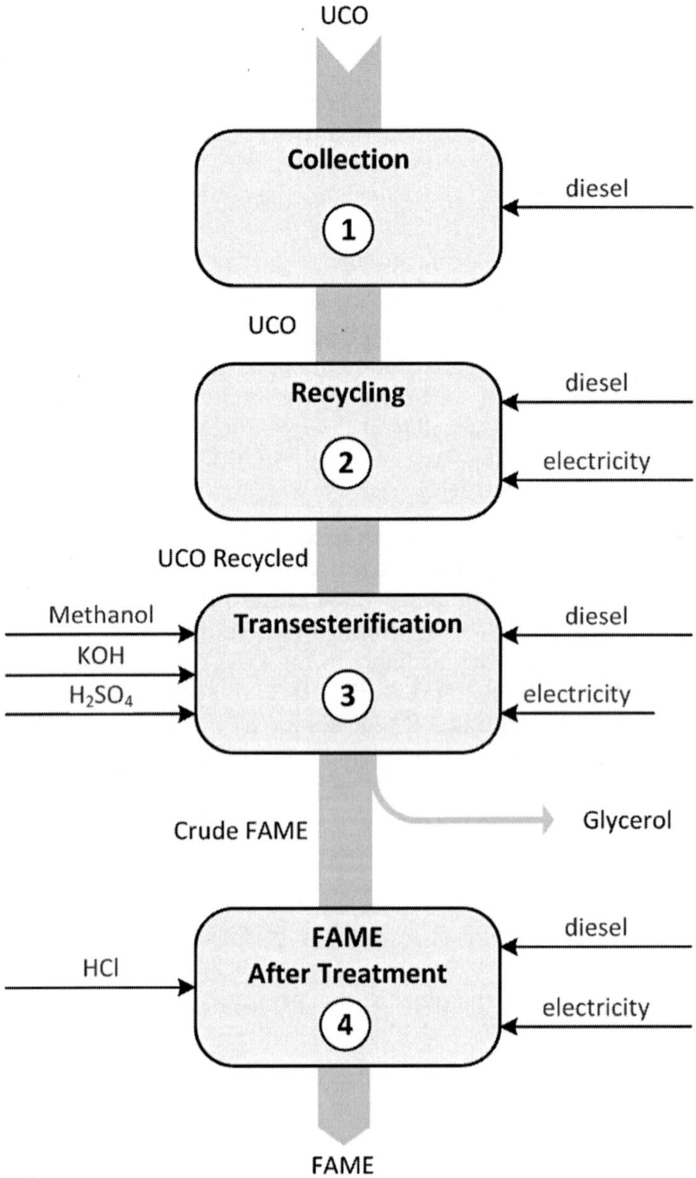

Figure 1. Flow diagram of a biodiesel plant producing with used cooking oil. Source: CIEMAT [12].

The UCO collection potential from households can be higher than the HORECA one. This is the case of Spain. According to GEREGRAS, the Spanish association of vegetable oil and fat residues management companies, the potential for UCO collection in Spain in 2020 is about 280,000 tons from which more than 120,000 tons are from the HORECA sector and 160,000 tons from the domestic sector. However, in 2011, the quantity of UCO collected was about 90,000 tons, which were mostly coming from the HORECA sector [13]. The main problems that the UCO collection from households is facing are non-technological. They are related to the lack of awareness of the possibility to collect UCO, the lack of promotion about the benefits to the environment and making the collection more attractive to the UCO holders, for example, by reducing the distances to the UCO collection points and having a strong hygiene maintenance of the containers.

The collection and transport of the UCO is done by an authorized agent for waste management. In this phase, the only input is diesel fuel needed to run the vans and trucks that are used to transport the containers of UCO to recycling, also called the refining unit.

The collection can be organized in different ways. In the HORECA sector and food industry, normally, the collection is done on-site by the authorized agent in medium-to-big containers that are transported via van or truck. In the case of households, there are different possibilities: door-to-door collection or distributed collection at points throughout an area with well-identified containers. These containers can be adapted to receive UCO in used bricks, bottles or reusable bottles designed for this purpose, or have a hole through which the oil can be poured. According to the results of IEE projects OILECO and RecOil, the best practices, showing higher acceptance, are those having collection points with adapted containers [9,10]. This system can also be the most sustainable if the transport of the oil to the collection point by the household consumer is not accounted for in the life cycle assessment.

In the recycling process, the inputs are normally electricity and gas oil or natural gas. The gas oil or natural gas is consumed to produce steam, which is used to clean the containers, so that they can be reused again.

As far as the transesterification is concerned, the UCO normally follows a different process compared to the case of crude vegetable oils, which affects the consumption of materials and energy sources. The main differences are:

- The pre-treatment is not so elaborated, because the oil comes already "clean" from the recycling unit. Normally, cleaning is only done via centrifugation and drying to eliminate possible small particles and excess water.
- Afterwards, the oil is esterified with an acid catalyzer to transform into esters the high quantity of free fatty acids that are normally present in UCO; this esterification is followed by a basic catalyzed transesterification. Some plants have the esterification and transesterification tanks in parallel. They treat in the esterification plant those oils having high FFA (more than 4%) and in the transesterification plant those oils with less FFA [12].
- The ester obtained follows a cleaning and drying process. Other plants have opted to distillate it to ensure the fulfillment of the quality standards, EN 14214 [14] in the EU.
- The glycerol does not follow an after-treatment process. The crude glycerol is either treated as a residue or sold as a low added-value co-product or to be used as fuel in cement kilns [15]. Some plants have after-treatment processes similar to the ones using vegetable oils to obtain a high-quality glycerin product; these after-treatment processes normally consist of adding an acid, e.g., HCl, to split the soaps into FFA and salts, which can then be separated by decantation, and distilling the remaining flow up to 130 °C to evaporate the remaining methanol [15].

Some biodiesel producers that do not have their plants adapted to work with UCO, if they aim to use UCO, normally add a small part (e.g., 10%) to their main streams, e.g., rapeseed oil. With this, they do not need to adapt their plants to deal with this lower quality resource, while producing a high-quality biodiesel.

The inputs and outputs of the transesterification phase can be seen in Table 1 together with the flow diagram of Figure 1. These data have been obtained from an LCA carried out by CIEMAT [12], based on a supply chain established in the north of Spain and with data of two real biodiesel production plants using this resource. It is important to note that in this case, for the production of steam, diesel fuel is considered instead of natural gas.

Table 1. Consumption data for processes considered in the life cycle analysis for UCO. Source: CIEMAT [12].

Process	I/O	Value	Units
Collection			
Diesel	Input	0.48316	MJ/kg UCO
Recycling			
Diesel	Input	0.036	MJ/kg oil
Electricity	Input	0.35	MJ/kg oil
Transesterification			
Recycled oil	Input	1.03	kg/kg FAME
Methanol	Input	0.155	kg/kg FAME
KOH	Input	0.02770	kg/kg FAME
H_2SO_4	Input	0.017	kg/kg FAME
Diesel	Input	0.455	kWh/kg FAME
Electricity	Input	0.137	kWh/kg FAME
FAME	Output	1.00	kg/kg FAME
Glycerol crude	Output	0.20	kg/kg FAME

4. THERMOECONOMIC ANALYSIS

The methodology applied in this paper has been introduced in previous works [16,17,18]. Thermoeconomic input-output analysis, formerly symbolic exergoeconomics [19], applies the mathematical framework of input-output analysis [20] to exergy cost theory [20]. It provides general relationships between the production demand and the resource cost of an energy system with the efficiency and irreversibilities of each individual process of that system. The distinguishing element in thermoeconomics from conventional energy and exergy analysis is purpose. Matter and energy flows entering and exiting a given system are classified into fuel and product. Fuel (F) refers to the resources that the component uses to achieve its purpose, and product (P) corresponds to the flows related to that purpose [21].

The system interactions are represented by means of the fuel-product table (see Table 2), which describes how the production processes are related.

Table 2. Fuel-product table.

		Final Product	Process Resources					Total
			1	...	j	...	n	
External Resources			E_{01}	...	E_{0j}	...	E_{0n}	P_0
Process Products	1	E_{10}	E_{11}	...	E_{1j}	...	E_{1n}	P_1
	⋮	⋮	⋮		⋮		⋮	⋮
	i	E_{i0}	E_{i1}	...	E_{ij}	...	E_{in}	P_i
	⋮	⋮	⋮		⋮		⋮	⋮
	n	E_{n0}	E_{n1}	...	E_{nj}	...	E_{nn}	P_n
Total		F_0	F_1	...	F_j	...	F_n	

Let us consider a system formed by, say n processes, and the environment named as process Number 0. Let P_i denote the exergy of the process i production, which is used in part to meet the intermediate requirement as input resources of other processes and in part to meet the final demand of the system. If E_{ij} denotes the exergy that process i uses as a resource for process j and E_{i0} denotes the final demand produced in process i, the following expression is formed:

$$P_i = E_{i0} + \sum_{j}^{n} E_{ij}, \quad i = 1, \ldots, n \tag{1}$$

Note that, in this model, the production of the environment P_0 is the total exergy of the external resources. On the other hand, the input resources of each process i, say F_i, is in part coming from external resources, say E_{0i}, and in part from the production of other processes, then we have the condition:

$$F_i = E_{0i} + \sum_{j}^{n} E_{ji}, \quad i = 1, \ldots, n \tag{2}$$

where F_0 is the total system demand.

The second law states that the difference between fuel and product is a positive value and equal to the irreversibility (I), accounting for the losses appearing in the process: $F_i - P_i = I_i \geq 0$. Furthermore, the unit exergy consumption is defined as the ratio between fuel and product: $k_i = F_i/P_i \geq 1$

The exergy cost of the products of a system refers to the exergy of the resources consumed to obtain such products. Therefore, the exergy cost of the production of a process, denoted as $C_{P,i}$, is the sum of the cost of the resources consumed by the process: system internals, i.e., products of other processes C_{ji} and the cumulative exergy of the external resources $C_{0,i}$.

$$C_{P,i} = C_{0,i} + \sum_j C_{ij} \tag{3}$$

To calculate the production exergy cost, the assumption that the costs of the products obtained in a process are proportional to its exergy is made; see [22]:

$$\frac{C_{ij}}{C_{P,i}} = \frac{E_{ij}}{P_i} = y_{ij} \tag{4}$$

where y_{ij} is the distribution exergy ratio, which means the portion of production of process i uses as resources of process j.

Combining Equations (3) and (4), the exergy cost can be calculated by solving the following set of linear equations:

$$C_{P,i} = C_{0i} + \sum_{j=1}^{n} y_{ji} C_{P,j} \qquad i = 1, \ldots, n \tag{5}$$

which could be written in matrix form as:

$$\mathbf{C}_P = \mathbf{C}_e + {}^{\mathrm{T}}\langle \mathbf{FP} \rangle \mathbf{C}_P \tag{6}$$

where⟨**FP**⟩ is a (n × n) matrix, whose entries are the distribution exergy ratios, and C_e is a (n × 1) vector, whose entries are the cumulative exergy of the external resource uses in each process.

Therefore, the production cost could be obtained as:

$$\mathbf{C}_P = {}^{\mathrm{T}}\langle \mathbf{P}^* | \, \mathbf{C}_e \quad \text{where} \quad \langle \mathbf{P}^* | \equiv (\mathbf{U} - \langle \mathbf{FP} \rangle)^{-1} \tag{7}$$

where ⟨**P***| is called the production cost matrix, and its elements $\pi * ij$ are the global distribution ratios, which represent the amount of external resources (measured in exergy) of process i required to obtain one exergy unit of the product of the process j.

Table 3 shows the exergy of flows and the cumulative exergy of external resources of the UCO biodiesel pathway. From this information, the F-P table of the UCO pathway model is built; see Table 4.

Table 3. Exergy and cost of flows of the UCO biodiesel pathway.

					Exergy		Cost	
Material	Process	I/O	Unit (u)	u/kg FAME	MJ/u	MJ/kg FAME	MJ/u	MJ/kg FAME
Diesel	1	Input	MJ	0.6459	1.0000	0.6459	1.1773	0.7604
UCO	1	Output	kg	1.3368	39.5994	52.9364	-	-
Diesel	2	Input	MJ	0.0371	1.0000	0.0371	1.1773	0.0437
Electricity	2	Input	MJ	0.4679	1.0000	0.4679	2.8690	1.3423
UCO recycled	2	Output	kg	1.0300	39.5994	40.7873	-	-
Methanol	3	Input	kg	0.1550	23.1540	3.5889	30.3074	4.6977
KOH	3	Input	kg	0.0277	0.8229	0.0228	13.3852	0.3708
H_2SO_4	3	Input	kg	0.0170	1.6417	0.0279	2.6295	0.0447
Diesel	3	Input	MJ	1.5651	1.0000	1.5651	1.1773	1.8426
Electricity	3	Input	MJ	0.4713	1.0000	0.4713	2.8690	1.3522
Crude FAME	3	Output	kg	1.0100	39.8548	40.2533	-	-
Crude Glycerol	3	Output	kg	0.2000	23.6779	4.7356	-	-
HCl	4	Input	kg	0.0406	1.3409	0.0544	5.2484	0.2131
Diesel	4	Input	MJ	0.0749	1	0.0749	1.1773	0.0882
Electricity	4	Input	MJ	0.0187	1.0000	0.0187	2.869	0.0535
FAME	4	Output	kg	1	39.8376	39.8376	-	-

Table 4. F-P (fuel-product) tables for UCO biodiesel (MJ/kg FAME).

	F_0	F_1	F_2	F_3	F_4	Total
P_0		53.58	0.51	5.68	0.15	59.91
P_1	0.00	0.00	52.94	0.00	0.00	52.94
P_2	0.00	0.00	0.00	40.79	0.00	40.79
P_3	4.74	0.00	0.00	0.00	40.25	44.99
P_4	39.84	0.00	0.00	0.00	0.00	39.84
Total	44.58	53.58	53.44	46.46	40.40	

The local and global distribution matrices, are obtained from Table 4 using Equations (4) and (7):

$$\langle \mathbf{FP} \rangle = \begin{bmatrix} 0.0000 & 1.0000 & 0.0000 & 0.0000 \\ 0.0000 & 0.0000 & 1.0000 & 0.0000 \\ 0.0000 & 0.0000 & 0.0000 & 0.8947 \\ 0.0000 & 0.0000 & 0.0000 & 0.0000 \end{bmatrix} \qquad \langle \mathbf{P}^* | = \begin{bmatrix} 1.0000 & 1.0000 & 1.0000 & 0.8947 \\ 0.0000 & 1.0000 & 1.0000 & 0.8947 \\ 0.0000 & 0.0000 & 1.0000 & 0.8947 \\ 0.0000 & 0.0000 & 0.0000 & 1.0000 \end{bmatrix}$$

The production exergy cost can be decomposed [18], considering the different types of resources required. The resources cost vector can be separated into two terms:

$$\mathbf{C}_e = \mathbf{C}_e^{\mathrm{rs}} + \mathbf{C}_e^{\mathrm{nrs}} \tag{8}$$

The first term represents the cumulative exergy of the renewable resources and the latter the non-renewable or fossil resources. Therefore, the exergoecologic cost can be broken down into the renewable and fossil parts:

$\mathbf{C}_P = \mathbf{C}_P^{\mathrm{rs}} + \mathbf{C}_P^{\mathrm{nrs}}$, where:

$$\mathbf{C}_P^{\mathrm{rs}} \equiv {}^{\mathrm{T}}\langle \mathbf{P}^* | \mathbf{C}_e^{\mathrm{rs}}$$
$$\mathbf{C}_P^{\mathrm{nrs}} \equiv {}^{\mathrm{T}}\langle \mathbf{P}^* | \mathbf{C}_e^{\mathrm{nrs}} \tag{9}$$

This distinction depending on the origin of the resources allows calculating the sustainability indicator. Font de Mora et al. [16] defined the concept of ExROI and renewability. ExROI measures the amount of product obtained with one unit of non-renewable resources:

$$\text{ExROI} = \frac{P}{C_P^{\text{nrs}}} \tag{10}$$

A production process is sustainable if ExROI > 1. The higher the ExROI value, the more sustainable a production process will be.

Font de Mora et al. [17] introduced the concept of the renewability ratio, defined as:

$$\rho = \frac{C_P^{\text{rs}}}{C_P} \quad 0 \leq \rho \leq 1 \tag{11}$$

the portion of renewable resources per total resources used or the weight of the renewable exergy cost with respect to the total cost.

The results of the cost analysis, obtained by applying Equation (9), are shown in Table 5. In this case, the primary UCO stream is considered as a renewable resource, and its cost is equal to its exergy. From these values, it is possible to compute the ExROI and renewability factor, which are 4.10 and 0.83, respectively. This means that from each unit of non-renewable sources, i.e., fossil fuels, it is possible to obtain 4.10 units of energy and that the percentage of renewable energy contained in the fuel is about 83%. These values can be increased if the conventional methanol of fossil origin is substituted by a bioproduct, either bioethanol or methanol from biomass.

Table 5. Cost analysis for UCO biodiesel (MJ/kg FAME).

No.	Process	C_e^{nrs} MJ/Kg	C_P MJ/kg	C_P^{rs} MJ/Kg	C_P^{nrs} MJ/kg	ρ %
1	Collection	0.76	53.697	52.936	0.760	98.58
2	Refining	1.39	55.083	52.936	2.146	96.10
3	Esterification	8.31	63.391	52.936	10.454	83.51
4	FAME after treatment	0.35	57.073	47.364	9.709	82.99

The values obtained for the UCO biodiesel can be compared to the ones obtained for biodiesels from vegetable oils in Figure 2. The values for vegetable oils were calculated in [15]. As can be seen, biodiesel production from UCO has the highest ExROI value and the second highest renewability factor, only exceeded by palm oil-based biodiesel. The high renewability of palm oil is due to the use of biomass resources to cover part of the energy needs in the cycle. In addition, UCO biodiesel has the lowest non-renewable exergy cost, which means that, overall, it consumes less non-renewable resources, i.e., mainly fossil fuels, than its counterparts. The left graphic, in Figure 2, shows also the renewable and non-renewable exergy cost per produced unit of the different biodiesels analyzed in the paper. This analysis could be extended to cover also non-edible vegetable oils, for example castor and Jatropha oils, and their blends with UCO [23].

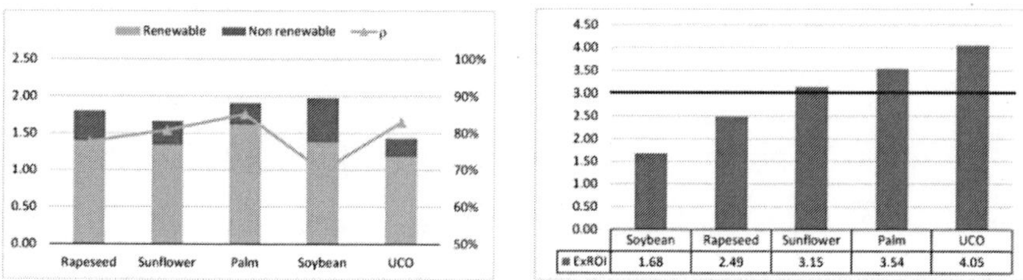

(a) (b)

Figure 2. Comparison of ExROI and renewability factors of different biodiesel life cycles.(**a**) Renewability ratio; (**b**) ExROI.

5. ANALYSIS OF POTENTIAL IMPROVEMENTS

As shown in the previous section, the exergy cost of biodiesel depends on the exergy cost values of the external resources. Therefore, depending on the boundary limits defined by the LCA and from the data provided by LCA databases, whose values may be based on general hypothesis and conversion factors, the results may vary considerably. In order to understand the effect that variations in the consumption of external resources may have on the production cost values, a sensitivity analysis must be performed.

According to Equation (7), the cost of the external resources does not depend on the production matrix; thus, the effect in the exergy cost due to the variation in the external cost could be calculated as:

$$\Delta \mathbf{C}_P = {}^{\mathrm{T}}\langle \mathbf{P}^* | \, \Delta \mathbf{C}_e \tag{12}$$

Introducing the elasticity coefficient $\%\Delta x = \Delta x / x$, Equation (12) could be rewritten as:

$$\%\Delta C_{P,i} = \sum_{j=1}^{n} \frac{C_{0,j}}{C_{P,i}} \pi_{ji}^{*} \, \%\Delta C_{0,j} \tag{13}$$

This expression is also valid to compute the variation of renewable and non-renewable costs, and Equation (13) could be written in terms of non-renewable costs, as:

$$\%\Delta \mathbf{C}_P^{\mathrm{nrs}} = \left[\mathbf{A}_P^{e}\right] \%\Delta \mathbf{C}_e^{\mathrm{nrs}} \tag{14}$$

where [A*eP*] is called the elasticity matrix, whose elements are defined as

$$a_{ij} = C_{0,j}\pi_{ji}^{*} / C_{P,i}.$$

Equation (14) could be used to analyze the potential improvements, in terms of sustainability, of replacing non-renewable resources by alternative renewable resources. In the case of the life cycle of UCO, its elasticity matrix is:

$$\left[\mathbf{A}_P^{e}\right] = \begin{bmatrix} 1.0000 & 0.0000 & 0.0000 & 0.0000 \\ 0.3543 & 0.6457 & 0.0000 & 0.0000 \\ 0.0727 & 0.1326 & 0.7974 & 0.0000 \\ 0.0702 & 0.1277 & 0.7656 & 0.0365 \end{bmatrix}$$

From the elasticity matrix, in combination with the information of external resource consumption provided in Table 3, it is possible to obtain the relative impact of the cost of each external resource in the production cost of biodiesel (FAME), which is shown in Table 6. Note, that the sum of each row in the elasticity matrix must be equal to one.

Table 6. Relative impact of external resources on the production cost of FAME from UCO.

Flow	Collection	Recycling	Esterification	FAME
Diesel	1.0000	0.3746	0.2532	0.2530
Methanol	0.0000	0.0000	0.4433	0.4273
Electricity	0.0000	0.6254	0.2594	0.2554
KOH	0.0000	0.0000	0.0350	0.0337
H_2SO_4	0.0000	0.0000	0.0042	0.0041
HCl	0.0000	0.0000	0.0000	0.0217

The non-renewable resource having higher impact in the production cost of FAME is methanol 42.73%. This means that, if we replace methanol by bio-methanol obtained from wood [3,15], the non-renewable production cost of FAME produced from UCO decreases by 42.73%, and the ExROI value is increased up to 7.07.

By recycling 7% biodiesel (82% renewable source) to produce the required electricity in the production of FAME, the production cost is reduced 0.2554 × 82.8% = 21.15%, and the ExROI is increased up to 5.10.

By using part of the produced biodiesel in the collection process, in order to substitute the consumption of diesel, the production cost will be reduced 0.0692 × 82.8% = 5.73%, and the ExROI will be increased up to 4.28.

Finally, if several measures are combined, like the use of organic methanol and the recirculation of 16% of the biodiesel (6.3 MJ/kg) to substitute part of the energy needs in the process, the ExROI would be increased to 69 and the renewability to 98.7%.

6. CONCLUSIONS

This paper applies the methodology and concepts introduced in [16,17,18] to the life cycle production of biodiesel produced from used cooking oils. Used cooking oils have the potential to replace 1.5% to 1.8% of the EU-27 diesel consumption with a GHG emission reduction potential of 88% compared to fossil diesel. Given the importance that this alternative fuel is gaining and the support that is given by the energy and transport policy frameworks worldwide, it is important to assess its life cycle from the thermoeconomic perspective, analyze how the process can be improved and compare the results with other biodiesel types. The results show that biodiesel from used cooking oil is the most sustainable biodiesel from the point of view of the use of non-renewable resources. The ExROI value is higher than three, which is, according to Hall et al. [24], the minimum ExROI value that society must attain from its energy exploitation to support continued economic activity and social function. The most important measure to improve the production cycle is substituting the fossil-derived methanol by an alcohol of biological origin, which provides an ExROI increase of 75%.

AUTHOR CONTRIBUTIONS

Emilio Font de Mora designed the research and written the sections related to UCO-Based Biodiesel. Thermoeconomic Analysis is written by César Torres. Antonio Valero reviewed the research. All authors have read and approved the final manuscript.

NOMENCLATURE

c Unit Exergy Cost (MJ/MJ)

n Number of processes

C Exergy Cost (MJ/kg FAME)

F Exergy of Fuel (MJ/kg FAME)

P Exergy of Product (MJ/kg FAME)

y Exergy distibution ratios (MJ/MJ)

ρ Renewability ratio (%)

π* Global distribution ratios (MJ/MJ)

MATRIX AND VECTORS

⟨FP⟩ Exergy distribution ratios matrix (n × 1)

C_e Vector of external resources (n × 1)

⟨**P***| Production cost operator matrix (n × n)

U_n Identity Matrix (n × n)

[A*eP*] Elasticity Matrix (n × n)

SUBSCRIPTS AND SUPERSCCRIPTS

0 Environment process

e External resources

F Fuel

nrs Non renewable resources

P Product

rs Renewable resources

T Transpose Matrix

ABBREVIATIONS

CIEMAT	Centro de Investigaciones Energéticas, Medioambientales y Tecnológicas
EC	European Commission
EROI	Energy Return of Energy Investment
EU	European Union
EU-27	European Union comprising 27 member states (excluding Croatia)
ExROI	Exergy Return of Exergy Investment
FAME	Fatty Acid Methyl ester (biodiesel)
FFA	Free Fatty Acid
GEREGRAS	Spanish association of vegetable oil and fat residues

	management companies
HORECA	Sector of the food service industry
LCA	Life Cycle Analysis
UCO	Used Cooking Oil

REFERENCES

1. LIFE Programme. Ecobus—Collecting Used Cooking Oils to Their Recycling as Biofuel for Diesel Engines; Technical Report; European Commission: Brussels, Belgium, 2012.
2. European Union. Directive 2009/28/EC of the European Parliament and the Council of 23 April 2009 on the promotion of the use of energy from renewable sources and amending and subsequently repealing Directives 2001/77/EC and 2003/30/EC. 2009. Available online: http://europa.eu/legislation_summaries/energy/renewable_energy/en0009_en.htm (accessed on 12 May 2015).
3. Bailey, R. The Trouble with Biofuels: Cost and Consequences of Expanding Biofuel Use in United Kingdom. Energy Environ. Resour. **2013**, 1, 1–24.
4. ePURE. Double Counting, Half Measures: Study on the Effectiveness of Double Counting as a Support for Advanced Biofuels; Technical Report; European Renewable Ethanol Association: Bruxelles, Belgium, 2013.
5. Policy toolkit to improve the support mechanisms for advanced biofuels. Availabe online: http://biofuelstp.eu/policy/policy-toolkit-ebtp-06-10-2011.pdf (accessed on 14 May 2015).
6. FEDIOL. Vegetable oils production, import, exports and consumption 2011 annual statistics. Available online: http://www.fediol.be/web/statistics2011/1011306087/list1187970179/f1.html (accessed on 14 May 2015).
7. United States Department of Agriculture Foreign Agricultural Service (USDA). PDS - Production. Available online: https://apps.fas.usda.gov/psdonline/ (accessed on 14 May 2015).

8. European Union. Directive 2008/98/EC of the European Parliament and of the Council of 19 November 2008 on waste and repealing certain Directives. Available online: http://faolex.fao.org /docs/pdf/eur83580.pdf (accessed on 12 May 2015).
9. OilECO Project Project OILECO, supported by the Intelligent Energy - Europe (IEE) Programme of the European Commision. 2013. Available online: http:/www.oileco.org (accessed on 15 May 2015).
10. Paraiba, O.; Tsoutsos, T.; Tournaki, S.; Antunes, D.; Lino, J.; Manning, E. Strategies for the optimization of the domestic used cooking oil to biodiesel chain, the European project RecOil. In Proceedings of the 20th European Biomass Conference and Exhibition, Milan, Italy, 18–22 June 2012.
11. BioDieNet Project. The Future of Small Scale, Localised Biodiesel Production from Used Cooking Oil and Its Use in Higher Blends; Technical Report; Project supported by the Intelligent Energy–Europe (IEE) Programme of the European Commission: Brussels, Belgium, 2009.
12. Lechon, Y.; Cabal, H.; de la Rua, C.; Lago, C.; Izquierdo, L.; Saez, R.M.; Fdez. San Miguel, M. Análisis de ciclo de vida de Combustibles Alternativos para el Transporte. Fase I. Análisis de ciclo de vida Comparativo del Biodiesel de Cereales y de la Gasolina; Technical Report; Ministerio de Educación y Ciencia y Ministerio de Medio Ambiente: Madrid, Spain, 2005.
13. Instituto para la Diversificación y Ahorro de la Energía (IDAE). Plan de Energías Renovables (PER) 2011–2020; IDAE: Madrid, Spain, 2011.
14. EN14214—Liquid petroleumproducts—Fatty acid methyl esters (FAME) for use in diesel engines and heating applications—Requirements and test methods. Available online: http://www.sis.se/en/petroleum-and-related-technologies/fuels/ liquid-fuels/ss-en-142142012a12014-ac2014 (accessed on 12 May 2015).
15. Font de Mora, E. Application of Thermoeconomics to Assess and Improve the Efficiency of Bioenergy Production Plants and Land-to-Tank Cycles. Ph.D. Thesis, Universidad de Zaragoza, Zaragoza, Spain, 2013.

16. Font de Mora, E.; Torres, C.; Valero, A. Assessment of biodiesel energy sustainability using the exergy return on investment concept. Energy **2012**, 45, 474–480.
17. De Mora, E.F.; Torres, C.; Valero, A.; Zambrana, D. Defossilisation assessment of biodiesel life cycle production using the ExROI indicator. Int. J. Thermodyn. **2013**, 16, 81.
18. Torres, C.; Valero, A.; Valero, A. Exergoecology as a tool for ecological modelling. The case of the US food production chain. Ecol. Modell. **2012**, 255, 21–28.
19. Torres, C. Symbolic thermoeconomic analysis of energy systems. In Exergy, Energy System Analysis and Optimization, from Encyclopedia of Life Support Systems (EOLSS); Frangopoulos, C., Ed.; EOLSS Publishers: Oxford, UK, 2006.
20. Miller, R.E.; Blair, P.D. Input-Output Analysis: Foundations and Extensions; Cambridge University Press: Cambridge, UK, 2009.
21. Tsatsaronis, G.; Winhold, M. Exergoeconomic analysis and evaluation of energy-conversion plants. Energy **1985**, 10, 69–80.
22. Lozano, M.; Valero, A. Theory of the exergetic cost. Energy **1993**, 18, 939–960.
23. Sankaranarayanan, S.; Kannan, S. Sonochemistry: A simple and efficient route for biodiesel production from neat and blended vegetable oils using LDH-derived base catalysts. Int. J. Adv. Eng. Sci. Appl. Math. **2013**, 5, 210–218.
24. Hall, C.A.S.; Balogh, S.; Murphy, D.J. What is the minimum EROI that a sustainable society must have? Energies **2009**, 2, 25–47.

7

CHAPTER

MOVING TOWARDS A COMPETITIVE FULLY ENZYMATIC BIODIESEL PROCESS

Silvia Cesarini [1,2], F. I. Javier Pastor [1], Per M. Nielsen [2] and Pilar Diaz [1,*]

[1] *Department of Microbiology, University of Barcelona, Av. Diagonal 643, 08028 Barcelona, Spain*

[2] *Department Bioenergy Opportunities, NOVOZYMES A/S, Krogshoejvej 36, 2880 Bagsvaerd, Denmark*

ABSTRACT

Enzymatic biodiesel synthesis can solve several problems posed by the alkaline-catalyzed transesterification but it has the drawback of being too expensive to be considered competitive. Costs can be reduced by lipase improvement, use of unrefined oils, evaluation of soluble/immobilized lipase preparations, and by combination of phospholipases with a soluble lipase for biodiesel production in a single step. As shown here, convenient natural tools have been developed that allow synthesis of high quality FAMEs (EN14214) from unrefined oils in a completely enzymatic single-step process, making it fully competitive.

Keywords: biodiesel; lipase; enzyme improvement; crude oil; enzyme immobilization; transesterification; phospholipase; degumming

1. INTRODUCTION

Bio-inspired processes or enzymatic reactions have a low environmental impact, reduce the amount of waste material and can minimize costs, thus serving the requirements to integrate environmental sustainability with economic growth and welfare. The global economic and energy crisis has given new impetus to the search for sustainable energy alternatives, and research in the area has significantly increased in the last years, including biodiesel as one of the targets. Biodiesel is defined as the mono alkyl ester of long chain fatty acids derived from vegetable oils or animal fats, for use in compression-ignition engines [1]. Essentially all industrial biodiesel is produced nowadays using chemical catalysts [2]. The reaction for the chemical synthesis of biodiesel (transesterification) is the conversion of vegetable oils or animal fats into methanol (FAME) or ethanol (FAEE) esters. This reaction proceeds very efficiently in an alkaline medium by addition of sodium methoxide, sodium or potassium hydroxide, or by acid catalysis, usually with sulfuric acid [2]. However, this process requires efficiently pre-treated (degummed or refined) raw materials, is affected by the presence of water in the sample, and cannot make use of the free fatty acids (FFAs) contained by most crude oils [3], being thus difficult to achieve sufficiently high degrees of conversion [2,4]. Moreover, the chemically catalyzed process releases low value glycerol, and generates large volumes of effluents containing high quantities of catalyst and other chemicals. Therefore, chemical biodiesel may not accomplish the requirements for a clean biofuel production.

By contrast, the use of lipases for biodiesel synthesis (Figure 1) has become increasingly important [5,6,7]. Enzymatic production of biodiesel with lipases has received great consideration in recent years and is undergoing a rapid development [8,9].

Lipase-mediated transesterification can be carried out under mild reaction conditions to produce high quality biodiesel free of contaminants, favoring also an easy recovery of glycerol, without need for purification or chemical waste release [2,10]. In addition, the enzymatic process tolerates the common water content of oil and increases the biodiesel yield by avoiding the typical soap formation due to alkaline transesterification [11]. However, the enzymatic method has only scarcely been industrially adopted, mainly due to the still high costs of both, the immobilized enzymes traditionally used, and the raw materials [2,12].

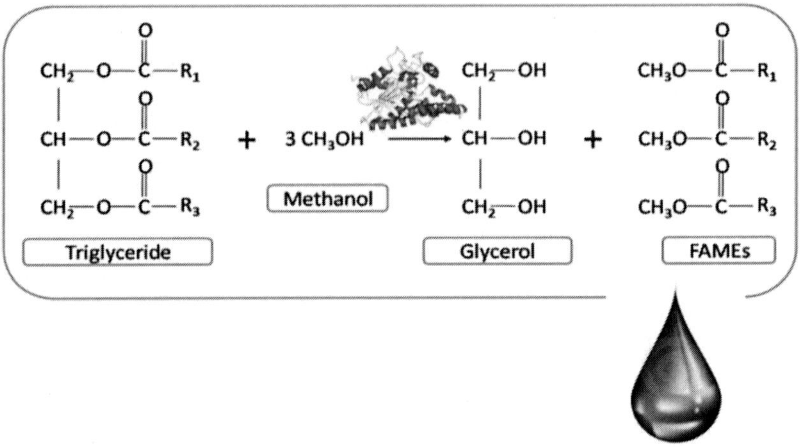

Figure 1. Schematic model of the enzymatic transesterification reaction for FAMEs synthesis from triacyl-glycerides.

2. THE ENZYMES

The enzymes directly involved in biodiesel synthesis are lipases (EC 3.1.1.1). These enzymes are versatile and powerful tools in biotechnology [13], capable of carrying out reactions involving insoluble lipid substrates [14,15], and preserving their catalytic activity in organic solvents [16]. Lipases can catalyze not only hydrolysis reactions, but also participate in various reverse reactions such as esterifications or transesterifications in low water systems or in anhydrous organic solvents [16,17]. Moreover, lipases can perform reactions with high specificity, regioselectivity and enantioselectivity, becoming the most used enzymes in synthetic organic chemistry [13,18,19,20]. Therefore, there is substantial interest in developing appropriate lipase preparations for their use in biodiesel synthesis [21,22]. As an initial approach, the study of biodiversity is a valuable tool for the discovery of new lipolytic enzymes useful for biocatalysis [20,23,24]. However, performance of an enzyme that is active in a given reaction is not always sufficient for its application in an industrial process [13,25]. Thus, new strategies such as process modification or enzyme improvement are required for optimal yield [26]. Various strategies may be run in parallel to obtain the required biocatalyst in a defined process. For example, medium engineering or enzyme immobilization can be assayed [13]. Alternatively, the enzyme can be modified by either rational protein design [27] or by directed evolution [28,29], or both [30]. In this context, a previously isolated and characterized lipase, LipC from P.

aeruginosa 42A2 [31] was improved for evaluation in biodiesel production, using rational design and saturation mutagenesis [32].

2.1. Improvement of LipC Thermoresistance

The cold-activity shown by LipC [31] provides this lipase a valuable property that could be used in low–medium temperature operating systems (4–25 °C). However, a higher thermostability of the enzyme would be required for long-term process in many applications, such as biodiesel production. For such purpose, a rational design strategy to improve LipC thermostability without compromising its cold-adapted properties was developed. In general, the structure of cold-adapted enzymes is more flexible compared to those of their thermostable homologues [33,34]. Thus, introduction of appropriate mutations at the structure sites displaying a high degree of flexibility [32,35] should provide higher compactness and thus more resistance to higher temperatures. As the 3D structure of LipC is not known, the choice of the appropriate sites on this lipase was performed by executing B-FITTER [36] on the known structure of homologous P. aeruginosa mesophilic LipA (55% protein identity) [37]. Most of the residues displaying a high degree of flexibility in the crystal structure of LipA [37] were also found in LipC, thus being good candidates for modification, and chosen to design focused saturation mutagenesis libraries. Following flexibility criteria, eight focused libraries (Figure 2) using different codon degeneracies were constructed involving one or two amino acid positions.

A total of 3256 recombinant clones bearing activity on tributyrin agar plates at 4 °C were obtained and screened for activity prior and after heat shock (15 min, 60 °C). Pseudomonas PABST7.1 pBBlipCH, bearing wild type LipC and used as a control, showed an average 7.6% residual activity after the heat shock. Among the clones screened, five potential variants were obtained with higher residual activity at 4 °C than wild type LipC after 15 min treatment at 60 °C. Among them, variant D2_H8 displayed a residual activity after heat shock of 49%, seven times higher than that shown by wild type LipC after the same treatment, and was selected as the best thermostable variant. Analysis of the DNA sequence of modified lipC gene in variant D2_H8 revealed two substitutions in residues Leu^{288} and Ser^{291}, which were replaced by an alanine and a glutamic acid, respectively. The 3D model structure of variant D2_H8 was constructed and overlapped with that of wild type LipC, and their flexibility parameters were compared. No evident structural changes were detected between them in the conformation of the loop involving residues Leu288Ala and Ser291Glu, but flexibility was found to be slightly lower for mutant D2_H8 than for wild type LipC, showing global B-Factors (obtained from B-Factor putty in PyMol) of 25 and 28.9, respectively.

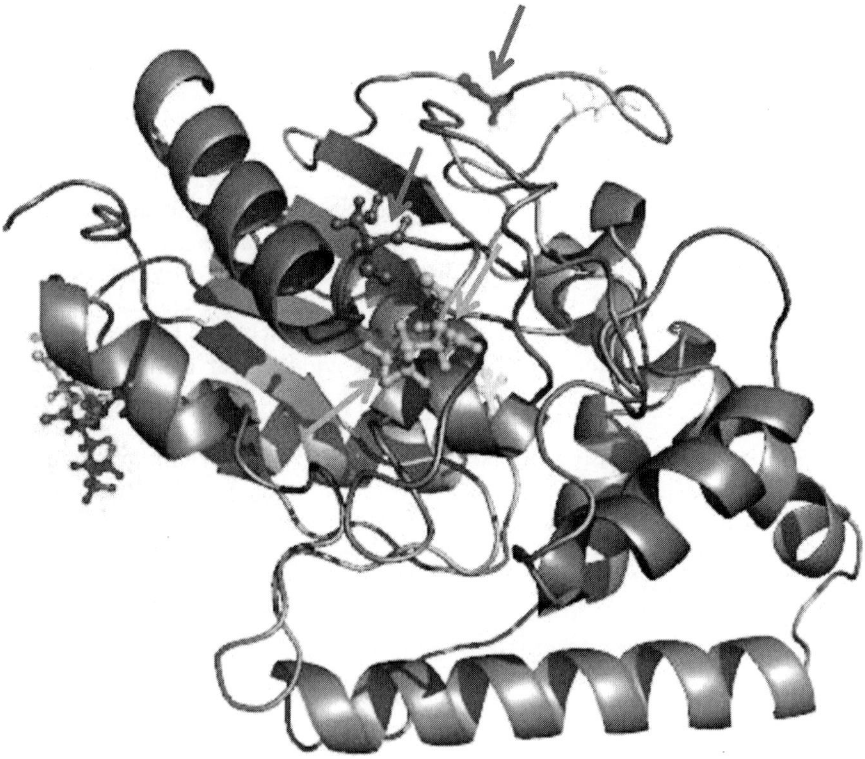

Figure 2. Structural 3D model of Pseudomonas sp. 42A2 LipC displaying the sites chosen for saturation mutagenesis. Site-directed saturation mutagenesis was performed on the highlighted residues after rational protein design. The two residues modified in thermostable mutant D2_H8 are depicted by arrows and colored in orange and purple.

Re-evaluation of variant D2_H8, bearing two amino acid changes and showing higher thermal stability, revealed no significant changes in optimum temperature, still found between 4–20 °C. The substitutions occurred in this variant produced the desired modifications to make the enzyme more compact and thermostable, probably due to replacement of a serine, a hydrophilic amino acid by a glutamic acid, capable of forming salt bridges and to bind metal ions, causing charge changes in the environment of the region and probably allowing interactions with neighboring residues

[38]. This mutant was the result of three levels of selection pressure applied during the screening: (i) selection of only active LipC mutants (tributyrin plates); (ii) selection of active mutants at 4 °C (temperature employed for screening assays); and (iii) selection of active mutants at 4 °C displaying an increased thermal stability. Taking into consideration that increased thermal stability and low optimum temperature may be regarded as apparently contradictory conditions for an enzyme [39], isolation of variant D2_H8 (named LipCmut), with a seven-fold thermostability increase over native LipC while maintaining the optimum temperature at 4–20 °C, allowed generation of a stable and biotechnologically well suited cold-active LipC variant bearing the expected properties, and also showing the capability of using longer carbon chain-length (C_{10}, C_{12}, C_{18}) substrates at higher temperatures, as previously reported [31]. To our knowledge, this was the first time that a cold adapted enzyme is converted into a thermostable variant maintaining its optimum activity at 4 °C.

2.2. Lipase Immobilization

Enzyme immobilization is a well-reported technology that allows application of enzymes in many biocatalyzed processes like in lipase-mediated biodiesel production [9,22,40]. In general, immobilization allows reuse of the biocatalyst, makes the product recovery easier, and frequently enhances enzyme resistance against inactivation by different denaturants, providing more stable and efficient catalysts [41,42]. Nevertheless, many immobilization procedures use sophisticated protocols for lipase entrapment on expensive supports [43], not always suitable for a real scaling-up, and causing an increase in the costs of industrial processes [44,45]. On the contrary, simple enzyme adsorption allows preparation and use of immobilized lipases under mild conditions without significant loss of activity [46], and the associated process is very simple, with the carrier being easily recovered for repeated immobilization rounds [22,46]. Therefore, four non-commercial lipases from Pseudomonas species—LipA and LipC from P. aeruginosa 42A2 [31], LipCmut, the thermostable mutant of LipC previously described [47], and LipI.3 from Pseudomonas CR611 [21], which were produced in either homologous or heterologous hosts—were prepared, immobilized on different cheap supports, and examined for potential biodiesel production. The enzymes were obtained directly from crude cell extracts or from growth culture supernatants, and further immobilized on different carriers: Polypropylene polymers like Accurel EP100 or Accurel MP1000 [48,49] and a silica powder, Celite®545 [50] by simple adsorption. In this case, best immobilization was obtained on polypropylene supports: Accurel EP100 for LipA and Accurel MP1000 for

LipC and LipCmut. Lip I.3, requiring a refolding step [21], was poorly immobilized on all supports tested (best results for Accurel MP1000). The behavior of all four immobilized lipases was tested for triolein transesterification, the best results being obtained for lipases immobilized on Accurel MP1000. Therefore, a new protocol was introduced for immobilized lipase production which does not require protein purification and uses crude enzymes immobilized by a fast adsorption technique on low-cost supports, making the method suitable for an eventual scaling up aimed at biodiesel synthesis. The low price of the supports tested and the simplicity of the procedure, skipping the tedious and expensive purification steps, will certainly contribute to cost reduction in biotechnological lipase-catalyzed processes.

3. THE FEEDSTOCKS

Nowadays, biodiesel is mainly produced by transesterification of edible oils such as soybean, rapeseed, sunflower or palm oil [51], which currently account for over 85% of production costs [52]. Selection of the appropriate raw material is of major importance for ensuring the feasibility of the process at industrial scale [53]. Thus, exploring the use of low-value or non-edible feedstocks (Figure 3) for biodiesel synthesis is a goal for innovation among biodiesel-producing partners. Besides waste cooking oils or animal fats, the biodiesel fuel market can also use unrefined oils [54,55] like crude soybean or canola oils (Figure 3).

Non-refined (crude) soybean oil is an alternative solution already investigated for enzymatic transesterification [55,56,57]. However, crude vegetable oils, also referred to as non-degummed oils, contain relatively high amounts of FFAs, plus certain impurities such as phosphorous compounds known as hydratable phospholipids or phosphatides, that may inhibit the transesterification reaction.

During an alkaline catalyzed chemical biodiesel production, FFAs react with the catalyst to produce soaps that induce formation of emulsions, making the final separation process very difficult. Furthermore, soaps can bind to the catalyst, and addition of extra catalyst is necessary. This is one of the main reasons why when waste oils are used as feedstocks, they require alternative and more complicated processing, since e.g., used cooking oil contains as much as 15% FFAs that would cause an important loss in the process yield [58]. Thus, the FFAs content in oil feedstocks should be as low as possible (less than 0.5%–2.0% wt) in alkaline-catalyzed transesterification reactions [59]. In contrast, enzyme catalysis is usually insensitive to high FFAs concentrations [60] as they can also be used as substrates and converted into fatty acid alkyl esters (FAAEs) through esterification reactions [61]. In this case, the enzyme-catalyzed esterification rate and yield depends on the lipase employed [62].

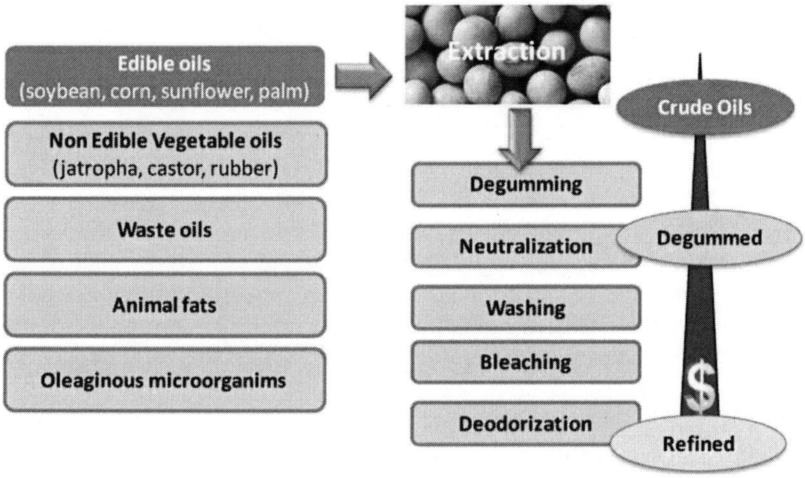

Figure 3. Putative feedstocks for biodiesel production. Edible oils must go through a refining process that significantly increases their costs. Refining is also a requirement for chemical biodiesel production but not for the enzymatic process, which can proceed even in the presence of FFAs and phosphatides. For other oils, only extraction would be required for the enzymatic process. Therefore, as shown here, the use of crude oils may contribute to high cost reductions, thus making the enzymatic process more economically competitive.

Concerning phosphatides, their content in e.g., crude soybean oils may vary from 0.5%–3% w/w [63], corresponding to a phosphorous content in the range 250–1200 ppm. These phospholipids can cause oil storage problems since they precipitate forming a gum, and produce water accumulation. Presence of phospholipids is thus a problem for alkaline biodiesel production, but not for the enzyme-catalyzed process, where, in general, it is not necessary to completely refine crude oils [64]. However, phospholipids have been described as inhibitory substances in transesterification reactions catalyzed by immobilized lipases. In that case, inhibition is due to formation of a film around the support of the enzyme that prevents proper interaction of the lipase with the substrate [56]. Therefore, for immobilized lipases, a degumming would be the only required pre-treatment for enzymatic transformation of vegetable oils into biodiesel, with no further refining steps being necessary [56]. The possibility of using unrefined, low price oils would result in a dramatic reduction of the global costs of enzyme-catalyzed biodiesel process because the yield loss from the refining/degumming is eliminated.

4. LIPASE-CATALYZED BIODIESEL SYNTHESIS

Many lipases from different sources have been studied, improved and tested for biodiesel synthesis [51]. In general, this process has been performed using immobilized lipases [9]. However, the latest results promote the use of soluble enzyme preparations to reduce the cost of the enzymatic preparations and to make the whole process more competitive with respect to the chemical systems. Some examples of immobilized and soluble lipase preparations evaluated for FAMEs synthesis are presented below.

4.1. Evaluation of Soluble and Immobilized Lipases

The soluble and immobilized preparations of LipA, LipI.3 and the cold-adapted lipases LipC and its thermo stable variant LipCmut described in Section 2.2 were evaluated for biodiesel synthesis. Although cold adapted enzymes can provide significant savings due to the possibility of performing processes at much milder conditions, the use of psychrophilic lipases is a rare phenomenon among previously described industrial processes [65]. For evaluation purposes, these lipases were over expressed and produced in recombinant hosts using a low-cost preparation protocol to further reduce costs of eventual scaling-up. The enzymes were obtained from crude cell extracts or from growth culture supernatants without any further purification step. When required, they were immobilized by simple adsorption on Accurel EP100, Accurel MP1000 and Celite®545 [66], and several conditions of transesterification were tested to evaluate FAMEs production using water-degummed soybean oil.

Prior to FAMEs synthesis, the immobilized and soluble forms of the lipases were assayed for activity, thermoresistance, and methanol tolerance. After 1 h incubation between 4 and 37 °C, the cold adapted lipases LipC and LipCmut maintained more than 65% activity [47], while, as expected, thermoresistant LipCmut displayed a significantly higher resistance at 50 °C, showing good tolerance to the methanol concentrations tested. Reactions for FAMEs synthesis were carried out at different ratios of water and methanol to set up the conditions for the highest yield. A great FAMEs increase with the amount of water content in the reaction was observed.

Immobilized LipCmut reached the highest production, with 25% FAMEs release using 7.5% MeOH and 10% water. For reactions carried out with the lipases in soluble form, lower FAMEs synthesis was achieved (5.4% for LipC and 11.6% for LipCmut) at their best reaction conditions. Nevertheless, when the loading of soluble enzymes in the reaction was increased to 10% (v/v), FAMEs production was comparable to that obtained for the immobilized forms. Thus, these results open up the possibility to perform transesterification reactions using soluble enzymes and at low temperatures, a procedure that could help to reduce the costs of enzymatic biodiesel processes. No matter the moderate efficiency of LipC and LipCmut preparations for FAMEs production, it must be highlighted that in the literature very few works can be found related to biodiesel synthesis using psychrophilic lipases [65]. Use of cold-adapted enzymes in biodiesel production would allow reducing the reaction temperature, with significant economical savings, and could contribute to the design of innovative synthesis systems for alternative chemicals production.

4.2. Testing Soluble Lipase Callera Trans L

Although the use of immobilized lipases for FAMEs production has been extensively reported [67,68], they show several drawbacks: high costs of the enzyme, use of expensive supports, and the immobilization process itself. For instance, the price of immobilized Novozym 435 is very high and would require many re-uses (typically, more than 100 cycles) to become cost-efficient for bulk biodiesel production. On the contrary, soluble lipases can be produced and sold at a much lower price (30–50 times lower; P. M. Nielsen, personal communication) and can also be re-used for several times after recovery from the water phase, which makes them cost-efficient compared to immobilized enzymes. Moreover, soluble lipases can also be used under mild reaction conditions, have a faster reaction time, display higher conversion rates than immobilized enzymes [69], and allow, without any FAMEs yield loss, the presence of water in the process, which is required for oil degumming or during downstream steps [57].Therefore, the use of soluble lipases on low cost, unrefined oils, will launch the enzymatic FAMEs synthesis process into the range of the most competitive chemical systems for fuel production [70].

Up to date, however, soluble lipases have scarcely been tested in transesterification [70,71,72] due to their water requirement. But in fact water would not inhibit transesterification if used at low concentrations, as it is essential to maintain the specific tridimensional structure of the lipase, especially if used in soluble form [58]. Accordingly, a new liquid lipase (Callera Trans L, Novozymes) was tested to obtain FAMEs from crude, non-degummed soybean oil. A successful conversion into FAMEs (96% w/w) was obtained after 24 h reaction at 35 °C in 200 rpm agitation in the presence of water (3%–15%). Moreover, a general good reduction of free fatty acids in the final product was produced, especially in the two lower (3%–5%) water systems, where Callera Trans L produced an exceptional 96.3% and 95.6% FAMEs release, a transformation never found before for non-degummed oils.

To evaluate the mode of action of Callera Trans L, variation of FFAs, FAMEs and acyl-glycerides during the reaction was determined by acid titration, GC and HPLC. A clear increase of FFAs during the first 3 h was observed, suggesting that the reaction takes place through a first step of enzyme-catalyzed triacyl-glycerides (TAGs) hydrolysis that releases FFAs (Figure 4). As the reaction continues, the concentration of FFAs decreases, suggesting that once the acyl-glycerides have been hydrolyzed, the released FFAs are further esterified with methanol by the enzyme into the corresponding FAMEs. In that case, FFAs would be considered the intermediates of the reaction [70], which probably proceeds through an initial hydrolysis reaction where TAGs are hydrolyzed to DAGs, MAGs and FFAs, followed by the subsequent esterification of the FFAs derived from the complete acyl-glycerides hydrolysis, as previously suggested [70]. Evidence for this mode of action was obtained through HPLC analysis where TAGs are hydrolyzed by Callera Trans L into DAGs, MAGs and FFAs during the first 5 h of reaction at all water concentrations. This hydrolysis is also favored by the low initial methanol concentration, added step-wise during the reaction. The FAMEs appeared during the first 5 h (Figure 4) are probably due to a true transesterification activity of the enzyme, whereas at longer reaction times, when TAGs have almost disappeared, esterification activity is predominant and FAMEs formation derives from the FFAs generated by the complete hydrolysis of TAGs, DAGs and MAGs. This effect was particularly evident at 3%–5% water concentrations, where FAMEs production by Callera Trans L was more effective (Figure 4).

These results show not only the ability of soluble Callera Trans L for FAMEs synthesis but also its double hydrolysis and esterification activities in low water systems. The use of a soluble instead of immobilized lipase resulted in a significant simplification of oil transesterification and is the background for the new, cost-effective process developed by Novozymes using liquid lipase Callera Trans L, which is now being used at several production plants for biodiesel synthesis from different feedstocks [73]. In the first quarter of 2014, both Blue Sun Biodiesel in St. Joseph, Mo., and Viesel Fuel LLC in Stuart, Fla., announced the full-scale production of biodiesel based on lipase as catalyst.

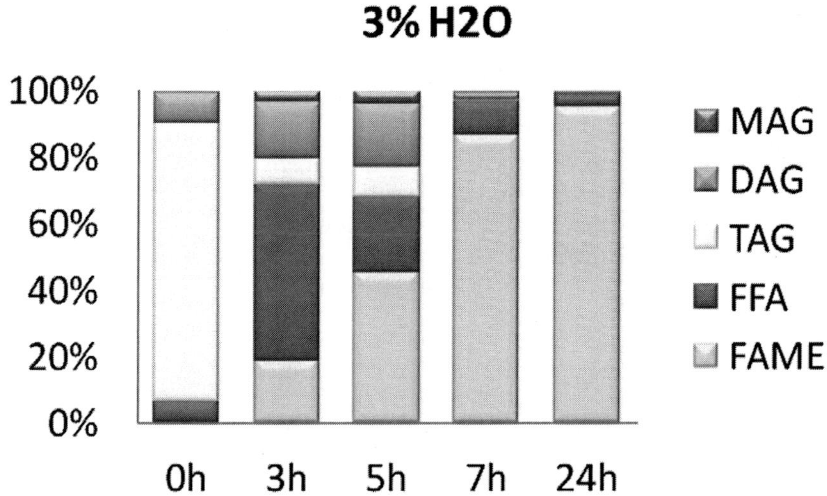

Figure 4. Mode of action of lipase Callera Trans L when used in a low water system (3% water). From the HPLC analysis of the reaction products, it can be concluded that an initial hydrolysis releases FFAs, MAGs and DAGs from the original oil (mostly TAGs). All these products are good substrates for Callera Trans L, which then proceeds to FAMEs formation through esterification of the FFAs with methanol. Although most of the reaction occurs through hydrolysis followed by esterification, the results indicate that some transesterification occurs as well since the beginning of the reaction, where the enzyme uses the TAGs present in the oil for FAMEs synthesis. Reaction conditions were: crude soybean oil (7.8% FFAs; 690 ppm P); 1.5 equivalents. MeOH added at t = 0, 3 h, 5 h, 7 h; 1% w/w Callera Trans L; T = 35 °C; mixing: 200 rpm; and reaction time: 24 h.

5. DEGUMMING

As mentioned in Section 3, crude, non-degummed oils contain impurities like phosphorous compounds (phosphatides), which have to be removed for efficient transesterification. Oil degumming is a requirement to obtain refined, edible oils, but it is also essential for biodiesel production using immobilized lipases [56]. Moreover, according to the legal specifications, phosphorus concentration in final biodiesel must be reduced to less than 10 ppm (EN14214:2003; ASTM D6751:2012). For biodiesel synthesis, a high content of phospholipids in the feedstocks means a concomitant loss of yield in FAMEs production, as the fatty acids enclosed into the phospholipid molecules are not accessible to the lipase for transesterification, and the gums entrap oil when it is separated. In fact, degumming represents a 2.5% loss of total oil; being the current market price \$1,100 per ton, this corresponds to a loss of \$27.5 per ton of oil. Depending on the raw materials used, degumming becomes a requirement for biodiesel synthesis to achieve phospholipids removal and to reduce the final phosphorus content below the specified limits. In addition to an extra tank, the degumming process involves the use of acids and high temperatures, all factors boosting the process costs and health risks.

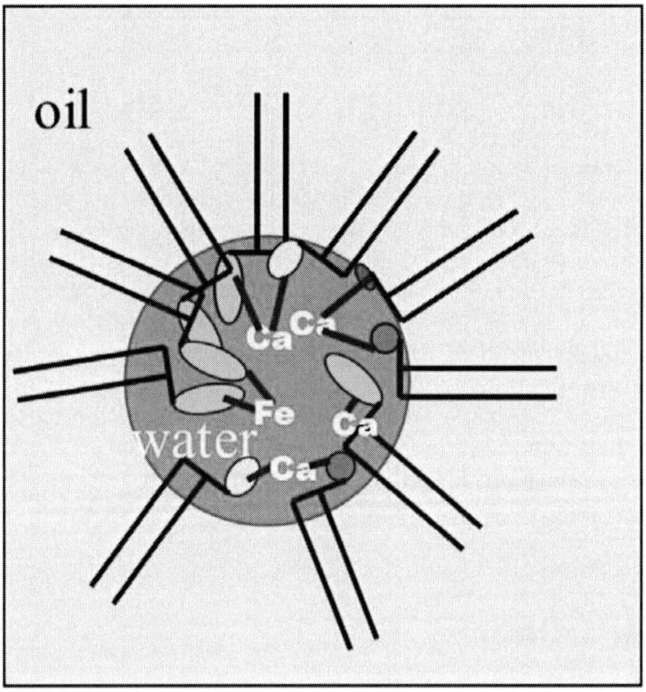

Figure 5. Schematic representation of phospholipids combined with metal cations as can be found in most unrefined, non-degummed oils. In order to disrupt the phospholipid micelles for gums removal, ions must be chelated, a process that involves acid treatment with high-shear mixing of the oil at high temperatures. Once the micelles are disrupted, the phosphorous group can be removed to produce DAGs with a more hydrophobic moiety, adequate for enzymatic FAMEs synthesis.

Phospholipids are commonly removed by previously developed refining/degumming processes of physical or chemical nature. In the 1990s, the first industrial enzymatic oil-degumming process was launched, the Lurgi's Enzymax®, based on phospholipase conversion of non-hydratable phospholipids (NHPs) to hydratable lyso-phospholipids [74]. NHPs, which consist on phospholipids combined with metal cations such as Fe, Ca or Mg (Figure 5), can be chemo-enzymatically converted into hydratable compounds and simply removed by water washing and centrifugation [75].

The easily hydratable phospholipids are phosphocholine (PC), phosphatidylinositol (PI) and lysophosphatidylcholine (LPC), whereas phosphatidilethanolamine (PE) and phosphatidic acid (PA) are generally considered as NHPs [76]. The proposed industrial enzymatic degumming [63] involves a first step of citric acid treatment with high-shear mixing at high temperatures to chelate the metals and to open the phospholipid micelles (Figure 5). After cooling down, degumming can be completed with a final phospholipase treatment. For this purpose, different types of phospholipases can be applied depending on the acyl ester bond they hydrolyze in the phospholipid molecule [77]: Phospholipase A_1 (PLA_1), which catalyzes the hydrolysis of the fatty acid at position sn-1 in the phospholipid molecule releasing a lyso-phospholipid (Lyso-PL) and a free fatty acid (FFA), has already been applied to the degumming process [63,77,78] also in combination with phospholipase C (PLC), which cleaves the phosphorus–oxygen bond between glycerol backbone and phosphate, releasing a diacylglycerol (DAG) and the phosphate ester group [79]. Since phospholipase-aided degumming requires the presence of water in the reaction mixture and Callera Trans L can efficiently be used for FAMEs synthesis in low-water containing systems, a more cost-effective biodiesel production process could be performed by combination of enzymatic degumming and transesterification of crude feedstocks using phospholipases and Callera Trans L (Section 4.2) in a single-pot reaction.

5.1. Combined Enzymatic Degumming and Transesterification

The possibility of using efficient soluble lipases like Callera Trans L allows combination of the required steps for enzyme-catalyzed synthesis of biodiesel from crude oils: Phospholipase-catalyzed oil degumming followed by lipase-mediated synthesis of FAMEs. The use of water, required in the degumming step, would not inhibit transesterification with Callera Trans L, as water is tolerated by this enzyme [57]. To search for such a more sustainable and competitive process, two phospholipases (PLA_1 and PLC), a lyso-phospholipase (LPLL-2) and their combinations with liquid lipase

Callera Trans L were tested to unify the enzymatic oil degumming with the transesterification (Figure 6), aimed at producing a final biodiesel that meets the phosphorus legal limits. The two processes were studied combined or separately using crude soybean oil to define the optimal conditions for each parameter and enzyme.

As stated in Section 5, citric acid is used as helper in degumming for disruption of phospholipid micelles due to its chelating action towards binding ions [63,78]. Therefore, influence of citric acid on lipase activity was studied. A clear negative effect of citric acid on transesterification was observed, with maximum FAMEs production of 67%, too low for a competitive biodiesel process. On the other hand, the effect of methanol on phospholipases was tested. For these assays, the degumming process included phospholipases and the conventional citric acid step, carried out on crude soybean oil with or without addition of methanol. Higher phosphorus values were found in the oil phase for all enzyme mixtures when assayed in the presence of methanol, and the same results were obtained when control samples containing only crude soybean oil were assayed with (575 ppm P) and without methanol (64 ppm P). This suggests a direct action of methanol on removal of the phospholipids from soybean oil and shows that phospholipases are not inhibited by this solvent. In fact, for crude soybean oil, when methanol was present, the oil phase was completely clear and no sediment (gums) could be observed, contrarily to what happened to acid degummed or untreated samples, which displayed some turbidity or a sediment, respectively [80]. Therefore, methanol can solubilize the phospholipids of crude soybean oil [81], constituted mainly by PC (and to a lower extent by PE), described as easily hydratable phospholipids and also soluble in short-chain alcohols like ethanol [82]. The evidence of solubilization of phospholipids by methanol is of great importance in the process. It means that gums can be removed with no requirement for a conventional acid treatment, thus making the released phospholipids more available for enzymatic treatment, that is, methanol seems to break apart the micelles, contributing to the release of extra FFAs (Figure 6). Therefore, presence of methanol may allow skipping the acid degumming step without any loss of performance and it can concomitantly provide more FFAs to the reaction for a higher FAMEs yield.

Since citric acid has a negative effect on transesterification, and taking into consideration the role of methanol in phospholipid solubilization, enzymatic degumming with phospholipases was further coupled to transesterification in the same batch, using Callera Trans L best operating conditions (24 h incubation, 35 °C, 250 rpm, 24 h), with no need for a conventional citric acid treatment. Performance of different types of phospholipases with Callera Trans L for FAMEs production and phosphorous decrease was tested (Table 1). The high P concentration (823 ± 56 ppm) found in the oil phase when only transesterification, used as a control, was run is remarkable. In this case, the phosphorus content was approximately the same as that of raw material (823 ppm), evidencing that Callera Trans L does not have any phospholipase activity. The transesterification yield of Callera Trans L reached >95% FAMEs production when phospholipases were added to the reaction mixture, where concomitantly a dramatic phosphorus content decrease, below 10 ppm, was found (Table 1). A higher FAMEs production was obtained in samples containing combinations of phospholipases, probably due to a synergic effect between phospholipases themselves and the lipase. Therefore, phospholipase activity leads to a gain of oil useful for lipase-mediated transesterification. This means that beyond the reduction of phosphorus in the final biodiesel, phospholipases contribute to the release of FFAs from phospholipid molecules, which can then be used by the lipase for increased FAMEs synthesis (Figure 6). When a lyso-phospholipase (LLPL-2) was added to help PLA_1 in reducing the final content of phosphorus, the P values in the oil phase were even lower (Table 1) due to the mode of action of LLPL-2, which releases the FFAs located at position sn-2 after PLA_1 hydrolysis on the phospholipid molecule. After the combined action of PLA_1 and LLPL-2, a glycerol-phosphatide would presumably be released which, being more polar than a lyso-phospholipid, would migrate to the glycerol phase, thus reducing the P content in the FAMEs phase. Higher reductions of phosphorus were observed when PLA_1 was combined with PLC, which directly removes the phosphodiester group, releasing a diacyl glyceride (DAG), known to be also a good substrate for Callera Trans L transesterification [57].

Figure 6 shows a schematic representation of the successful, completely enzymatic process that has been developed, resulting in a more economic and eco-friendly biodiesel production. Combination of crude soybean oil degumming and transesterification in a unique step is possible by using phospholipases and liquid lipase Callera Trans L. In the combined process, an important cost reduction can be achieved: In addition to the $27.5 per ton savings in the case of crude soybean oil, costs can also be substantially lowered by avoiding the extra tank commonly required for oil degumming pre-treatments, by using mild temperatures (35 °C or below) and a soluble lipase. Moreover, citric acid treatment can be skipped, thus increasing the savings of the whole process. Therefore, the developed method bears the conditions for being easily scaled-up and meets the requirements of a sustainable and competitive process.

Table 1. FAMEs release (%) and P (ppm) content of reactions performed in a single batch with the enzyme combinations stated. Crude oils were incubated with the enzymes for 24 h at 35 °C and 200 rpm agitation. All enzymes were added in soluble form providing an average 3% water in the final mixture. Methanol was added following a continuous gradient of 0.4 mL/h for 10 h, to a total amount of 1.5 molar equivalents.

	Soybean Oil (1% FFA; 900 ppm P)		Corn Oil (6% FFA; 62 ppm P)		Canola Oil (1% FFA; 250 ppm P)	
Enzyme Combination *	FAMEs	P	FAMEs	P	FAMEs	P
Callera	85.2	823	93.4	7.5	98.7	145
PLA_1 + Callera	98.2	8.0	94.5	7.3	98.3	142
PLC + Callera	90.8	12.8	95.7	9.2	98.2	130
PLA_1 + LLPL-2 + Callera	97.8	6.0	96.2	9.3	98.9	145
PLA_1 + PLC + Callera	96.6	4.6	95.5	7.5	99.1	130

* Callera: Lipase Callera Trans L® (Novozymes A/S); PLA_1: Phospholipase A_1 (Lecitase® Ultra, Novozymes A/S), releases fatty acids from position sn-1; PLC: Phospholipase C (Purifine, Verenium), removes the phosphate group; LLPL-2: Lysophospholipase (Novozymes A/S), releases fatty acids from position sn-2.

5.2. Other Feedstocks Evaluation

To evaluate performance of the degumming-transesterification combined process, Callera Trans L and phospholipases were tested in a single step procedure for FAMEs production and P removal starting from other oils commonly used for chemical biodiesel production but more restrictive for an enzymatic process, like corn oil and canola oil. The corn oil here evaluated was obtained after dry corn milling from a biorefinery–bioethanol plant. When tested for combined enzymatic biodiesel synthesis it showed a slightly lower FAMEs production compared to soybean oil. Optimum operational conditions were found with the combination PLA_1 + LLPL-2 + Callera, with a 96.2% production of FAMEs (Table 1). Transesterification was not complete, probably due to the composition of corn oil, rich in sterols and waxes, which are hardly processed by the lipase for methyl esterification. Like for soybean oil, phospholipases helped to increase the transesterification yield, with final phosphorus values below 10 ppm, even when phospholipases were not present in the reaction mixture. This was probably due to the low initial content of phosphorus in corn oil (62 ppm), mostly constituted by hydratable phospholipids, which can be easily removed by simple washing with the water present in the system. However, it must be highlighted that when a phospholipase is present in the reaction mixture, phospholipids are hydrolyzed releasing FFAs, and more useful oil is available for lipase methylester formation.

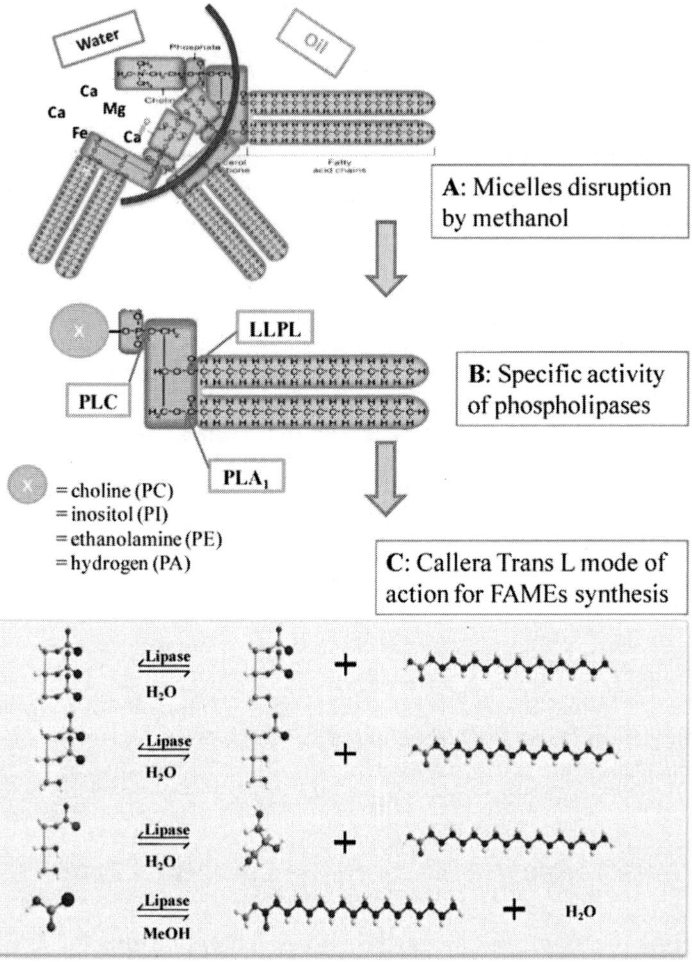

Figure 6. Schematic representation of the combined enzymatic degumming-transesterification reaction performed using phospholipases and lipase Callera Trans L for a more competitive FAMEs synthesis from soybean oil. As shown in the combined process developed here for soybean oil, the acid treatment can be skipped as methanol, one of the substrates of the reaction, can disrupt the micelles and dissolve the gums (**A**). Further phospholipase treatment contributes to remove the phosphate group (PLC), whereas the fatty acids at positions 1 and 2 can be released by a PLA_1 and LLPL-2, respectively, (**B**) to provide extra FFAs, which will then be converted into FAMEs by Callera Trans L esterification activity (**C**).

When crude canola oil was tested, a high transesterification rate was achieved. However, the final phosphorus content was not reduced below 130 ppm. In this particular batch of canola oil phospholipids have a very high content of Ca^{2+} and Mg^{2+}, which make it especially difficult to degum. In some cases, crude canola oil is considered one of the most difficult oils for degumming because it is mostly constituted by NHPs, barely attacked by phospholipases without the aid of an acid treatment. Moreover, it requires a strong chelating agent such as citric acid at higher concentrations (0.1%) to achieve a suitable phosphorus reduction. Such conditions can hardly be reached with milder citric acid concentrations or even with phosphoric acid [83], thus making it difficult to reach the required phosphorus content in the final biodiesel.

In order to confirm the predicted activity of the phospholipases tested in the combined degumming-transesterification process, and to exclude the possibility that the phospholipids contained in the original raw material had migrated to the glycerin phase or to the interphase as entire molecules without any gain of oil, the phospholipids content of both, the oil and glycerin phases obtained from the combined reactions were analyzed by ultra-performance liquid chromatography-tandem mass spectrometry (UPLC/MS-MS). Among the compounds putatively released by phospholipases, glycerol and other polar molecules containing the remaining phosphorus, like the glycerolphosphatide and the phosphodiester group should be found in the glycerin phase, thus justifying the good reduction of P achieved in the final FAMEs phase for samples treated with the enzyme combinations assayed.

5.3. Phospholipase Activity in the Combined Process

Comparing the oil phase of control samples containing only Callera Trans L, where only transesterification occurs, with those resulting from reactions containing both, the lipase and PLA_1 (PLA_1 + Callera), intact phospholipids containing phosphocholine completely disappeared, indicating that phospholipase A_1 activity almost completely removed all phosphocholine from the oil phase (Figure 6). Interestingly, also lysophosphatidylcholine, apparently contained in the raw oil from the beginning, disappeared from the oil phase after PLA_1 + Callera treatment, suggesting that PLA_1 might display an extra activity causing removal of the fatty acid of the LPC molecule and not only the one at position sn-1, as would be expected. This extra activity, not described before for a PLA_1, could be due to the long reaction time used in the combined process: Generally phospholipases are used for 2–6 h, whereas they were acting for 24 h in the developed process.

When the final glycerine phase of all combined reaction samples was analyzed for phosphorous compounds, the predicted mode of action of the phospholipases could be confirmed (Figure 6). As expected, phosphocholine was found only in the glycerin phase of the reactions where PLC was present (PLC + Callera and PLA_1 + PLC + Callera), proving the specific activity of this phospholipase for the phosphodiester bond. Instead, as a result of PLA_1 activity, lyso-phosphocholine could only be detected in very small amounts in the glycerine phase, suggesting a kind of instability of the LPC molecule. In fact, glycerol-phosphocholine, resulting from the combined activity of PLA_1 + LLPL-2 + Callera, should be present in the glycerin phase due to its polarity. But it was also found in the oil phase, supporting the hypothesis of an extra activity of PLA_1 causing additional removal of the fatty acid at position sn-2, or being due to a kind of instability of the LPC molecule. LLPL-2 activity was confirmed in sample PLA_1 + LLPL-2 + Callera by the finding of glycerol-phosphocholine in the glycerine phase. Parallel analysis by UPLC/MSMS of the phosphatides found in the oil and glycerin phases resulting from the combined reactions performed turned out to be a powerful tool for the study of such a complex system, allowing dissection of the activity of each enzyme, and confirming the importance of phospholipases in the combined degumming-transesterification process.

6. CONCLUSIONS

In order to reduce the costs of the industrial enzymatic biodiesel production process, several recombinant and improved lipases were prepared and tested in their immobilized and soluble forms for FAMEs synthesis. Cost reduction was also achieved by using crude (non-degummed) soybean oil as a cheaper raw material. Use of liquid lipase Callera Trans L in its soluble form instead of the common immobilized preparations in biodiesel production can provide numerous and sustainable benefits: besides reducing the costs derived from enzyme immobilization, the lipase can be used in combination with other enzymes such as phospholipases for gums removal, thus allowing the use of much cheaper, non-refined oils. The possibility to perform degumming and transesterification in a single tank denotes a great efficiency increase in the new era of enzymatic biodiesel production at industrial scale.

ACKNOWLEDGMENTS

We thank Novozymes (Denmark) for kindly providing soluble lipase Callera Trans L and the phospholipases, and research facilities. Mathias Nordblad from Technical University of Denmark and Rune F. Haller from the Department of Analytical Development of Novozymes, Denmark are also acknowledged for help in HPLC and UPLC/MS-MS analysis, respectively. This work was financed by the Scientific and Technological Research Council (MINECO, Spain), grant CTQ2010-21183-C02-02/PPQ, by the IV Pla de Recerca de Catalunya, grant 2009SGR-819, by PCI-AECID project A203563511, and by the Generalitat de Catalunya to the "Xarxa de Referència en Biotecnologia" (XRB). Silvia Cesarini acknowledges a doctoral fellowship from the Spanish Ministry of Science and Education (AP2008-04579).

AUTHOR CONTRIBUTIONS

Silvia Cesarini participated in the design of the study, carried out the experiments, organized and interpreted the data, and drafted the manuscript. F. I. Javier Pastor contributed to the manuscript draft discussion. Per M. Nielsen coordinated and designed part of the study, critically interpreted the data and revised the manuscript. Pilar Diaz coordinated the project, contributed to the critical discussion and data interpretation, and prepared, revised and corrected the manuscript.

REFERENCES

1. Al-Zuhair, S. Production of biodiesel: Possibilities and challenges. Biofuels Bioprod. Biorefin. **2007**, 1, 57–66.
2. Fjerbaek, L.; Christensen, K.V.; Norddahl, B. A review of the current state of biodiesel production using enzymatic transesterification. Biotechnol. Bioeng. **2009**, 102, 1298–1315.
3. Liu, L.; Lampert, D. Monitoring chemical interesterification. J. Am. Oil Chem. Soc. **1999**, 76, 783–787.
4. Shimada, Y.; Watanabe, Y.; Sugihara, A.; Tominaga, Y. Enzymatic alcoholysis for biodiesel fuel production and application of the reaction to oil processing. J. Mol. Catalysis B **2002**, 17, 133–142.
5. Marchetti, M.; Miguela, V.U.; Errazu, A.F. Possible methods for biodiesel production. Renew. Sustain. Energy Rev. **2007**, 11, 1300–1311.
6. Marchetti, J.M.; Miguel, V.U.; Errazu, A.F. Techno-economic study of different alternatives for biodiesel production. Fuel Process. Technol. **2008**, 89, 740–748.

7. Ranganathan, S.V.; Narasimhan, S.L.; Muthukumar, K. An overview of enzymatic production of biodiesel. Bioresour. Technol. **2008**, 99, 3975–3981.
8. Ghaly, A.E.; Dave, D.; Brooks, M.S.; Budge, S. Production of biodiesel by enzymatic transesterification: Review. Am. J. Biochem. Biotechnol. **2010**, 6, 54–76.
9. Zhao, X.; Qi, F.; Yuan, C.; Du, W.; Liu, D. Lipase-catalyzed process for biodiesel production: Enzyme immobilization, process simulation and optimization. Renew. Sustain. Energy Rev. **2015**, 44, 182–197.
10. Szczesna Antczak, M.; Kubiak, A.; Antczak, T.; Bielecki, S. Enzymatic biodiesel synthesis—Key factors affecting efficiency of the process. Renew. Energy **2009**, 34, 1185–1194.
11. Dizge, N.; Aydiner, C.; Imer, D.Y.; Bayramoglu, M.; Tanriseven, A.; Keskinler, B. Biodiesel production from sunflower, soybean, and waste cooking oils by transesterification using lipase immobilized onto a novel microporous polymer. Bioresour. Technol. **2009**, 100, 1983–1991.
12. Vyas, A.P.; Verma, J.L.; Subrahmanyam, N. A review on fame production processes. Fuel **2010**, 89, 1–9.
13. Bornscheuer, U.T.; Bessler, C.; Srinivas, R.; Krishna, S.H. Optimizing lipases and related enzymes for efficient application. Trends Biotechnol. **2002**, 20, 433–437.
14. Jaeger, K.E.; Eggert, T. Lipases for biotechnology. Curr. Opin. Biotechnol. **2002**, 13, 390–397.
15. Jaeger, K.E.; Dijkstra, B.W.; Reetz, M.T. Bacterial biocatalysts: Molecular biology, three-dimensional structures, and biotechnological applications of lipases. Annu. Rev. Microbiol. **1999**, 53, 315–351.
16. Klibanov, A.M. Asymmetric transformations catalyzed by enzymes in organic-solvents. Acc. Chem. Res. **1990**, 23, 114–120.
17. Gupta, R.; Gupta, N.; Rathi, P. Bacterial lipases: An overview of production, purification and biochemical properties. Appl. Microbiol. Biotechnol. **2004**, 64, 763–781.
18. Reetz, M.T. Directed evolution as a means to create enantioselective enzymes. Abstr. Pap. Am. Chem. Soc. **2002**.

19. Fillat, A.; Romea, P.; Urpi, F.; Javier Pastor, F.I.; Diaz, P. Improving enantioselectivity towards tertiary alcohols using mutants of Bacillus sp bp-7 esterase estbp7 holding a rare ggg(x)-oxyanion hole. Appl. Microbiol. Biotechnol. **2014**, 98, 4479–4490.
20. Bassegoda, A.; Fillat, A.; Pastor, F.I.J.; Diaz, P. Special rhodococcus sp cr-53 esterase est4 contains a ggg(a)x-oxyanion hole conferring activity for the kinetic resolution of tertiary alcohols. Appl. Microbiol. Biotechnol. **2013**, 97, 8559–8568.
21. Alves, J.S.; Vieira, N.S.; Cunha, A.S.; Silva, A.M.; Zachia Ayub, M.A.; Fernandez-Lafuente, R.; Rodrigues, R.C. Combi-lipase for heterogeneous substrates: A new approach for hydrolysis of soybean oil using mixtures of biocatalysts. RSC Adv. **2014**, 4, 6863–6868.
22. Poppe, J.K.; Matte, C.R.; Ruaro Peralba, M.D.C.; Fernandez-Lafuente, R.; Rodrigues, R.C.; Zachia Ayub, M.A. Optimization of ethyl ester production from olive and palm oils using mixtures of immobilized lipases. Appl. Catal. A **2015**, 490, 50–56.
23. Ruiz, C.; Pastor, F.I.J.; Diaz, P. Isolation of lipid- and polysaccharide-degrading micro-organisms from subtropical forest soil, and analysis of lipolytic strain bacillus sp cr-179. Lett. Appl. Microbiol. **2005**, 40, 218–227.
24. Bassegoda, A.; Pastor, F.I.J.; Diaz, P. Rhodococcus sp strain cr-53 lipr, the first member of a new bacterial lipase family (family x) displaying an unusual y-type oxyanion hole, similar to the Candida antarctica lipase clan. Appl. Environ. Microbiol. **2012**, 78, 1724–1732.
25. Bommarius, A.; Bommarius-Riebel, B. Fundamentals of Biocatalysis; Wiley-VCH: Weinheim, Germany, 2005.
26. Lutz, S.; Patrick, W.M. Novel methods for directed evolution of enzymes: Quality, not quantity. Curr. Opin. Biotechnol. **2004**, 15, 291–297.
27. Bornscheuer, U.T.; Pohl, M. Improved biocatalysts by directed evolution and rational protein design. Curr. Opin. Chem. Biol. **2001**, 5, 137–143.
28. Arnold, G. Methods in molecular biology. In Methods in Molecular Biology; Humana Press Inc.: New York, NY, USA, 2003.

29. Reetz, M.T.; Hobenreich, H.; Soni, P.; Fernandez, L. A genetic selection system for evolving enantioselectivity of enzymes. Chem. Commun. **2008**, 5502–5504.
30. Bassegoda, A.; Cesarini, S.; Diaz, P. Lipase improvement: Goals and strategies: Minireview. Comput. Struct. Biotechnol. J. **2012**, 2, 1–8.
31. Bofill, C.; Prim, N.; Mormeneo, M.; Manresa, A.; Javier Pastor, F.I.; Diaz, P. Differential behaviour of pseudomonas sp 42a2 lipc, a lipase showing greater versatility than its counterpart lipa. Biochimie **2010**, 92, 307–316.
32. Reetz, M.T.; Carballeira, J.D. Iterative saturation mutagenesis (ISM) for rapid directed evolution of functional enzymes. Nat. Protoc. **2007**, 2, 891–903.
33. Gerday, C.; Aittaleb, M.; Bentahir, M.; Chessa, J.P.; Claverie, P.; Collins, T.; D'Amico, S.; Dumont, J.; Garsoux, G.; Georlette, D.; et al. Cold-adapted enzymes: From fundamentals to biotechnology. Trends Biotechnol. **2000**, 18, 103–107.
34. Siddiqui, K.S.; Cavicchioli, R. Cold-adapted enzymes. Annu. Rev. Biochem. **2006**, 75, 403–433.
35. Radivojac, P.; Obradovic, Z.; Smith, D.K.; Zhu, G.; Vucetic, S.; Brown, C.J.; Lawson, J.D.; Dunker, A.K. Protein flexibility and intrinsic disorder. Protein Sci. **2004**, 13, 71–80.
36. Reetz, M.T.; Soni, P.; Fernandez, L.; Gumulya, Y.; Carballeira, J.D. Increasing the stability of an enzyme toward hostile organic solvents by directed evolution based on iterative saturation mutagenesis using the b-fit method. Chem. Commun. **2010**, 46, 8657–8658.
37. Nardini, M.; Lang, D.A.; Liebeton, K.; Jaeger, K.E.; Dijkstra, B.M. Crystal structure of Pseudomonas aeruginosa lipase in the open conformation—The prototype for family i.1 of bacterial lipases. J. Biol. Chem. **2000**, 275, 31219–31225.
38. Daniel, R.M.; Dunn, R.V.; Finney, J.L.; Smith, J.C. The role of dynamics in enzyme activity. Annu. Rev. Biophys. Biomol. Struct. **2003**, 32, 69–92.
39. Feller, G. Protein stability and enzyme activity at extreme biological temperatures. J. Phys. Condes. Matter **2010**, 22, 17.
40. Adlercreutz, P. Immobilisation and application of lipases in organic media. Chem. Soc. Rev. **2013**, 42, 6406–6436.

41. Tischer, W.; Kasche, V. Immobilized enzymes: Crystals or carriers? Trends Biotechnol. **1999**, 17, 326–335.
42. Bezerra, C.S.; de Farias Lemos, C.M.G.; de Sousa, M.; Barros Goncalves, L.R. Enzyme immobilization onto renewable polymeric matrixes: Past, present, and future trends. J. Appl. Polym. Sci. **2015**,
43. Ferrer, M.; Plou, F.J.; Fuentes, G.; Cruces, M.A.; Andersen, L.; Kirk, O.; Christensen, M.; Ballesteros, A. Effect of the immobilization method of lipase from Thermomyces lanuginosus on sucrose acylation. Biocatal. Biotransform. **2002**, 20, 63–71.
44. Bezbradica, D.I.; Mateo, C.; Guisan, J.M. Novel support for enzyme immobilization prepared by chemical activation with cysteine and glutaraldehyde. J. Mol. Catal. B Enzym. **2014**, 102, 218–224.
45. Godoy, C.A.; de las Rivas, B.; Guisan, J.M. Site-directing an intense multipoint covalent attachment (MCA) of mutants of the geobacillus thermocatenulatus lipase 2 (btl2): Genetic and chemical amination plus immobilization on a tailor-made support. Process Biochem. **2014**, 49, 1324–1331.
46. Jesionowski, T.; Zdarta, J.; Krajewska, B. Enzyme immobilization by adsorption: A review. Adsorption **2014**, 20, 801–821.
47. Cesarini, S.; Bofill, C.; Pastor, F.I.J.; Reetz, M.T.; Diaz, P. A thermostable variant of P. aeruginosa cold-adapted lipc obtained by rational design and saturation mutagenesis. Process Biochem. **2012**, 47, 2064–2071.
48. Gangoiti, J.; Santos, M.; Llama, M.J.; Serra, J.L. Production of chiral (r)-3-hydroxyoctanoic acid monomers, catalyzed by Pseudomonas fluorescens gk13 poly(3-hydroxyoctanoic acid) depolymerase. Appl. Environ. Microbiol. **2010**, 76, 3554–3560.
49. Al-Duri, B.; Goddard, R.; Bosley, J. Characterisation of a novel support for biocatalysis in supercritical carbon dioxide. J. Mol. Catal. B Enzym. **2001**, 11, 825–834.
50. Chang, S.F.; Chang, S.W.; Yen, Y.H.; Shieh, C.J. Optimum immobilization of Candida rugosa lipase on celite by rsm. Appl. Clay Sci. **2007**, 37, 67–73.
51. Hama, S.; Kondo, A. Enzymatic biodiesel production: An overview of potential feedstocks and process development. Bioresour. Technol. **2013**, 135, 386–395.

52. Moser, B.R. Influence of blending canola, palm, soybean, and sunflower oil methyl esters on fuel properties of biodiesel. Energy Fuels **2008**, 22, 4301–4306.
53. Haas, M.J.; McAloon, A.J.; Yee, W.C.; Foglia, T.A. A process model to estimate biodiesel production costs. Bioresour. Technol. **2006**, 97, 671–678.
54. Encinar, J.M.; González, J.F.; Rodríguez-Reinares, A. Ethanolysis of used frying oil: Biodiesel preparation and characterization. Fuel Process. Technol. **2007**, 88, 513–522.
55. Encinar, J.M.; Sanchez, N.; Martinez, G.; Garcia, L. Study of biodiesel production from animal fats with high free fatty acid content. Bioresour. Technol. **2011**, 102, 10907–10914.
56. Watanabe, Y.; Shimada, Y.; Sugihara, A.; Tominaga, Y. Conversion of degummed soybean oil to biodiesel fuel with immobilized Candida antarctica lipase. J. Mol. Catal. B Enzym. **2002**, 17, 151–155.
57. Cesarini, S.; Diaz, P.; Nielsen, P. Exploring a new, soluble lipase for fames production in water-containing systems using crude soybean oil as a feedstock. Process Biochem. **2013**, 48, 484–487.
58. Nielsen, P.M.; Brask, J.; Fjerbaek, L. Enzymatic biodiesel production: Technical and economical considerations. Eur. J. Lipid Sci. Technol. **2008**, 110, 692–700.
59. Lam, M.K.; Lee, M.T.; Mohamed, A.R. Homogeneous, heterogeneous and enzymatic catalysis for transesterification of high free fatty acid oil (waste cooking oil) to biodiesel: A review. Biotechnol. Adv. **2010**, 28, 500–518.
60. Halim, S.F.A.; Kamaruddin, A.H. Catalytic studies of lipase on fame production from waste cooking palm oil in a tert-butanol system. Process Biochem. **2008**, 43, 1436–1439.
61. Brask, J.; Damstrup, M.L.; Nielsen, P.M.; Holm, H.C.; Maes, J.; de Greyt, W. Combining enzymatic esterification with conventional alkaline transesterification in an integrated biodiesel process. Appl. Biochem. Biotechnol. **2011**, 163, 918–927.
62. Christensen, M.W.; Abo, M.; Du, W.; Liu, D. Enzymes in biodiesel production. In Proceedings of the World Congress on Industrial Biotechnology and Bioprocessing, Toronto, ON, Canada, 11–14 July 2006.

63. Cowan, D.; Nielsen, P.M. Enzymatic degumming of edible oils and fats. In Bleaching and Purifying Fats and Oils: Theory and Practice; Patterson, H.B.W., Ed.; AOCS Press: Urbana, IL, USA, 2009; pp. 216–235.
64. Watanabe, Y.; Nagao, T.; Nishida, Y.; Takagi, Y.; Shimada, Y. Enzymatic production of fatty acid methyl esters by hydrolysis of acid oil followed by esterification. J. Am. Oil Chem. Soc. **2007**, 84, 1015–1021.
65. Luo, Y.; Zheng, Y.T.; Jiang, Z.B.; Ma, Y.S.; Wei, D.X. A novel psychrophilic lipase from Pseudomonas fluorescens with unique property in chiral resolution and biodiesel production via transesterification. Appl. Microbiol. Biotechnol. **2006**, 73, 349–355.
66. Cesarini, S.; Infanzon, B.; Pastor, F.I.J.; Diaz, P. Fast and economic immobilization methods described for non-commercial pseudomonas lipases. BMC Biotechnol. **2014**.
67. Fukuda, H.; Hama, S.; Tamalampudi, S.; Noda, H. Whole-cell biocatalysts for biodiesel fuel production. Trends Biotechnol. **2008**, 26, 668–673.
68. Noureddini, H.; Gao, X.; Philkana, R.S. Immobilized Pseudomonas cepacia lipase for biodiesel fuel production from soybean oil. Bioresour. Technol. **2005**, 96, 769–777.
69. Chen, X.; Du, W.; Liu, D.; Ding, F. Lipase-mediated methanolysis of soybean oils for biodiesel production. J. Chem. Technol. Biotechnol. **2008**, 83, 71–76.
70. Lv, D.; Du, W.; Zhang, G.; Liu, D. Mechanism study on ns81006-mediated methanolysis of triglyceride in oil/water biphasic system for biodiesel production. Process Biochem. **2010**, 45, 446–450.
71. Chen, X.; Du, W.; Liu, D. Effect of several factors on soluble lipase-mediated biodiesel preparation in the biphasic aqueous-oil systems. World J. Microbiol. Biotechnol. **2008**, 24, 2097–2102.
72. Tan, T.; Xu, J.; Asano, Y. Biocatalysis in biorefinery: A green and highly efficient way to convert renewables preface. J. Mol. Catal. B **2009**, 56, 77–77.
73. Nielsen, P.M. Industrial Biofames Production with Lipase Callera Trans L, Proceedings of the 104th AOCS Annual Meeting and Expo, Montréal, QC, Canada, 28 April–1 May 2013.

74. Aalrust, E.; Beyer, W.; Ottofrickenstein, H.; Penk, G.; Plainer, H.; Reiner, R. Enzymatic Treatment of Edible Oils. U.S. Patent 5264367, 23 November 1993.
75. Dijkstra, A. Continuous Removal of Gum Phase from Tri: Glyceride Oil by Mixing Alkali with the Oil. In Proceedings of the World Conference on Oilseed Technology and Utilization, Champaign, IL, USA, 1 November 1993.
76. Nieuwenhuyzen, W.V.; Tomás, M.C. Update on vegetable lecithin and phospholipid technologies. Eur. J. Lipid Sci. Technol. **2008**, 110, 472–486.
77. Clausen, K. Enzymatic oil-degumming by a novel microbial phospholipase. Eur. J. Lipid Sci. Technol. **2001**, 103, 333–340.
78. Holm, H.C.; Nielsen, P.M.; Christensen, M.W. Production of Fatty Acid Alkyl Esters. U.S. Patent 20080199924 A1, 21 August 2008.
79. Dayton, C.; Galhardo, F. Enzymatic Degumming Utilizing a Mixture of Pla and Plc Phospholipases. U.S. Patent 20080182322 A1, 31 July 2008.
80. Cesarini, S.; Haller, R.F.; Diaz, P.; Nielsen, P.M. Combining phospholipases and a liquid lipase for one-step biodiesel production using crude oils. Biotechnol. Biofuels **2014**, 7.
81. Pinisetty, D.; Moldovan, D.; Devireddy, R. The effect of methanol on lipid bilayers: An atomistic investigation. Ann. Biomed. Eng. **2006**, 34, 1442–1451.
82. Daicheng, L.; Fucui, M. Soybean phospholipids. In Recent Trends for Enhancing the Diversity and Quality of Soybean Products; Krezhova, D., Ed.; InTech: Rijeka, Croatia, 2011; Volume 22.
83. Cowan, D. Chapter 15. Lipases for the Production of Food Components; Wiley-Blackwell: Chichester, UK, 2009; pp. 332–359.

8

CHAPTER

BIODIESEL PRODUCTION FROM NON-EDIBLE BEAUTY LEAF (CALOPHYLLUM INOPHYLLUM) OIL: PROCESS OPTIMIZATION USING RESPONSE SURFACE METHODOLOGY (RSM)

Mohammad I. Jahirul [1,*], Wenyong Koh [1], Richard J. Brown [1], Wijitha Senadeera [1], Ian O'Hara [2] and Lalehvash Moghaddam [2]

[1] *Biofuel Engine Research Facility (BERF), Science and Engineering Faculty, Queensland University of Technology (QUT), Brisbane 4000, Australia*

[2]*Centre for Tropical Crops and Biocommodities (CTCB), Queensland University of Technology (QUT), Brisbane 4000, Australia*

ABSTRACT

In recent years, the beauty leaf plant (Calophyllum Inophyllum) is being considered as a potential 2nd generation biodiesel source due to high seed oil content, high fruit production rate, simple cultivation and ability to grow in a wide range of climate conditions. However, however, due to the high free fatty acid (FFA) content in this oil, the potential of this biodiesel feedstock is still unrealized, and little research has been undertaken on it. In this study, transesterification of beauty leaf oil to produce biodiesel has been investigated. A two-step biodiesel conversion method consisting of acid catalysed pre-esterification and alkali catalysed transesterification has been utilized. The three main factors that drive the biodiesel (fatty acid methyl ester (FAME)) conversion from vegetable oil (triglycerides) were studied using response surface methodology (RSM) based on a Box-Behnken experimental design. The factors considered in this study were catalyst concentration, methanol to oil molar ratio and reaction temperature. Linear and full quadratic regression models were developed to predict FFA and FAME concentration and to optimize the reaction conditions. The significance of these factors and their interaction in both stages was determined using analysis of variance (ANOVA). The reaction conditions for the largest reduction in FFA concentration for acid catalysed pre-esterification was 30:1 methanol to oil molar ratio, 10% (w/w) sulfuric acid catalyst loading and 75 °C reaction temperature. In the alkali catalysed transesterification process 7.5:1 methanol to oil molar ratio, 1% (w/w) sodium methoxide catalyst loading and 55 °C reaction temperature were found to result in the highest FAME conversion. The good agreement between model outputs and experimental results demonstrated that this methodology may be useful for industrial process optimization for biodiesel production from beauty leaf oil and possibly other industrial processes as well.

Keywords:biodiesel; beauty leaf; transesterification; response surface methodology (RSM)

1. INTRODUCTION

The current global energy supply is heavily dependent on finite reserves of fossil fuels (oil, natural gas, coal) which represent 88% of total global energy consumption. Based on current production scenarios, it is expected that the peak of global oil production will occur between 2015 and 2030 [1]. Therefore, fossil resources have practical limitations in their capacity to supply future global energy requirements, and there are currently few large scale alternatives available. Moreover, combustion of fossil fuels results in greenhouse gas emissions and contributes to anthropogenic climate change. Despite global measures such as the Kyoto Protocol and scientific innovation, atmospheric CO_2 concentration continues to increase and is exceeding benchmark levels much earlier than had previously been predicted [2].

With a growing world population, increasing energy consumption per capita, and the impacts of global warming resulting from greenhouse gas emissions, the need for long-term alternative energy source is acute [3,4,5]. Over the past few decades, biodiesel produced from oilseed crops and animal fat is receiving much attention as a renewable and sustainable alternative for automobile engine fuels, particularly for petroleum diesel [6]. It is currently produced in commercial quantities from edible oil feedstocks such as soybean, palm, rapeseed and canola oil. Biodiesels produced from these feedstocks are generally referred to as first generation biodiesels [7]. Although biodiesels from these feedstocks offer reductions in greenhouse gas emissions (GHG) and improve domestic energy security, first generation biodiesels are unlikely to be sustainable in the longer term due to land use impacts and the price and social impacts associated with using a food-based feedstock. Second generation biodiesels produced from non-edible feedstocks have the potential to overcome the disadvantages associated with first generation feedstocks, while addressing many of the climate change and energy availability challenges [8].

Vegetable oils are extremely viscous, ranging from 10 to 17 times higher viscosity than that of petroleum diesel [9,10]. This makes the raw oils unsuitable for direct use as a fuel in a modern diesel engine. As a consequence, researchers and scientists have developed various methods to reduce the viscosity of bio-oils to make them suitable for diesel engine use. Some of these methods include dilution with other fuels, transesterification, microemulsification, pyrolysis and catalytic cracking [11]. Among these techniques, transesesterification is the most widely used solution due to its high conversion efficiency, simplicity, low conversion cost and the good fuel qualities of the product [12].

Transesterification is a chemical reaction in which oils (triglycerides) react with alcohols (e.g., methanol, ethanol) under acid or alkali catalysed conditions, producing fatty acid alkyl esters and glycerol. A catalyst is used to improve the reaction rate and ester yield. Because the transesterification reaction is reversible, excess alcohol is used to shift the equilibrium to favour production of esters. After the reaction is completed, glycerol is removed as a by-product and the esters are purified into biodiesel [12].

One limitation with the alkali catalysed transesterification process is that this process is not suitable for vegetable oils containing high levels of free fatty acids (FFA). This is because FFAs react with the catalyst to form soaps, resulting in emulsification and separation problems [13]. In addition excessive soap formation reduces biodiesel yield and obstructs subsequent purification processes including glycerol separation and water washing [14]. However, the maximum limit of FFA in vegetable oil for alkali catalysed transesterification is still uncertain with different benchmarks being reported. For example, Van Gerpan reported that vegetable oils containing up to 5% FFA can be transesterified using an alkali catalyst while Dorodo et al. [15] and Ramadhas et al. [16] reported that FFA content should not be greater than 3% and 2%, respectively. Many researchers have also reported that FFA should be kept less than 1% for alkali catalysed transesterification [17,18,19]. In order to overcome the difficulties related to transesterifying high FFA oils, a pre-esterification process can be used in which a homogeneous acid catalysed process is used prior to transesterification [14,20].

The yield and quality of biodiesel are affected by several pre-esterification and transesterification reaction parameters such as the quantity of alcohol, reaction temperature, FFA content of the oil and the type and concentration of catalyst [21,22]. For the stoichiometric transesterification reaction, three moles of methanol are required per mole of triglyceride to yield three moles of methyl esters and one mole of glycerol. The theoretical molar ratio of methanol to triglyceride should, therefore, be 3:1 [18]. However, the ratio of alcohol to oil used in the reaction is much higher than this to promote complete conversion of oils to FAME and varies with oil quality and the type of catalyst used. For example, the molar ratio of alcohol to oil for alkali catalysed reactions is typically 6:1, and for acid catalysed reactions it may be 15:1 or higher. An increase in the concentration of catalyst generally increases the conversion of triglycerides into fatty acid esters [18]. Insufficient catalyst leads to an incomplete conversion reaction and lower levels of fatty acid esters, whereas excess catalyst has a negative impact on end product yield, because of the formation of soaps.

On the other hand, a higher reaction temperature increases the reaction rate and decreases the reaction time due to the reduction in viscosity of the oils. High reaction temperatures above optimal levels, however, leads to a decrease in biodiesel yield, as higher reaction temperatures accelerate the saponification of triglycerides [23]. Therefore, researches seek to optimise the important reaction parameters for different biodiesel feedstock in order to achieve an efficient and economical biodiesel production process.

The availability and price of feedstock are significant factors as feedstock cost represents approximately 75%–88% of the total biodiesel production cost [24,25]. However, there are vast areas of grazing (e.g., cleared) and degraded (e.g., mined) land on which biodiesel crops can be successfully established for complementing fuel supplies. In a recent study, a number of species have been found suitable for growth on degraded land which has the capacity for producing a considerable amount of non-edible oil for biodiesel production [26]. Among these species, beauty leaf (Calophyllum Inophyllum) has been identified as the most suitable feedstock for future generation biodiesel [26,27]. It is a moderately sized tree that grows between 8–20 m tall and is most notable for its decorative leaves and fragrant flowers. The tree grows in tropical and sub-tropical climates close to sea level. It is a moderately quick growing tree reaching up to 1 m tall within a year. It has also been seen to flourish even with the presence of weeds and other species, so the plant can be grown in mixed cultures. The Beauty leaf tree has the ability to produce about 4800 kg of non-edible oil per year per hectare [27]. However, the potential of beauty leaf as a source of future generation biodiesel is yet to be established in part due to a lack of knowledge of its optimum production process.

Response surface methodology (RSM) is a collection of mathematical and statistical techniques that are useful for modelling, analysis and optimisation problems in which the response of interest is influenced by several factors [28,29]. In this technique, a well-designed experiment can substantially reduce the number of tests, and yet provide the essential information required for process optimisation. RSM uses statistical methods for experimental design to identify important factors by characterizing the response surface using a polynomial model [30]. In the practical application of RMS it is necessary to develop a statistically valid approximating model for the true response surface. The relationship of the response variable Y and the dependent variables X_1, X_2, ... X_k, in the RSM application is generally expressed as in the following equation [31]:

$$Y = f(X_1, X_2, \dots\dots\dots X_k) + \varepsilon \tag{1}$$

where ε is the noise or error observed in the response Y; and f is the response surface.

This study aims to investigate the effect of several reaction parameters on the production of biodiesel from high FFA vegetable oil obtained from the beauty leaf seed. This study also implemented RSM in developing linear and full quadratic polynomial equations for predicting FFA and FAME content and predicting the optimum reaction condition for pre-esterification and transesterification processes.

2. MATERIALS AND METHOD

2.1. Beauty Leaf Oil Extraction Method

The beauty leaf oil used in this study was obtained through a chemical oil extraction methods using n-hexane as a solvent. In this process, dried seed kernels were ground using a blender and coffee grinder to obtain a fine consistency to maximise particle surface area. The ground kernels were then put into conical flasks into which n-hexane was added at a ratio of 1.6:1 by weight (n-hexane:seed kernels). The mixture was given an initial stir to ensure that all kernels were wetted with hexane. The conical flask openings were covered with aluminium foil and placed on laboratory scale orbital mixer in a fume hood, and the samples were extracted for at least 8 h with 150 rpm shaking speed. Following extraction, the hexane/oil mixtures were collected, filtered and decanted into aluminium foil containers for solvent evaporation, and placed into the fume hood for 8 to 10 h. Hexane was again added to the conical flask of kernels, but at a ratio of 8:1 (weight) for the second extraction, and a similar procedure was followed for recovery of the oil. When it was determined that the hexane had been fully evaporated, the raw beauty leaf oil was collected.

2.2. Analysis Methods

FFA content of the beauty leaf oil was analysed using D5555-95 (2011) standard test method. Ester content of the FAME was analysed by gas chromatography and flame ionization detection (GC-FID) in accordance with EN 14103 standards. The gas chromatograph (GC) was a Hewlett-Packard 6890 System fitted with Varian Select™ 30 m × 0.32 mm × 0.25 μm column. Oil density and surface tension were analysed following ASTM D1298 and ASTM D971-12 standard test methods using a KSV Sigma 702 Tensiometer Viscosity was measured using Brookfield DV-III Rheometer and following the ASTM D445 standard test method. The fatty acid compositions of the oils were analysed using a Hewlett Packard Plus 6890 series GC-FID and a capillary column of acidified polyethylene glycol (HP-INNOWax 19091N-133, 30 m × 250 μm × 0.25 μm).

2.3. Pre-Esterification and Transesterification Methods

Both acid-catalysed pre-esterification and base-catalysed transesterification were conducted in a 500 mL triple neck bottom flask reactor (Figure 1a). An oil quantity of 40 g was used for the acid-catalysed pre-esterification experiments, and 30 g was used for each base-catalysed transesterification trial. For each experiment, oil was carefully transferred into the reaction flask and preheated in an oil bath to the reaction temperature. For acid-catalysed esterification trials, sulfuric acid (H_2SO_4) was used as catalyst. The sulphuric acid and methanol solution were freshly prepared and added to the preheated oil, and the mixture was agitated for two hours. At the completion of the two hours, the mixture was centrifuge in a self-sanding tube for 7 min to separate the methanol-water and esterified oil phases as shown in Figure 1b. The majority of the excess methanol, sulfuric acid and impurities were separated into the top phase. The bottom phase containing the oil was collected for base-catalysed transesterification.

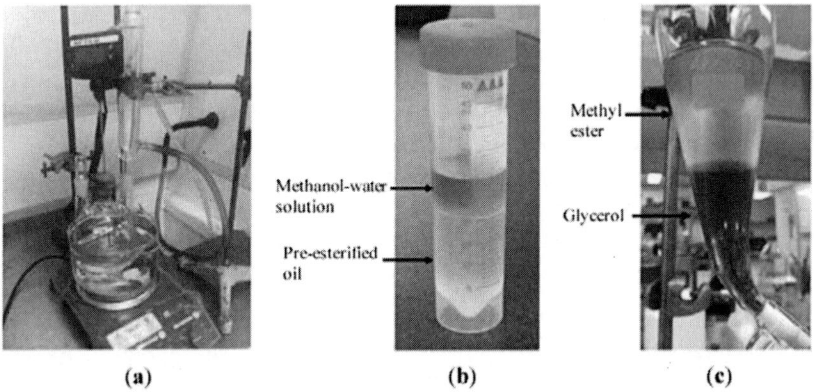

Figure 1. (**a**) Esterification and transesterification reactor; (**b**) Layer of Methanol-Water (top) and oil (bottom) after acid-catalysed pre-esterification; (**c**) Layer of beauty leaf oil methyl ester (top) and glycerol (bottom) after base-catalysed Transesterification.

In the based-catalysed transesterification trials, sodium methoxide ($NaOCH_3$) was used as a catalyst with a reaction time of 1.5 h. Similarly to the acid-catalysed pre-esterification trials, the phases of the transesterification product were separated using a centrifuge and the bottom layer drained using a separation funnel as shown in the Figure 1c. The top layer containing beauty leaf methyl ester was collected for analysis.

Experiments were carried out according to a Box-Behnken response surface design which involves 3 factors and requires 3 levels and a total of 15 runs. The factors and the ranges and levels used in this study are shown in Table 1. The Minitab 16 statistical software package was used to randomly generate runs orders of the experiments which are shown in Table 2.

Table 1. Experimental range and levels of independent variables.

Variables	Unit	Symbol coded	Range & levels		
			−1	0	1
Acid-catalyzed pre-esterification					
MeOH: Oil	mole	M	10:1	20:1	30:1
H_2SO_4	wt% oil	C	5	10	15
Temperature	°C	T	45	60	75
Base-catalyzed transesterification					
MeOH: Oil	mole	M'	4:1	6:1	8:1
CH_3ONa	wt% oil	C'	0.6	0.8	1
Temperature	°C	T'	45	60	75

Table 2. Coded experimental design.

Run	Acid-catalysed pre-esterification			Base-catalysed transesterification		
	M	C	T	M'	C'	T'
1	0	−1	1	0	0	0
2	−1	−1	0	1	0	−1
3	−1	0	−1	−1	−1	0
4	1	−1	0	−1	1	0
5	0	0	0	−1	0	−1
6	1	1	0	−1	0	1
7	0	1	1	1	1	0
8	0	0	0	0	1	1
9	0	0	0	0	−1	−1
10	0	−1	−1	0	−1	1
11	0	1	−1	1	0	1
12	−1	0	1	1	−1	0
13	1	0	−1	0	0	0
14	−1	1	0	0	0	0
15	1	0	1	0	1	-1

Since the key focus of acid-catalysed esterification reaction is on reducing the free fatty acid content to be <3–5 wt%, a full quadratic model was used for statistical analysis in order to correlate the %FFA with the operating variables. The form of the full quadratic model for the first step is as shown in Equation (2):

$$Y = \beta_0 + \beta_1 X_1 + \beta_2 X_2 + \beta_3 X_3 + \beta_{1,2} X_1 X_2 + \beta_{1,3} X_1 X_3 + \beta_{2,3} X_2 X_3 + \beta_{1,1} X_1^2 + \beta_{2,2} X_2^2 + \beta_{3,3} X_3^2 \quad (2)$$

where Y is the %FFA; β_0 is a constant; β_1, β_2, β_3 are regression coefficients and X_1, X_2, X_3 are independent variables. For the base-catalysed transesterification reaction, the same form of the full quadratic model is used; however, in this case Y is the ester content of the biodiesel (%).

3. RESULTS, ANALYSIS AND DISCUSSION

3.1. Beauty Leaf Oil Characterizations

The quality of beauty leaf oil has been characterised in terms of chemical composition and physical properties in order to identify its suitability as a feedstock for biodiesel for diesel engine combustion. The results of analyses of the chemical composition and properties of the crude beauty leaf oil used in this study are shown in Table 3 and Table 4. The compositional analysis shows that beauty leaf oil contains high levels of stearic (C18:0), oleic (C18:1) and linoleic (C18:2) acids (Table 3). This indicates the potential for high combustion quality and hence suitability of beauty leaf oil as a fuel [32]. However, kinematic viscosity and surface tension are significantly higher than other oils which may lead to poor atomisation and volatility characteristics. Therefore, beauty leaf oil may not be suitable as fuel for direct use in conventional diesel engines [21].

Table 3. Fatty acid composition of beauty leaf oil.

Fatty Acid	Weight percentage
Palmitic, C16:0	13.66
Palmitoleic, C16:1	0.24
Heptadecanoic, C17:0	0.15
Heptadecanoic, C17:1	0.06
Stearic, C18:0	16.55
Oleic, C18:1	42.48
Linoleic, C18:2	25.56
Linolenic, C18:3	0.20
Arachidic, C20:0	0.87
Arachidonic, C20:1	0.23

Table 4. Properties of beauty leaf oil.

Properties	Values
Density, kg/m^3	936
Surface tension (mN/s)	35.6
Kinematic Viscosity @40 °C, cSt	40.05
Free fatty acid (wt%)	12

On the other hand, the FFA content of beauty leaf oil was 12% (w/w) of oil (Table 4) which is much higher than the recommended FFA content of vegetable oil for the base catalyzed transesterification [15,16]. To overcome this high FFA level, biodiesel production was conducted in two processing steps as described. For both pre-esterification and transesterification, final acid value and production yield has been optimized, and a statistical production model has been developed.

3.2. Acid-Catalysed Pre-Esterification

Table 5 summarizes the experimental conditions and results from each pre-esterification experimental run. The results indicate a significant reduction in FFA content of the beauty leaf oil following acid-catalysed pre-esterification. FFA content of the pre-esterified samples ranged from 3.25 to 1.83. The minimum FFA content resulted from the esterification condition with 30:1 MeOH to oil molar ratio, 10% weight concentration of catalyst and 75 °C reaction temperature (Test 15). An FFA content of less than 2% was also achieved for Test 6 (30:1 MeOH to oil molar ratio, 15% weight concentration of catalyst and 60 °C reaction temperature). Based on the experimental results, a linear and a quadratic equations have been developed using Minitab 16 software in order to predict the FFA percentages as a function of methanol to oil molar ratio, catalyst concentration and reaction temperature in acid catalyst esterification. The developed quadratic models equations are shown in Equations (3) and (4):

$$FFA(\%) = 3.97617 - 0.0528M - 0.0055C - 0.00542T \quad (3)$$

$$FA(\%) = 1.493 - 0.00498M - 0.02487C + 0.06596T + 0.000085MC - 0.0007083MT$$
$$-0.0002233CT - 0.0001554M^2 + 0.001558C^2 - 0.000458T^2 \quad (4)$$

Table 5. Experimental conditions and results for acid-catalysed pre-esterification.

Test	MeOH: Oil Molar Ratio	H_2SO_4 Conc. (wt%)	Temp (°C)	FFA (wt%)
1	20	5	75	2.39
2	10	5	60	3.25
3	10	10	45	2.88
4	30	5	60	2.10
5	20	10	60	2.61
6	30	15	60	1.97
7	20	15	75	2.39
8	20	10	60	2.61
9	20	10	60	2.53
10	20	5	45	2.61
11	20	15	45	2.67
12	10	10	75	3.02
13	30	10	45	2.12
14	10	15	60	3.10
15	30	10	75	1.83

Table 6 summarizes the regression coefficient generated using Minitab 16 software. The significance of each coefficient in this equation was evaluated by ANOVA using Minitab software in terms of the p-value. Low p-value indicates that the corresponding coefficient is significant. In the linear model shows that the methanol to oil ratio (MeOH:Oil) was the most significant with p-value of 0 followed by temperature and catalyst concentration (H_2SO_4) with the p-value of 0.07 and 0.511 respectively. For the full quadratic model, temperature was the most significant with p-value of 0.069. Following this, the interaction effect between methanol to oil molar ratio and temperature was most significant with p-value of 0.08. Finally, the quadratic effect of temperature was most significant with p-value of 0.097. Table 6 also shows that all p-values were fairly high, considerably above 0.05 which infer that the coefficients in full quadratic model are not statistically insignificant in 95% confidence interval.

The accuracy of the prediction model obtained by the regression analysis was verified by a scattered diagram (Figure 2), where experimental results for FFA were compared to predicted values from the model. In linear model, the regression coefficient (R^2) and the adjusted regression coefficient (R^2 (adj)) of 0.983 and 0.949 shows a good fit between actual and predicted results whereas in full quadratic model, those parameters were 0.981 and

0.949 respectively. Therefore, it is apparent that the linear model is statistically more appropriate than that of full quadratic model. Moreover, the full quadratic model is over-specified as none of the coefficients are significant in 95% confident interval.

Table 6. Regression coefficients for %FFA prediction.

Predictor	Linear		Full quadratic	
	Coefficient	*p*-value	Coefficient	*p*-value
Constant	3.97617	0	1.493	0.212
MeOH:Oil (M)	−0.0528	0	−0.00498	0.874
H_2SO_4 (C)	−0.0055	0.511	−0.02487	0.695
Temp (T)	−0.00542	0.07	0.06596	0.069
MeOH:Oil × H_2SO_4 (MC)			0.000085	0.934
MeOH:Oil × Temp (MT)			−0.0007083	0.080
H_2SO_4 × Temp (CT)			−0.0002233	0.744
MeOH:Oil × MeOH:Oil (M^2)			−0.0001554	0.771
H_2SO_4 × H_2SO_4 (C^2)			0.001558	0.476
Temp × Temp (T^2)			−0.000458	0.097

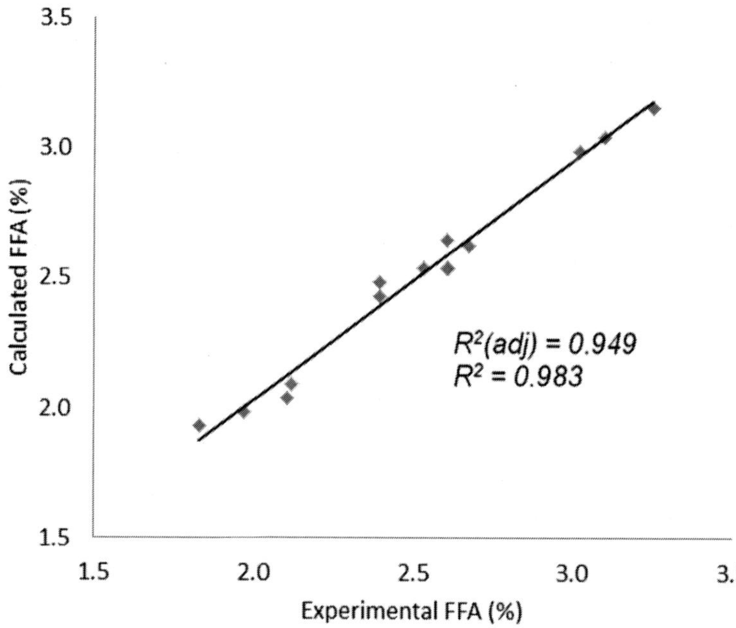

Figure 2. Scatter diagram of experimental FFA (%) and predicted FFA (%) of a linear model.

Figure 3a,b show the surface plots generated using the model equation on the effect of each variable on FFA content. Methanol to oil molar ratio has a strong effect on FFA reduction which is evident in Figure 3a. One the other hand, H_2SO_4 concentration has only a minimal impact on FFA reduction within the range of catalyst concentrations used, suggesting low linear effect on FFA. In addition, FFA content increases correspondingly after 10 wt% H_2SO_4, implying that further increase in H_2SO_4 concentration will have adverse effects. This might be the result of oil decomposition at high acid concentrations. Reaction temperature has a small effect on FFA content at lower values and effect on FFA reduction is significant after 65 °C, which was more evident in Figure 3b.

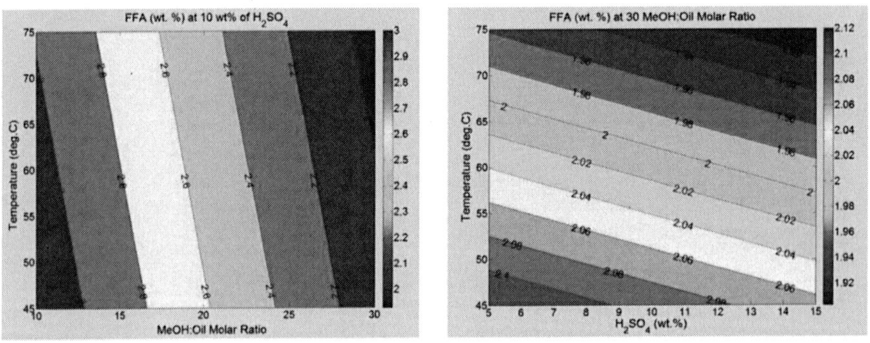

(a) (b)

Figure 3. Response surface of FFA content against (**a**) methanol to oil molar ratio and reaction temperature at 10% acid catalyst (H_2SO_4); (**b**) against catalyst concentration and reaction temperature at 30:1 methanol to oil molar ratio.

3.3. Base-Catalysed Transesterification of Pre-Esterified Beauty Leaf Oil

All samples produced from the acid-catalyzed esterification process were thoroughly mixed to produce a homogenous feedstock for transesterification. The FFA content of the mixture was found to be 2.46% (w/w). Similarly to the acid-catalysed pre-esterification trials, 15 experimental runs were undertaken based on Box-Behnken design as shown in Table 2. The results obtained from those experiments are shown in Table 7.

Table 7. Experimental data for base-catalysed transesterification.

Run	MeOH: oil molar ratio	$NaOCH_3$ (wt%)	Temp (°C)	FAME (%)
1	6	0.8	60	89.21
2	8	0.8	45	88.51
3	4	0.6	60	74.15
4	4	1.0	60	85.51
5	4	0.8	45	81.92
6	4	0.8	75	78.02
7	8	1.0	60	89.41
8	6	1.0	75	87.70
9	6	0.6	45	64.18
10	6	0.6	75	75.76
11	8	0.8	75	84.78
12	8	0.6	60	63.02
13	6	0.8	60	87.72
14	6	0.8	60	87.20
15	6	1.0	45	90.76

In the transesterification experiments, the ester content of the FAME ranged from 63.02% to 90.76% with the highest content resulting from reaction conditions with 6:1 methanol to oil molar ratio, 1 wt% $NaOCH_3$ and 45 °C temperature.

As from the experimental data processed using Minitab 16 software to generate the linear and quadratic model for statistical prediction of ester content as a function of methanol to oil molar ratio, catalyst concentration and reaction temperature. The linear and quadratic model equation resulting from this is shown in Equations (5) and (6) respectively and Table 8 summarises the resulting regression coefficients and corresponding p-value:

$$\%\text{FAME} = 40.9817 + 0.3825M' + 47.6688C' + 0.00742T \tag{5}$$

$$\begin{aligned}\%\text{FAME} = &-129.28 + 2.256M' + 339C' + 1.817T + 9.392M'C' + 0.0014M'T' \\ &-1.2203C'T' - 0.7892M'^2 - 171.57C'^2 - 0.007011T'^2\end{aligned} \tag{6}$$

Table 8. Regression coefficients for FAME (%) prediction.

Predictor	Linear		Full quadratic	
	Coefficient	*p*-value	Coefficient	*p*-value
Constant	40.9817	0.008	−129.28	0.141
MeOH:Oil (M')	0.3825	0.727	2.256	0.809
$NaOCH_3$ (C')	47.6688	0.001	339	0.021
Temp (T')	0.00742	0.959	1.817	0.239
MeOH:Oil × $NaOCH_3$ (M'C')			9.392	0.133
MeOH:Oil × Temp (M'T')			0.0014	0.985
$NaOCH_3$ × Temp (C'T')			−1.2203	0.141
MeOH:Oil × MeOH:Oil (M'^2)			−0.7892	0.0207
$NaOCH_3$ × $NaOCH_3$ (C'^2)			−171.57	0.025
Temp × Temp (T'^2)			−0.007011	0.502

The regression coefficient (R^2) and the adjusted regression coefficient (R^2 (adj)) of linear model were 0.6465 and 0.55 demonstrated that the linear model may not be suitable for estimate FAME in given reaction condition. Whereas full quadratic model with 0.9224 of regression coefficient (R^2) and 0.8413 of adjusted regression coefficient (R^2 (adj)) shows better model for FAME estimation. Figure 4 shows the accuracy of the prediction model in a scattered plot between experimental and predicted ester contents. All points are close to straight line demonstrate a good agreement between experimental results and those ones calculated by the model.

More detailed analysis of the effect of base-catalysed transesterification reaction parameters on beauty leaf FAME ester content are shown in Figure 5a,b. These figures predict that an optimal methanol to oil molar ratio would be 7.5:1, however, further increase in methanol would not have a positive effect on ester content. On the other hand, $NaOCH_3$ concentration has a strong effect on ester content of the FAME with corresponding increment with agreement in terms of linear and quadratic effects. Temperature had a less significant effect than methanol to oil molar ratio and catalyst concentration. Figure 5 show that the optimum temperature of transesterification was 65 °C. These figures illustrated that although all parameters are not statistically significant at 95% confident level but the relationship still contains useful information for some biodiesel production purposes.

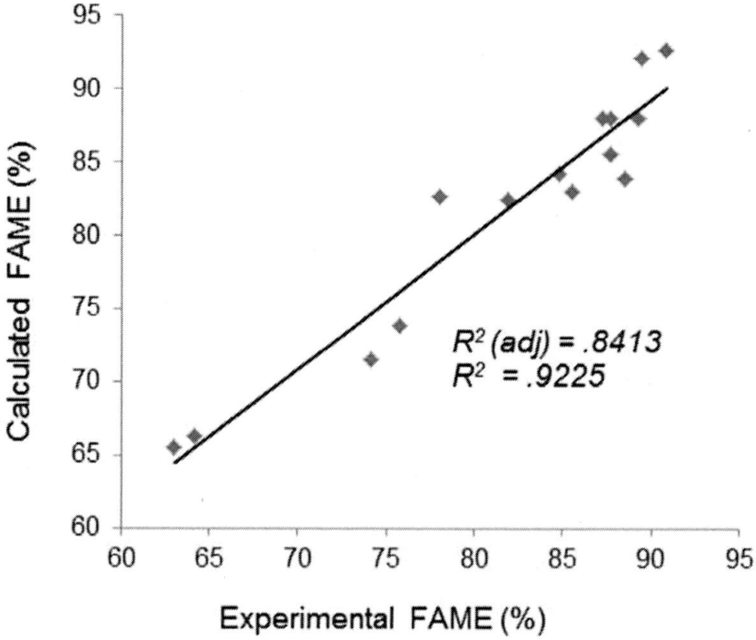

Figure 4. Scatter diagram of experimental and calculated FAME (%) of full quadratic model.

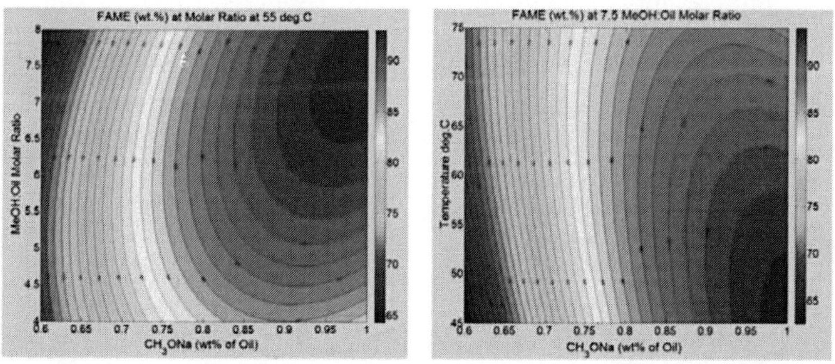

(a) (b)

Figure 5. Response surface ester content against catalyst concentration vs. (**a**) methanol to oil molar ratio at 55 °C; (**b**) reaction temperature at 7.5:1 methanol to oil molar ratio.

4. CONCLUSIONS

A response surface method based a Box-Behnken design was employed to determine a feasible experimental plan to optimize the beauty leaf oil to biodiesel conversion procedure. Due to the high FFA content of beauty leaf oil (12 wt%), a two-step process was employed utilizing sulphuric acid catalyzed pre-esterification followed by sodium methoxide catalyzed transesterification. Effects of reaction parameters such as methanol to oil molar ratio, catalyst loading and reaction temperature were statistically investigated on the reduction of FFA content in pre-esterification and ester content in transesterification. The optimal conditions for pre-esterification were 30:1 methanol to oil molar ratio, 10 wt% sulphuric acid catalyst and 75 °C reaction temperature which reduced the FFA content to 1.8 wt%. With the aid of statistical modelling, the predicted optimal conditions for transesterification methanol to oil molar ratio, catalyst concentration and reaction temperature were 7.5:1, 1% and 55 °C respectively. Based on these conditions, the highest achievable ester content of FAME predicted by the model was found to be approximately 93%. However a higher result may be achievable by future lowering FFA content of beauty leaf oil. In terms of a linear effect on FFA reduction for the first step, methanol to oil molar ratio was found to be highly significant and reaction temperature moderately significant. For transesterification, catalyst concentration was found be the most dominant variable in achieving high ester contents. The limitation of the developed response surface model is that all the p-values are greater than 0.05. Therefore, the developed models might be over-specified and some that terms can be omitted. However, the information contained in the model and experiment in this study is very significant in industrial biodiesel production.

ACKNOWLEDGMENTS

The authors wish to express their thanks to QUTPRA scholarship for providing fund for conducting this research. The authors would also like to thanks the Biofuel Engine Research Facility (BERF), QUT and Centre for Tropical Crops and Biocommodities (CTCB), QUT for providing experimental facilities.

AUTHOR CONTRIBUTIONS

All of the authors have contributed toward developing the ideas, establishing research design, method and analysis. Mohammad I. Jahirul and Wenyong Koh have been involved in sample preparation, experimental design and develop the model. Lalehvash Moghaddam has been involved in conducting experiment. Richard J. Brown, Wijitha Senadeera and Ian O'Hara have been involved in result analysis and editing. All of the authors have been involved in preparing the manuscript. All of authors are responsible for the integrity of the work as a whole.

REFERENCES

1. Jahirul, M.I.; Brown, R.J.; Senadeera, W.; O'Hara, I.M.; Ristovski, Z.D. The use of artificial neural networks for identifying sustainable biodiesel feedstocks. Energies **2013**, 6, 3764–3806.
2. Weitzman, M.L. A review of the stern review on the economics of climate change. J. Econ. Lit. **2007**, 45, 703–724.
3. Jahirul, M.I.R.; Rasul, M.G.; Chowdhury, A.A.; Ashwath, N. Biofuels production through biomass pyrolysis—A technological review. Energies **2012**, 5, 4952–5001.
4. Jahirul, M.I.; Hasanuzzaman, R.S.M.; Masjuki, H.H.; Kalam, M.A. A comparison of the air pollution of gasoline and CNG driven car for Malaysia. Int. J. Mech. Mater. Eng. **2009**, 2, 130–138.
5. Jahirul, M.I.; Masjuki, H.H.; Saidur, R.; Kalam, M.A.; Jayed, M.H.; Wazed, M.A. Comparative engine performance and emission analysis of CNG and gasoline in a retrofitted car engine. Appl. Therm. Eng. **2010**, 30, 2219–2226.
6. Reijnders, L. Conditions for the sustainability of biomass based fuel use. Energy Policy **2006**, 34, 863–876.
7. Rashid, U.; Anwar, F. Production of biodiesel through optimized alkaline-catalyzed transesterification of rapeseed oil. Fuel **2008**, 87, 265–273.
8. Posten, C.; Schaub, G. Microalgae and industrial biomass as a source for fuel—A process view. J. Biotechnol. **2009**, 142, 64–69.
9. Sinha, S.; Agarwal, A.K.; Garg, S. Biodiesel development from rice bran oil: Transesterification process optimization and fuel characterization. Energy Convers. Manag. **2008**, 49, 1248–1257.
10. Haseeb, A.S.M.A.; Fazal, M.A.; Jahirul, M.I.; Masjuki, H.H. Compatibility of automotive materials in biodiesel: A review. Fuel **2011**, 90, 922–931.

11. Lin, L.; Zhou, C.; Vittayapadung, S.; Shen, X.; Dong, M. Opportunities and challenges for biodiesel fuel. Appl. Energy **2011**, 88, 1020–1031.
12. Fernandoa, S.; Karraa, P.; Hernandezb, R.; Jhaa, S.K. Effect of incompletely converted soybean oil on biodiesel quality. Energy **2007**, 32, 844–851.
13. Rajendra, M.; Jena, P.C.; Raheman, H. Prediction of optimized pretreatment process parameters for biodiesel production using ANN and GA. Fuel **2009**, 88, 868–875.
14. Lam, M.K.; Lee, K.T.; Mohamed, A.R. Homogeneous, heterogeneous and enzymatic catalysis for transesterification of high free fatty acid oil (waste cooking oil) to biodiesel: A review. Biotechnol. Adv. **2010**, 28, 500–518.
15. Dorado, M.P.; Ballesteros, E.; Almeida, J.A.; Schellert, C.; Löhrlein, H.P.; Krause, R. An alkali-catalyzed transesterification process for high free fatty acid waste oils. Trans. ASABE **2002**, 45, 525–529.
16. Ramadhas, A.S.; Jayaraj, S.; Muraleedharan, C. Biodiesel production from high FFA rubber seed oil. Fuel **2005**, 84, 335–340.
17. Kumar, T.A.; Kumar, A.; Raheman, H. Biodiesel production from jatropha oil (Jatropha curcas) with high free fatty acids: An optimized process. Biomass Bioenergy **2007**, 31, 569–575.
18. Ma, F.; Hanna, M.A. Biodiesel production: A review. Bioresour. Technol. **1999**, 70, 1–15.
19. Zhang, Y.; Dubé, M.A.; McLean, D.D.; Kates, M. Biodiesel production from waste cooking oil: 1. Process design and technological assessment. Bioresour. Technol. **2003**, 89, 1–16.
20. Zhang, J.; Jiang, L. Acid-catalyzed esterification of Zanthoxylum bungeanum seed oil with high free fatty acids for biodiesel production. Bioresour. Technol. **2008**, 99, 8995–8998.
21. Balat, M.; Balat, H. A critical review of bio-diesel as a vehicular fuel. Energy Convers. Manag. **2008**, 49, 2727–2741.
22. Demirbas, A. Comparison of transesterification methods for production of biodiesel from vegetable oils and fats. Energy Convers. Manag. **2008**, 49, 125–130.
23. Leung, D.Y.C.; Guo, Y. Transesterification of neat and used frying oil: Optimization for biodiesel production. Fuel Proc. Technol. **2006**, 87, 883–890.

24. Bozbas, K. Biodiesel as an alternative motor fuel: Production and policies in the European Union. Renew. Sustain. Energy Rev. **2008**, 12, 542–552.
25. Haas, M.J.; McAloon, A.J.; Yee, W.C.; Foglia, T.A. A process model to estimate biodiesel production costs. Bioresour. Technol. **2006**, 97, 671–678.
26. Ashwath, N. Evaluating Biodiesel Potential of Australian Native and Naturalised Plant Species. Available online: https://rirdcinfoservicescomau/downloads/10-216 (accessed on 24 August 2012).
27. Jahirul, M.I.; Brown, J.R.; Senadeera, W.; Ashwath, N.; Laing, C.; Leski-Taylor, J.; Rasul, M.G. optimisation of bio-oil extraction process from beauty leaf (Calophyllum inophyllum) oil seed as a second generation biodiesel source. Proced. Eng. **2013**, 56, 619–624.
28. Jeong, G.; Don-Hee Park, D.H. Optimization of biodiesel production from castor oil using response surface methodology. Appl. Biochem. Biotechnol. **2009**, 156, 1–11.
29. Vicente, G.; Coteron, A.; Martinez, M.; Aracil, J. Application of the factorial design of experiments and response surface methodology to optimize biodiesel production. Ind. Crop. Prod. **1998**, 8, 29–35.
30. Ferella, F.; Mazziotti di Celso, G.; de Michelis, I.; Stanisci, V.; Vegliò, F. Optimization of the transesterification reaction in biodiesel production. Fuel **2010**, 89, 36–42.
31. Saidur, R.; Jahirul, M.I.; Hasanuzzaman, M.; Masjuki, H.H. Analysis of Exhaust Emissions of Natural Gas Engine by Using Response Surface methodology. J. Appl. Sci. **2008**, 8, 3328–3339.
32. Refaat, A.A. Correlation between the chemical structure of biodiesel and its physical properties. Int. J. Environ. Sci. Technol. **2009**, 6, 677–694.

9

CHAPTER

OPTIMIZED PRODUCTION OF BIODIESEL FROM WASTE COOKING OIL BY LIPASE IMMOBILIZED ON MAGNETIC NANOPARTICLES

Chi-Yang Yu, Liang-Yu Huang, I-Ching Kuan and Shiow-Ling Lee *

Department of Bioengineering, Tatung University, 40 Zhongshan N. Rd. Sec. 3, Taipei 10452, Taiwan

ABSTRACT

Biodiesel, a non-toxic and biodegradable fuel, has recently become a major source of renewable alternative fuels. Utilization of lipase as a biocatalyst to produce biodiesel has advantages over common alkaline catalysts such as mild reaction conditions, easy product separation, and use of waste cooking oil as raw material. In this study, Pseudomonas cepacia lipase immobilized onto magnetic nanoparticles (MNP) was used for biodiesel production from waste cooking oil. The optimal dosage of lipase-bound MNP was 40% (w/w of oil) and there was little difference between stepwise addition of methanol at 12 h- and 24 h-intervals. Reaction temperature, substrate molar ratio (methanol/oil), and water content (w/w of oil) were optimized using response surface methodology (RSM). The optimal reaction conditions were 44.2 °C, substrate molar ratio of 5.2, and water content of 12.5%. The predicted and experimental molar conversions of fatty acid methyl esters (FAME) were 80% and 79%, respectively.

Keywords: biodiesel; lipase; magnetic nanoparticles; response surface methodology; waste cooking oil

1. INTRODUCTION

Biodiesel is defined as the fatty acid alkyl monoesters derived from renewable feedstocks such as vegetable oils and animal fats [1]. It draws much attention as an alternative fuel because it is biodegradable, non-toxic, and can be used directly or blended with conventional petrodiesel in unmodified diesel engines. As compared to petrodiesel, biodiesel has a higher cetane number, no aromatics, almost no sulfur, and contains 10%–11% oxygen by weight [2], thus reducing the emission of carbon monoxide, hydrocarbon, and particulate matter in the exhaust gas. Biodiesel is commonly produced by transesterification of virgin vegetable oils with short chain alcohols using alkaline catalysts. However, the process requires high quality food-grade vegetable oils with low level of free fatty acids (FFA) to avoid saponification, which leads to low biodiesel conversion and causes difficulties in the separation of glycerol. A major hurdle in the commercialization of biodiesel is its high manufacturing cost, primarily from virgin vegetable oils. Therefore, waste cooking oil (WCO) has become a promising feedstock for biodiesel production. WCO is much less expensive than pure vegetable oils from soybean, sunflower or canola, and it is currently used as animal feed or is simply discarded. However, the harmful compounds of WCO may return into the food chain when used as animal feed and the disposal of WCO often leads to contamination of recipient waters [3]. In addition to the advantage of lower cost, utilization of WCO as a feedstock for biodiesel production partly solves the problem of disposing WCO. Biodiesel attained out of animal fat and WCO has a lower price than those derived from refined vegetable oils and fossil diesel [4].

Lipases (E.C.3.1.1.3) are capable of catalyzing a variety of reactions such as hydrolysis, alcoholysis, esterification, transesterification, and hence are widely used in industry [5]. Biodiesel can also be synthesized via lipase-catalyzed transesterification; the process produces high purity products and enables easy separation of the glycerol byproduct [6]. The enzymatic process is compatible with low quality feedstocks with high levels of FFA. It also requires less energy input due to lower reaction temperature than the akali-catalyzed process. However, there has been very limited commercial success due to the high cost of lipases. One common strategy for reducing the cost of lipases is to recycle the biocatalyst through immobilization [7].

Various immobilization techniques have been applied to lipases for the production of biodiesel. Support materials such as Celite [8], acrylic resin

[9], and ion exchange resins [10,11] were used for attaching lipases through adsorption. The adsorption technique is easy to perform, but it suffers from desorption of enzyme molecules. Immobilization through covalent bonds has the advantage of minimal enzyme leakage, and support like electrospun polyacrylonitrile was reported [12]. Cross-linked enzyme aggregates can be formed by means of bifunctional or multifunctional reagents such as glutaraldehyde [13]. Lipases can be also entrapped in polymeric matrix like phyllosilicate sol-gel [14] and silica gels [15] or encapsulated in silica aerogel [16].

Magnetic nanoparticles (MNP) has been widely applied in many fields of life science, for instance, magnetic resonance imaging contrast enhancement, tissue repair, immunoassay, hyperthermia, and drug delivery [17], owing to possessing unique properties of nontoxicity, biocompatibility, injectability, and high level of accumulation in the target tissue and organ [18]. It has been used as carriers for immobilizing drugs, proteins, enzymes, antibodies, and nucleotides [17]. Use of MNP as support for enzyme immobilization allows for large surface area for potential high enzyme loading, selective separation from the reaction mixture under magnetic field, and low mass transfer resistance due to small particle size [19]. We have previously immobilized lipase from Pseudomonas cepacia onto MNP for the synthesis of fatty acid methyl esters (FAME) with soybean oil as a feedstock [20,21].

Response surface methodology (RSM), a collection of mathematical and statistical techniques useful for modeling and analysis of problems in which a response of interest is influenced by several variables [22], has been widely applied to the synthesis of biodiesel. For instance, temperature, substrate molar ratio, and n-hexane content were optimized for the production of biodiesel using lipase entrapped in biomimetic silica [23], or temperature, flow rate, and substrate molar ratio were optimized for continuous production of biodiesel in a lipase-catalyzed packed-bed reactor [9,24].

The production of biodiesel using immobilized lipase has been studied extensively; however, most of related studies utilized pure vegetable oils as feedstocks [7]. In this work, we evaluated the potential of WCO as a feedstock for the synthesis of FAME using lipase immobilized on MNP as a catalyst. The crucial reaction variables (temperature, substrate molar ratio, and water content) were optimized with RSM. In addition, reusability and storage stability of immobilized enzyme were also evaluated.

2. RESULTS AND DISCUSSION

2.1. Effects of Amount of Added Lipase on Immobilization Efficiency and Activity Recovery

Effects of amount of added lipase on immobilization efficiency (ratio of amount of immobilized to added lipase) and activity recovery (ratio of specific activities for immobilized to free lipase) are shown in Figure 1. Maximal activity recovery of 60% with the corresponding immobilization efficiency of 92.3% was observed when 5 mg of lipase was added. The activity recovery decreased as the amount of lipase increased; however, the immobilization efficiency showed little change. The decrease in activity recovery could be attributed to greater steric hindrance generated and less accessible active sites as more enzyme molecules were attached to the support.

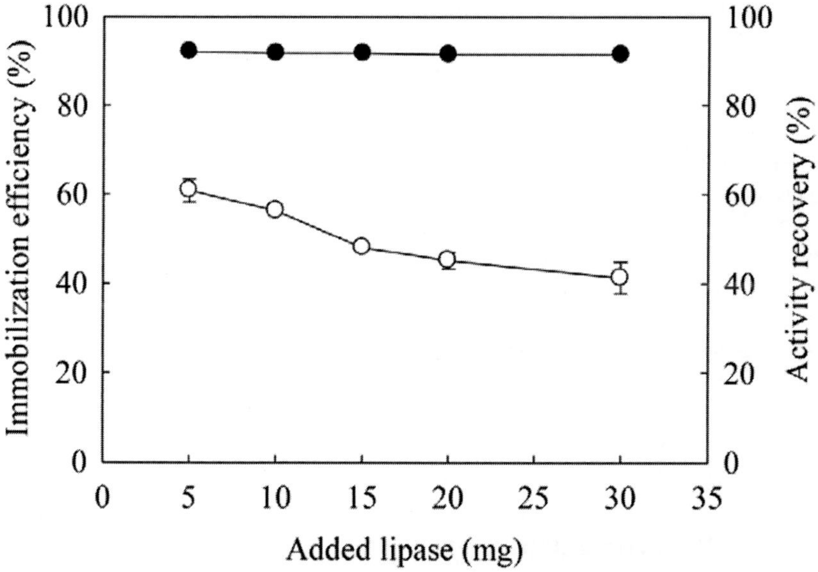

Figure 1. Effects of amount of lipase added during immobilization on immobilization efficiency (●) and activity recovery (○).

We also evaluated the external mass transfer resistance with Mears' criterion [25]:

$$\frac{-r'_A \rho_b R n}{k_c C_{Ab}} < 0.15 \tag{1}$$

where $-r'A$ is the rate of reaction; ρ_b is the bulk density of catalyst; R is the radius of catalyst particle; n is the reaction order; k_c is the mass transfer coefficient; C_{Ab} is the bulk concentration of the substrate. External mass transfer resistance can be neglected if Inequality (1) is satisfied. The value of Mears' criterion for lipase-bound MNP was 3.8×10^{-3} when calculated with $-r'A$=0.02 kmol/(kg catalysts), $\rho_b = 5.15 \times 10^3$ kg/m^3 for magnetite [26], $R = 8 \times 10^{-9}$ m, n = 1, $k_c = 0.29$ m/s, and $C_{Ab} = 0.74 \times 10^{-3}$ kmol/m^3. The result was significantly smaller than the critical value of 0.15, indicating that external mass transfer resistance can be neglected. The k_c was estimated with the correlation proposed for free convection around a solid sphere [27]. The diffusion coefficient of p-nitrophenyl palmitate substrate, 2.3×10^{-5} cm^2/s, required for the determination of k_c was determined using the modified Stokes-Einstein equation [28]. The actual reaction order (n) was between 0 and 1 because the substrate concentration of 0.74 mM was close to the K_m value of 1.5 mM for immobilized lipase [20], and hence the Mears' criterion was over-estimated with the assumption of n = 1. As more lipase was added, the radius of the lipase-bound MNP was increased with the attachment of more lipase to MNP and a porous outer layer of lipase may form. The external mass transfer resistance, however, was not limiting because the calculated Mears' criterion was much smaller than 0.15 even if the radius was increased by several folds. Nevertheless, the formation of an outer layer of lipase possibly caused a decrease in the effective diffusion coefficient within the layer [29]. The internal mass transfer may become limiting. Because the characteristics of the lipase layer are not known, the extent of internal mass transfer resistance is difficult to estimate.

2.2. Effects of Dosage of Lipase-Bound MNP and Stepwise Addition of Methanol on the Conversion of FAME

As shown in Figure 2, the conversion of FAME increased linearly with dosage of lipase-bound MNP in the range examined. Maximal FAME conversion of 55.6% was observed with 40% (w/w of oil) lipase-bound MNP, which was then used in later experiments.

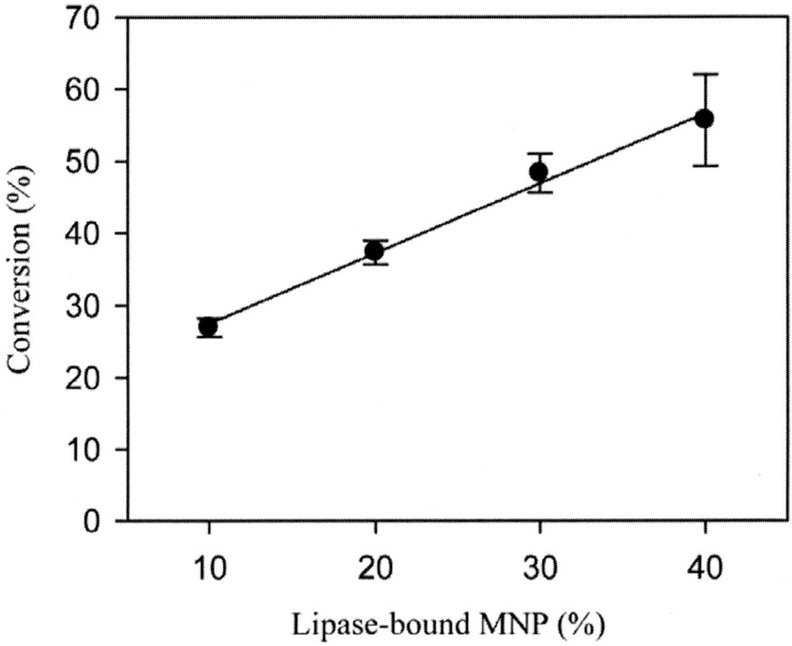

Figure 2. Effects of dosage of lipase-bound magnetic nanoparticles (MNP) on the conversion of fatty acid methyl esters (FAME). The dosage was expressed in weight percentage of waste cooking oil. The reaction was carried out at 40 °C for 72 h with water content of 10% (w/w of oil). The molar ratio of methanol to oil was 3:1; three separate additions at 0, 24 and 48 h, one third of the total amount each time. The equation of the fitted line is $y = x + 17.8$ with $R^2 = 0.99$.

Alcoholysis with relatively long-chain and branched alcohols proceeds efficiently even in organic-solvent free systems, but not in the case of methanolysis due to the inactivation of lipase by methanol [30]. One common strategy to avoid such inhibition is to add methanol in a stepwise fashion [30,31]. We examined the effect of stepwise addition at 0, 12 and 24 h or 0, 24 and 48 h (Figure 3). Stepwise addition at 12 h-intervals resulted in a higher rate of FAME conversion in comparison with the 24 h-interval addition. However, the conversions after 72 h were similar regardless of the interval used. For later experiments, stepwise addition of methanol at 24 h-intervals was selected.

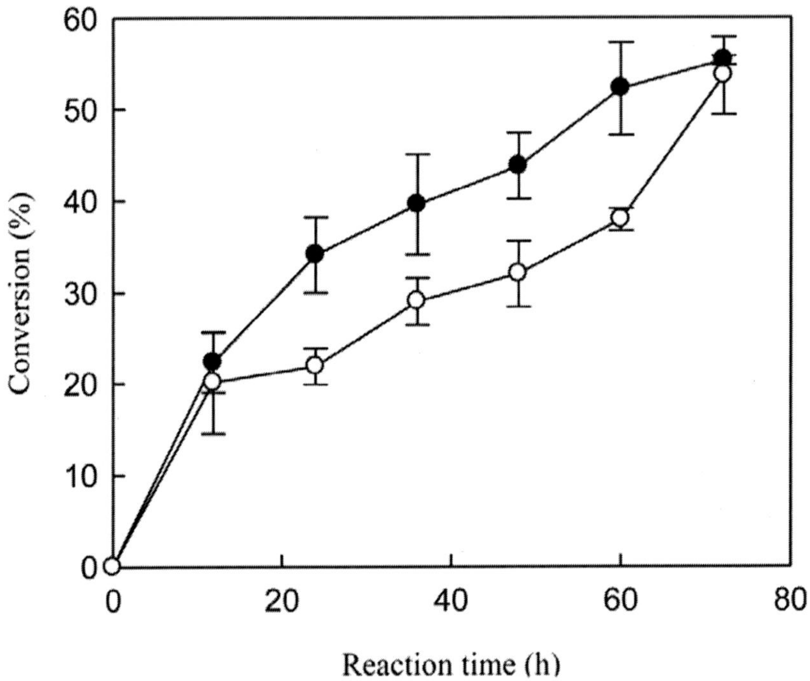

Figure 3. Time course of the conversion of FAME when methanol was added at different intervals. The reaction was carried out at 40 °C for 72 h with water content of 10% (w/w of oil). The molar ratio of methanol to oil was 3:1; three separate additions at 0, 12 and 24 h (•) or at 0, 24 and 48 h (○), one third of the total amount each time.

2.3. Model Fitting and Analysis of Variance

In addition to temperature and amount of methanol, the amount of water is also critical for the synthesis of FAME. Lipase possesses the unique feature of acting at the interface between an organic and an aqueous phase. The addition of water facilitates the formation of interfacial area; however, excess water may stimulate competitive hydrolysis reactions [7]. The optimal water content is a compromise between minimizing hydrolysis and maximizing enzyme activity for the transesterification reaction [15].

Based on previous reports using P. cepacia lipase immobilized on various supports for the transesterification of triglyceride to biodiesel [15,32,33], the variables selected for optimization and the corresponding ranges were temperature from 35 to 50 °C, water content of 1% to 20% (w/w of oil), and the molar ratio of methanol to oil from 3:1 to 8:1. The design of experiments and the corresponding data are given in Table 1. After fitting the data with various models followed by analysis of variance (ANOVA), the following quadratic polynomial most suitably described the correlation between conversion and the tested variables:

Table 1. Three-level-three-factor Box-Behnken design of experiments and the corresponding conversions.

Treatment No. [a]	Variable [b]			Conversion (%)
	Temperature (°C)	Molar ratio (methanol/oil)	Water content (%, w/w of oil)	
1	35 (−1)	3 (−1)	10.5 (0)	48 ± 1
2	50 (1)	3 (−1)	10.5 (0)	54.1 ± 0.3
3	35 (−1)	8 (1)	10.5 (0)	42.70 ± 0.01
4	50 (1)	8 (1)	10.5 (0)	52.6 ± 0.4
5	35 (−1)	5.5 (0)	1 (−1)	13.5 ± 0.1
6	50 (1)	5.5 (0)	1 (−1)	8 ± 1
7	35 (−1)	5.5 (0)	20 (1)	44.6 ± 0.9
8	50 (1)	5.5 (0)	20 (1)	58.13 ± 0.06
9	42.5 (0)	3 (−1)	1 (−1)	14.0 ± 0.9
10	42.5 (0)	8 (1)	1 (−1)	5.49 ± 0.02
11	42.5 (0)	3 (−1)	20 (1)	38.2 ± 0.5
12	42.5 (0)	8 (1)	20 (1)	20.0 ± 0.9
13	42.5 (0)	5.5 (0)	10.5 (0)	75.1 ± 0.8
14	42.5 (0)	5.5 (0)	10.5 (0)	76 ± 1
15	42.5 (0)	5.5 (0)	10.5 (0)	79.0 ± 0.2
16	42.5 (0)	5.5 (0)	10.5 (0)	82.1 ± 0.3
17	42.5 (0)	5.5 (0)	10.5 (0)	84.0 ± 0.8

[a] The treatments were performed in random order; [b] The values of 1, −1 and 0 in parentheses were coded levels.

$$Y = -347.13 + 13.1A + 33.48B + 8.38C - 0.16A^2 - 3.31B^2 - 0.43C^2 + 0.06AB + 0.07AC - 0.1BC \quad (2)$$

where Y, A, B and C were conversion of FAME, temperature, substrate molar ratio (methanol/oil), and water content (%, w/w of oil), respectively. The F-value of 25.92 for the model was higher than $F_{0.01,9,7}$ of 6.72, indicating the model was significant at confidence level of 99%. The F-value for lack of fit was 6.42, much lower than $F_{0.01,3,4}$ of 16.69, indicating lack of fit was insignificant. Overall, the model had a small p-value of 0.0001 and a suitable coefficient of determination (R^2 = 0.97), clearly indicating that the model was highly significant and sufficient to describe the correlation between the conversion of FAME and the tested variables. The high value of adjusted determination coefficient (Adj. $R^2 = 0.93$) also supported the significance of the model. The value of adequate precision (a measure of signal to noise ratio) of the model was 14.29, which is greater than four, thus providing adequate model discrimination [22]. Water content and all the square terms were significant for the process with p-values smaller than 0.05 (Table 2).

Table 2. Analysis of variance (ANOVA) for the regression model and respective model terms.

Source	Sum of squares	Degree of freedom	Mean square	*F*-value	Prob > *F* [a,b]
Model	11398.83	9	1266.537	25.92057	0.0001
Temperature (A)	69.32531	1	69.32531	1.418791	0.2724
Substrate molar ratio (B)	142.1298	1	142.1298	2.908786	0.1319
Water content (C)	1799.7	1	1799.7	36.83213	0.0005
A^2	346.665	1	346.665	7.094744	0.0323
B^2	1805.258	1	1805.258	36.94588	0.0005
C^2	6430.1	1	6430.1	131.5965	<0.0001
AB	4.2849	1	4.2849	0.087693	0.7757
AC	91.48922	1	91.48922	1.872391	0.2135
BC	23.3289	1	23.3289	0.477442	0.5118
Residual	342.0356	7	48.86223		
Lack of fit	283.2356	3	94.41188	6.422577	0.0521
Pure error	58.8	4	14.7		
Cor total	11740.87	16			

[a] Significant at "Prob > *F*" lower than 0.05; [b] Insignificant at "Prob > *F*" higher than 0.1.

2.4. Effects of Variables and Their Optimization

The correlation between the conversion of FAME and tested variables can be better understood by examining the contour plots. In Figure 4, the conversion increased significantly when water content increased from 1% to 10.5% (Figure 4avs. 4b), suggesting that certain amount of water was required to activate the immobilized lipase. Nevertheless, when increasing the water content further to 20%, the conversion decreased, indicating that the hydrolysis may start to compete with methanolysis. As the temperature was elevated from 35 to around 42.5 °C (center point), the conversion increased. However, when increasing the temperature further to 50 °C, the conversion started to decrease, showing that part of the immobilized lipase may start to inactivate. The conversion increased with the substrate molar ratio up to 5–6, but decreased when increasing the substrate molar ratio up to 8, which could be explained by the inactivation of lipase caused by excessive methanol.

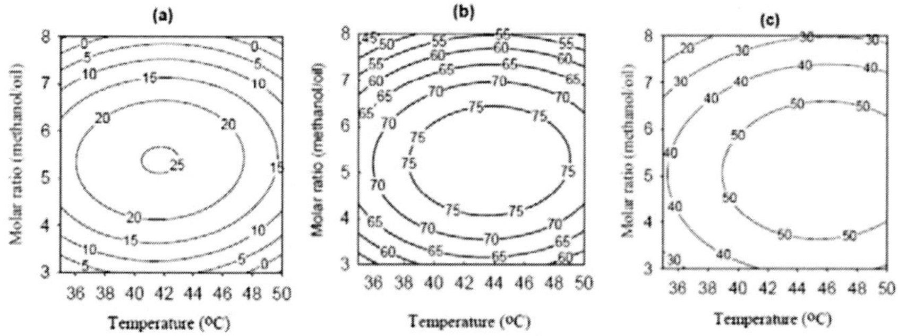

Figure 4. Contour plots of the molar conversion of FAME at different water content. (**a**) 1%; (**b**) 10.5%; and (**c**) 20%. The water content was expressed in weight percentage of waste cooking oil.

The optimal reaction conditions were 44.2 °C, substrate molar ratio of 5.2, and water content of 12.5%; the predicted and experimental values of conversion were 80% and 79%, respectively. The optimized conversion of FAME was higher than those using lipase from Thermomyces lanuginosa or Candida antarctica immobilized on granulated silica or C. antarctica lipase immobilized on macroporous acrylic resin (SP435) with restaurant grease as a feedstock [34]; but lower than those using P. cepacia lipase entrapped within a phyllosilictae sol-gel matrix (PS-30) with tallow and grease as feedstocks [14,34]. Our previous study using soybean oil as a feedstock showed 93% conversion of FAME with the same preparation of lipase-bound MNP [20]. The lower conversion with WCO could be explained by the presence of oxidized compounds such as aldehydes, epoxides, and polymers which were unrecognized as substrates by lipase [3].

2.5. Storage Stability and Reusability of Immobilized Lipase

The storage stability of immobilized lipase at 4 °C and room temperature was examined (Figure 5). Immobilized lipase stored at room temperature decayed at a faster rate as compared to that stored at 4 °C. The conversion of FAME for lipase stored at room temperature and 4 °C after 10 days were 31.1% and 69.1%, respectively, clearly indicating better storage stability at 4 °C.

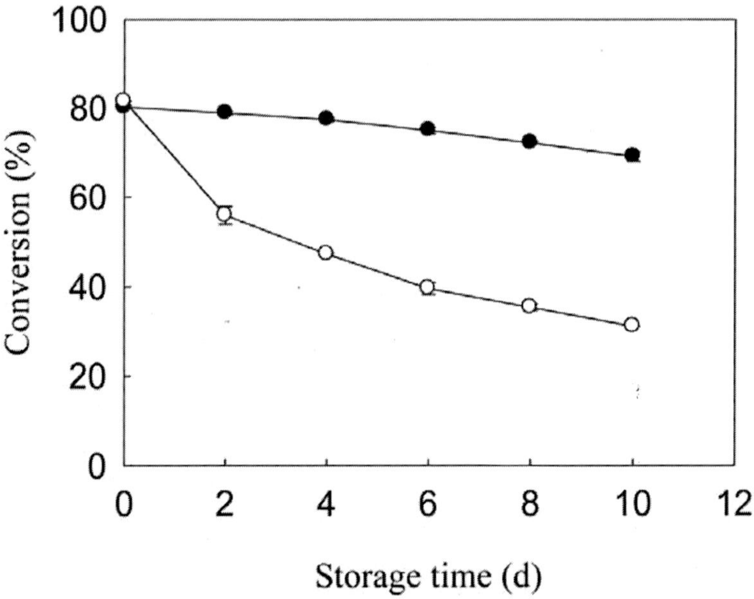

Figure 5. Stability of Pseudomonas cepacia lipase immobilized on magnetic nanoparticles after stored at 4 °C (•) and room temperature (○) at the time indicated. 40% (w/w of oil) immobilized lipase was used to catalyze transesterification using 4.8 g waste cooking oil under optimal reaction conditions for 72 h.

The reusability of immobilized lipase after washing with different solvent is shown in Figure 6. After three repeated uses, immobilized lipase recycled by washing with tert-butanol retained most of its initial conversion. tert-Butanol was reported being effective in the regeneration of immobilized lipase [35], perhaps due to its ability to alleviate the negative effects of both methanol and glycerol on activity [36]. After five cycles, lipase recycled without washing had the lowest relative conversion; however, the conversions showed little difference regardless of the solvent used. The decrease in FAME conversion after recycling can be partially attributed to the loss of lipase-bound MNP. In our previous work, lipase-bound MNP exhibited 89% of the initial activity after incubation at 40 °C for 30 min [20]. This implicated that thermal inactivation of immobilized lipase also contributed to the decrease in the conversion of FAME during reuse.

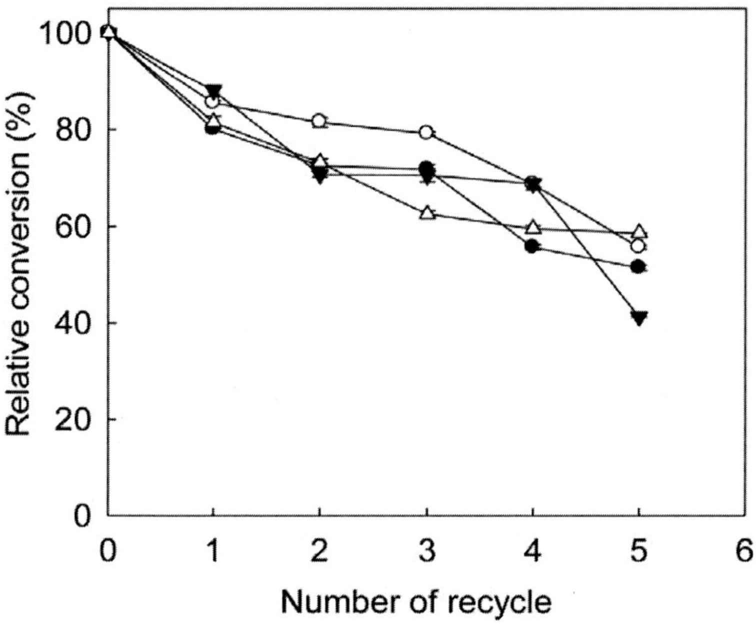

Figure 6. Reusability of Pseudomonas cepacia lipase immobilized on magnetic nanoparticles. Immobilized lipase was recycled without washing (▼) or after washing with tert-butanol (○); n-hexane (△); and deionized water (●). The initial conversion was defined as 100%. 40% (w/w of oil) immobilized lipase was used to catalyze transesterification using 4.8 g waste cooking oil under optimal reaction conditions for 72 h.

3. EXPERIMENTAL SECTION

3.1. Preparation of MNP

All reagents were purchased from Wako (Osaka, Japan) unless otherwise specified. MNP was prepared by dissolving 0.4 g of $FeCl_2{\cdot}4H_2O$ and 1.08 g of $FeCl_3{\cdot}6H_2O$ in 20 mL deionized water (final concentrations of Fe^{2+} and Fe^{3+} were 0.1 and 0.2 M, respectively), followed by addition of 15 mL of 29% (v/v) NH_4OH under vigorous stirring at room temperature. The precipitate was heated at 80 °C for 30 min before washing with 40 mL of deionized water twice followed by 40 mL of ethanol twice. The precipitate was finally resuspended in 40 mL of deionized water and then lyophilized. The untreated MNP were close to spherical with an average diameter of 16 nm by examining with high resolution TEM (JEOL, Akishima, Japan), and the XRD (MAC Science, Yokohama, Japan) pattern confirmed the synthesized MNP was pure Fe_3O_4 with a spinel structure [20].

3.2. Immobilization of Lipase

The procedure used was the same as previous report with minor modifications [19]. One hundred and fifty milligrams of MNP was added to 10 mL of binding buffer (3 mM sodium phosphate buffer, pH 6, containing 0.1 M NaCl) followed by sonication for 10 min. After removing the binding buffer, MNP was activated with 10 mL of 18.75 mg/mL carbodiimide prepared in the binding buffer for 15 min under sonication. MNP was then washed with 10 mL binding buffer three times, followed by incubation with 10 mL of 0.5 to 3 mg/mL Amano lipase PS (from P. cepacia; Sigma-Aldrich, St. Louis, MO, USA) solution prepared in the binding buffer at 4 °C for 30 min under sonication. After separation with a magnet, the lipase-bound MNP was washed with binding buffer several times and ready for use. The residual protein concentration in the supernatant was determined with BCA assay [37]. The immobilization efficiency was defined as follows:

$$\text{Immobilization efficiency (\%)} = [(\text{amount of added lipase} - \text{residual lipase in the supernatant})/\text{amount of added lipase}] \times 100 \quad (3)$$

3.3. Assay for Lipase Activity

The assay was modified from that described by Pencreac'h et al. [38]. The assay mixture contained 90 μL of 8.25 mM p-nitrophenyl palmitate (Sigma-Aldrich) in isopropanol and 810 μL of 50 mM Tris-HCl, pH 8.0, with 0.5% (w/v) Triton X-100 and 0.12% (w/v) arabic gum preheated to 40 °C. To initiate the reaction, 100 μL of lipase solution or suspension of immobilized lipase diluted to appropriate concentration with 1% (w/v) bovine serum albumin was added. The change in absorbance at 410 nm was monitored for 5 min at 40 °C using a thermostated spectrophotometer (UV-1800, Shimadzu, Kyoto, Japan). The activity was calculated from the difference in absorbance between 2 and 5 min with a standard curve for the hydrolysis product, p-nitrophenol. One activity unit was defined as the production of 1 μmol of p-nitrophenol per min at 40 °C. Measurements were performed in triplicates.

3.4. Experimental Design

Although the central composite design is often used for fitting second-order models, the Box-Behnken design is also efficient and needs less number of runs [22]. In our study, a 3-level-3-factor Box-Behnken design with five replicates at the center was applied and required 17 experiments. The variables selected for the synthesis of biodiesel and the corresponding ranges were reaction temperature (35–50 °C), substrate molar ratio (methanol: oil = 3:1–8:1), and water content (1%–20%, w/w of oil). Levels of the factors, in terms of coded and uncoded forms, are shown in Table 1. Treatments were performed in a fully random order to avoid bias.

3.5. Transesterification of Oil to Biodiesel

In a typical reaction, 1.92 g of lyophilized lipase-bound MNP (40%, w/w of oil) and 4.8 g of waste cooking oil were mixed in a 25 mL Erlenmeyer flask, followed by three separate additions of methanol, at 0, 24 and 48 h with one third of the total amount (methanol:oil = 3:1–8:1) each time. The mixture also contained deionized water (1%–20%, w/w of oil). The reaction mixture was incubated in the temperature range of 35–50 °C using a water bath with orbital shaking at 200 rpm for 72 h.

3.6. Analysis of FAME

The oil phase of the reaction mixture was first treated with 30 mg of sodium sulfate followed by centrifugation at 7500× g for 5 min. Fifty microliters of treated sample was mixed with 1 mL of 10 mg/mL methyl heptadecanoate (TCI, Portland, OR, USA) in hexane as an internal standard. The conversion of FAME was determined by injecting 1 μL of the sample into a gas chromatograph (Shimadzu GC-14A, Kyoto, Japan) equipped with a flame-ionization detector (FID, Shimadzu). A BPX70 capillary column (30 m × 0.25 mm i.d.; SGE Analytical Science, Ringwood, Australia) with hydrogen as carrier gas at a constant pressure of 6 kg/cm^2 was used. The injector and FID temperatures were set at 250 and 280 °C, respectively. The oven temperature was initially held at 150 °C for 30 s and then increased to 180 °C at 10 °C/min, finally to 198 °C at 1.5 °C/min. The amount of FAME in sample was determined from standard curves and measurements were performed in triplicates. The conversion was defined as follows:

$$Conversion\ (\%) = \frac{moles\ of\ FAME\ produced}{3 \times moles\ of\ oil} \times 100 \quad (4)$$

3.7. Data Analysis

The experimental data (Table 1) were fitted to the following second-order polynomial equation using the analysis procedure of the Design Expert software version 6.01 (Stat-Ease, Minneapolis, MN, USA):

$$Y = \beta_0 + \sum_{i=1}^{3} \beta_i x_i + \sum_{i=1}^{3} \beta_{ii} x_i^2 + \sum_{i=1}^{2} \sum_{j=i+1}^{3} \beta_{ij} x_i x_j \quad (5)$$

where Y is response (conversion of FAME); β_0, β_i, β_{ii}, and β_{ij} are constant coefficients; and x_i and x_j are the uncoded independent variables. All analytical steps including analysis of variance (ANOVA), regression analysis, optimization of the variables, and plotting of response surfaces were performed using the same software.

4. CONCLUSIONS

In this work, we demonstrated the potential of P. cepacia lipase immobilized on MNP as a biocatalyst for the synthesis of FAME using WCO as a feedstock, and the conversion of FAME reached 79% under optimal reaction conditions, which was comparable to those using other lipases in immobilized form. The proposed process may lower the production cost of biodiesel and facilitate the disposal of WCO. The immobilized lipase exhibited good storage stability at 4 °C and can be easily recovered by magnetic field for repeated use. Approximately 80% of the initial FAME conversion was retained after three repeated uses when lipase-bound MNP was washed with tert-butanol. Nevertheless, the reusability and storage stability at room temperature require further improvement for the immobilized lipase to be practical for industrial applications. Thermal inactivation is critical for both reusability and storage stability. One possible approach for improvement is to use thermally stable lipases [39,40]. Because large amount of lipase-bound MNP was used for the transesterification, those away from the magnetic field were easily washed off during recycling. Such loss of the biocatalyst could be reduced if stronger magnetic field is applied. Alternatively, the loss of lipase-bound MNP during recycling could be improved by using a packed-bed reactor, which also allows for continuous removal of products and protection of the enzyme from mechanical shear.

ACKNOWLEDGMENTS

Financial supports from National Science Council (NSC 100-2221-E-036-034) and Tatung University (B96-S03-059) are gratefully acknowledged.

REFERENCES

1. Canakci, M.; Sanli, H. Biodiesel production from various feedstocks and their effects on the fuel properties. J. Ind. Microbiol. Biotechnol **2008**, 35, 431–441.
2. Canakci, M.; Gerpen, J.V. Biodiesel production from oils and fats with high free fatty acids. Trans. ASAE **2001**, 44, 1429–1436.
3. Kulkarni, M.G.; Dalai, A.K. Waste cooking oil-an economical source for biodiesel: A review. Ind. Eng. Chem. Res **2006**, 45, 2901–2913.
4. Escobar, J.C.; Lora, E.S.; Venturini, O.J.; Yáñez, E.E.; Castillo, E.F.; Almazan, O. Biofuels: Environment, technology and food security. Renew. Sustain. Energy Rev **2009**, 13, 1275–1287.
5. Hasan, F.; Shah, A.A.; Hameed, A. Industrial applications of microbial lipases. Enzyme Microbial. Technol **2006**, 39, 235–251.
6. Bisen, P.; Sanodiya, B.; Thakur, G.; Baghel, R.; Prasad, G. Biodiesel production with special emphasis on lipase-catalyzed transesterification. Biotechnol. Lett **2010**, 32, 1019–1030.
7. Jegannathan, K.R.; Abang, S.; Poncelet, D.; Chan, E.S.; Ravindra, P. Production of biodiesel using immobilized lipase—A critical review. Crit. Rev. Biotechnol **2008**, 28, 253–264.
8. Shah, S.; Sharma, S.; Gupta, M.N. Biodiesel preparation by lipase-catalyzed transesterification of jatropha oil. Energy Fuels **2004**, 18, 154–159.
9. Shaw, J.F.; Chang, S.W.; Lin, S.C.; Wu, T.T.; Ju, H.Y.; Akoh, C.C.; Chang, R.H.; Shieh, C.J. Continuous enzymatic synthesis of biodiesel with Novozym 435. Energy Fuels **2008**, 22, 840–844.
10. Oliveira, D.; Oliveira, J.V. Enzymatic alcoholysis of palm kernel oil in n-hexane and $SCCO_2$. J. Supercrit. Fluids **2001**, 19, 141–148.
11. Mittelbach, M. Lipase catalyzed alcoholysis of sunflower oil. J. Am. Oil Chem. Soc **1990**, 67, 168–170.

12. Li, S.-F.; Fan, Y.-H.; Hu, R.-F.; Wu, W.-T. Pseudomonas cepacia lipase immobilized onto the electrospun PAN nanofibrous membranes for biodiesel production from soybean oil. J. Mol. Catal. B **2011**, 72, 40–45.
13. Kumari, V.; Shah, S.; Gupta, M.N. Preparation of biodiesel by lipase-catalyzed transesterification of high free fatty acid containing oil from Madhuca indica. Energy Fuels **2006**, 21, 368–372.
14. Hsu, A.-F.; Jones, K.; Marmer, W.; Foglia, T. Production of alkyl esters from tallow and grease using lipase immobilized in a phyllosilicate sol-gel. J. Am. Oil Chem. Soc **2001**, 78, 585–588.
15. Noureddini, H.; Gao, X.; Philkana, R.S. Immobilized Pseudomonas cepacia lipase for biodiesel fuel production from soybean oil. Bioresour. Technol **2005**, 96, 769–777.
16. Orçaire, O.; Buisson, P.; Pierre, A.C. Application of silica aerogel encapsulated lipases in the synthesis of biodiesel by transesterification reactions. J. Mol. Catal. B **2006**, 42, 106–113.
17. Gupta, A.K.; Gupta, M. Synthesis and surface engineering of iron oxide nanoparticles for biomedical applications. Biomaterials **2005**, 26, 3995–4021.
18. Ito, A.; Shinkai, M.; Honda, H.; Kobayashi, T. Medical application of functionalized magnetic nanoparticles. J. Biosci. Bioeng **2005**, 100, 1–11.
19. Huang, S.-H.; Liao, M.-H.; Chen, D.-H. Direct binding and characterization of lipase onto magnetic nanoparticles. Biotechnol. Prog **2003**, 19, 1095–1100.
20. Mak, K.-H. Immobilization of Lipase from Pseudomonas cepacia onto Magnetic Nanoparticles. Master's Thesis, Tatung University, Taipei, Taiwan, June 2008.
21. Mak, K.-H.; Yu, C.-Y.; Kuan, I.-C.; Lee, S.-L. Immobilization of Pseudomonas cepecia Lipase onto Magnetic Nanoparticles for Biodiesel Production. In Progress in Development and Applications of Renewable Energy; Yang, S.-S., Sayigh, A.A.M., Lai, C.-M., Chen, S., Eds.; National Taiwan University: Taipei, Taiwan, 2009; pp. 51–58.
22. Montgomery, D.C. Response Surface Methods and Designs. In Design and Analysis of Experiments, 6th ed; John Wiley & Sons: Hoboken, NJ, USA, 2005; pp. 405–463.

23. Kuan, I.-C.; Lee, C.-C.; Tsai, B.-H.; Lee, S.-L.; Lee, W.-T.; Yu, C.-Y. Optimizing the production of biodiesel using lipase entrapped in biomimetic silica. Energies **2013**, 6, 2052–2064.
24. Chen, H.-C.; Ju, H.-Y.; Wu, T.-T.; Liu, Y.-C.; Lee, C.-C.; Chang, C.; Chung, Y.-L.; Shieh, C.-J. Continuous production of lipase-catalyzed biodiesel in a packed-bed reactor: Optimization and enzyme reuse study. J. Biomed. Biotechnol **2011**, 2011, 1–6.
25. Mears, D.E. Tests for transport limitations in experimental catalytic reactors. Ind. Eng. Chem. Process Des. Dev **1971**, 10, 541–547.
26. Mineralogy Database, Available online: http://webmineral.com/ (accessed on 19 November 2013).
27. Cussler, E.L. Fundamentals of Mass Transfer. In Diffusion: Mass Transfer in Fluid Systems, 3rd ed; Cambridge University Press: New York, NY, USA, 2009; pp. 237–273.
28. Wilke, C.R.; Chang, P. Correlation of diffusion coefficients in dilute solutions. AIChE J **1955**, 1, 264–270.
29. Bailey, J.E.; Ollis, D.F. The Kinetics of Enzyme-Catalyzed Reactions. In Biochemical Engineering Fundamentals, 2nd ed; McGraw-Hill Inc: Columbus, OH, USA, 1986; pp. 86–156.
30. Watanabe, Y.; Shimada, Y.; Sugihara, A.; Tominaga, Y. Enzymatic conversion of waste edible oil to biodiesel fuel in a fixed-bed bioreactor. J. Am. Oil Chem. Soc **2001**, 78, 703–707.
31. Shimada, Y.; Watanabe, Y.; Sugihara, A.; Tominaga, Y. Enzymatic alcoholysis for biodiesel fuel production and application of the reaction to oil processing. J. Mol. Catal. B **2002**, 17, 133–142.
32. Shah, S.; Gupta, M.N. Lipase catalyzed preparation of biodiesel from Jatropha oil in a solvent free system. Process Biochem **2007**, 42, 409–414.
33. Tran, D.-T.; Yeh, K.-L.; Chen, C.-L.; Chang, J.-S. Enzymatic transesterification of microalgal oil from Chlorella vulgaris ESP-31 for biodiesel synthesis using immobilized Burkholderia lipase. Bioresour. Technol **2012**, 108, 119–127.
34. Hsu, A.-F.; Jones, K.; Foglia, T.A.; Marmer, W.N. Immobilized lipase-catalysed production of alkyl esters of restaurant grease as biodiesel. Biotechnol. Appl. Biochem **2002**, 36, 181–186.
35. Chen, J.-W.; Wu, W.-T. Regeneration of immobilized Candida antarctica lipase for transesterification. J. Biosci. Bioeng **2003**, 95, 466–469.

36. Li, L.; Du, W.; Liu, D.; Wang, L.; Li, Z. Lipase-catalyzed transesterification of rapeseed oils for biodiesel production with a novel organic solvent as the reaction medium. J. Mol. Catal. B **2006**, 43, 58–62.
37. Smith, P.K.; Krohn, R.I.; Hermanson, G.T.; Mallia, A.K.; Gartner, F.H.; Provenzano, M.D.; Fujimoto, E.K.; Goeke, N.M.; Olson, B.J.; Klenk, D.C. Measurement of protein using bicinchoninic acid. Anal. Biochem **1985**, 150, 76–85.
38. Pencreac'h, G.; Leullier, M.; Baratti, J.C. Properties of free and immobilized lipase from Pseudomonas cepacia. Biotechnol. Bioeng **1997**, 56, 181–189.
39. Palomo, J.M.; Segura, R.L.; Fernández-Lorente, G.; Pernas, M.; Rua, M.L.; Guisán, J.M.; Fernández-Lafuente, R. Purification, immobilization, and stabilization of a lipase from Bacillus thermocatenulatus by interfacial adsorption on hydrophobic supports. Biotechnol. Prog **2004**, 20, 630–635.
40. Hosseini, M.; Karkhane, A.; Yakhchali, B.; Shamsara, M.; Aminzadeh, S.; Morshedi, D.; Haghbeen, K.; Torktaz, I.; Karimi, E.; Safari, Z. In silico and experimental characterization of chimeric Bacillus thermocatenulatus lipase with the complete conserved pentapeptide of Candida rugosa lipase. Appl. Biochem. Biotechnol **2013**, 169, 773–785.

10
CHAPTER

LIPASE APPLICATIONS IN BIODIESEL PRODUCTION

Sevil Yücel[1], Pınar Terzioğlu[1, 2] and Didem Özçimen[1]

[1] *Yıldız Technical University, Faculty of Chemical and Metallurgical Engineering, Bioengineering Department, Istanbul, Turkey*

[2] *Mugla Sıtkı Koçman University, Faculty of Sciences, Chemistry Department, Mugla, Turkey*

1. INTRODUCTION

Because of the global warming and depletion of fossil fuels, in recent years, intensive investigations are carried on for providing the greater use of sustainable biofuels instead of fossil fuels. Biomass, which various biofuels are produced from, has an important role among other alternative energy sources including wind energy, solar energy, geothermal energy, etc.

Biodiesel is one of the important biofuels and a clean energy source as an alternative to petroleum-based diesel fuels. Biodiesel has some advantages and disadvantages. Transportability, high combustion efficiency, low sulphur and aromatic content, high cetane number and biodegradability are advantages of the biodiesel [1]. Disadvantages of biodiesel are high viscosity, lower energy content, high cloud and pour point, high nitrogen oxide emission, lower engine speed and power, injector cooking, high price and engine erosion [2].

The flash point of biodiesel is higher than diesel fuel. This feature is important for fuel storage and transportation in the way of safety. Cetane number of biodiesel (~50) is higher than diesel fuel [3]. Biodiesel does not include aromatic and sulphur content and contains oxygen at the rate of 10-11% by mass [4]. Cetane number is an important factor to determine the quality of diesel fuel, especially ignition quality of diesel fuel. In other words, it determines the ignition tendency of fuel when being injected into engine. Ignition quality of biodiesel is determined by the structure of methyl ester [5].Viscosity is also an important factor for biodiesel. Viscosity affects mostly fuel injection equipment and the increase of fuel viscosity changes the viscosity at low temperatures. High viscosity has an negative effect on fuel spray atomization [6]. Amounts of elements and compounds in biodiesel and diesel fuel are present in Table 1 [7]. Biodiesel has more polar structure than diesel fuel because of the oxygen, which is an electronegative element present in its structure, and therefore biodiesel has higher viscosity comparing with diesel fuel. In addition, elemental oxygen content is responsible for lower heating value of biodiesel when compared with diesel fuel. [7-9]. Biodiesel can be used in its pure form or when mixed with diesel fuel in certain proportions. Most common biodiesel blends are B2 (2 % biodiesel, 98 % diesel), B5 (5 % biodiesel, 95% diesel), B20 (20 % biodiesel, 80 % diesel) [10].

Table 1. The comparison of elemental and chemical content of diesel and biodiesel [7]

	Biodiesel Content (%)	Diesel Content (%)
Carbon	79.6	86.4
Hydrogen	10.5	13.6
Oxygen	8.6	-
Nitrogen	1.3	-
C/H	7.6	6.5
n-Aliphatics	15.2	67.4
Olephenics	84.7	3.4
Aromatics	-	20.1
Naphtens	-	9.1

The transesterification reaction can be influenced by several factors including molar ratio of alcohol, catalyst, presence of water, free fatty acid in oil samples, temperature, time and agitation speed. In this context, an understanding of the factors affecting the process is very important to make economically and environmentally biodiesel production [11].

To accelerate reaction rate, transesterification process is carried out in the presence of catalysts. So, biodiesel production is made by using chemical or enzymatic catalysts. Compared to chemical, enzymatic reaction is more attractive because of ability of make a high quality product, simplify the separation of products, mild reaction conditions, the reuse of the catalyst and especially environmental impact, although high conversion and reaction rate are obtained with chemical catalysts [11-14]. Lipase is important enzyme catalyst that catalyzes esterification and transesterification reaction to produce methyl esters (biodiesel). Figure 1 presents the enzymatic transesterification reaction [15].

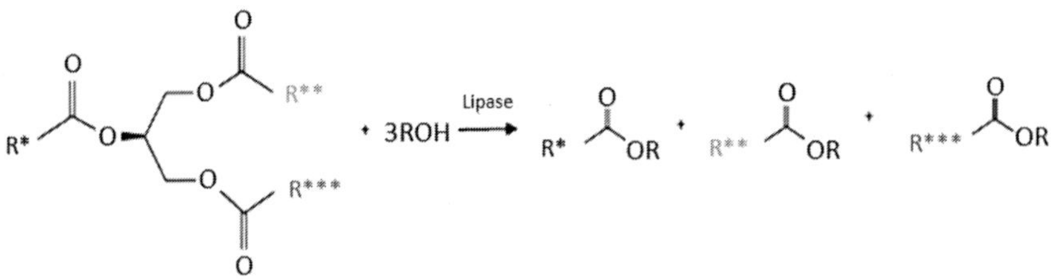

Figure 1. Enzymatic transesterification reaction [15].

In this study, enzymatic approach for biodiesel production was reviewed, and especially the usage of lipases in biodiesel production and factors affecting the effectiveness of lipase in reaction were explained in detail.

2. LIPASES IN BIODIESEL PRODUCTION

Biocatalyst based biotechnological applications are receiving increasing attention. Lipases (triacylglycerol acylhydrolases, EC 3.1.1.3) are the important biocatalysts because of their excellent biochemical and physiological properties. Lipases are the hydrolytic enzymes that can be used in various industrial applications for alcoholysis, acidolysis, amynolysis and hydrolysis reactions. Biodiesel production is one of the stunning applications of lipase. Lipase catalyzed biodiesel production was reported first by Mittelbach [16]. Lipase-catalyzed transesterification takes place in two steps, which involves hydrolysis of the ester bond and esterification with the second substrate [15]. A ping-ping bi bi mechanism generally used for kinetic studies of enzyme catalyzed transesterification.

Lipases can be isolated from many species of plants (papaya latex, oat seed lipase, and castor seed lipase), animals (pig's and human pancreatic lipases), bacteria, filamentous fungi and yeast [17-19]. For industrial enzyme production generally microorganisms are preferred because of their shortest generation time [20]. The other advantages of microorganisms can be listed as high yield of conversion of substrate into product, great versatility to environmental conditions and, simplicity in genetic manipulation and in cultivation conditions [20]. Although lipases from different sources are able to catalyze the same reaction, bacterial and fungal lipases are mostly used in biodiesel production such as *Aspergillus niger,Candida antarctica, Candida rugosa, Chromobacterium Viscosum, Mucor miehei, Pseudomonas cepacia, Pseudomonas fluorescens, Photobacterium lipolyticum, Rhizopus oryzae, Streptomyces sp., and Thermomyces lanuginose* [21]. *Candida rugosa*, obtained from yeast, is the most used microorganism for lipase production [22]. Recently, *Streptomyces sp.* was investigated as a potent lipase producing microbe for biodiesel production and found applicable in the field of biodiesel [23].

Specificity of lipases has a great importance in the selection of the usage area of lipases. Lipases can be divided into three groups due to their specificity as 1,3-specific lipases, fatty acid-specific lipases and nonspecific lipases. Especially, 1,3-specific lipases which release fatty acids from positions 1 and 3 of a glyceride and hydrolyze ester bonds in these positions such as *Aspergillus niger, Rhizopus oryzae and Mucor miehei* catalyze transesterification reactions efficiently [20,24]. The study of Du et al. [25], showed that higher yield (90%) was achieved for biodiesel production by using a sn-1,3-specific lipase, *Thermomyces lanuginosa* immobilized on silica gel (Lipozyme TL IM). Thus, the use of sn-1,3- specific lipases can give rise to biodiesel yield of above 90% under appropriate conditions [24]. Substrate specificity of lipases is also a crucial factor towards the biodiesel production which acts on the choice of the proper enzyme based on the composition of raw materials by consisting in the capability of distinguishing structural features of acyl chains [20,24]. Lipases from *Pseudomonas fluorescens, Pseudomonas cepacia, Candida rugosa, Candida antarctica and Candida cylindracea* are suitable for transesterification reaction by displaying both wide substrate specificity and regiospecificity [24].

2.1. Immobilization of Lipases

The immobilization of enzymes, which is attracting worldwide attention, was firstly reported in 1971 at Enzyme Engineering Conference [26]. During the past decade, chemical modification, physical modification, and gene expression techniques have been developed to obtain more economic, active, selective, or stable lipases. Immobilization is a modification method that can be defined as attaching the enzyme onto an insoluble solid support material [18]. By immobilization more operational and temperature stable lipases can be obtained and also lipases can be reused in the reactions. In addition, reusability of lipases will be a possible solution to the high cost of the enzymes and make them suitable for applications in industrial scale. The comparison of free enzymes and immobilized enzymes is given in Table 2. Methods for enzyme immobilization can be classified as adsorption, covalent bonding, entrapment, and cross-linking. The selection of method and support material is a prominent factor for obtaining an efficient lipase. The results of comparative studies revealed that the same lipase molecule can show very different catalytic activities after immobilization onto different supports [27].

Table 2. The comparison of free enzyme and immobilized enzyme [19]

Characteristics	Free Enzyme	Immobilized Enzyme
Price	High	Low
Efficiency	Low	High
Activity	Unstable	Stable
Reusability and recovery	Not possible	Possible
Tolerance to temperature, pH, etc.	Low	High
To separate from the substrate	Difficult	Easy
To separate from the product	Difficult	Easy

2.1.1. Adsorption technique

Adsorption is the adhesion of lipase on the surface of the adsorbent by weak forces, such as van der Walls, ionic and hydrophobic interactions, or dispersion forces [28]. Immobilization via adsorption method is the simply mixing of an aqueous solution of enzyme with the carrier material for a period and washing away the excess enzyme from the immobilized enzyme on the carrier after a time [29]. The level of adsorption is strictly related to the pH, temperature and ionic strength. Adsorption is the most widely employed method besides other methods because of its special commercial advantages and simplicity. Adsorption is the only reversible enzyme immobilization method. The advantages of adsorption is mild and easy preparing conditions, low cost, no need for chemical additives, the carrier can be recovered for repeated use, and high activity [30].

Various types of carriers used in immobilization of lipases. Acrylic resin, celite, polypropylene and textile membrane are broadly used carriers. Some of the reported results of adsorption technique based immobilized enzymes used in biodiesel production are summarized in Table 3. As can be seen from table generally the biodiesel yields using the enzymes obtained by adsorption method are higher than 85%. Novozym 435 is a commercial lipase, which is obtained by immobilization of *Candida antartica* lipase on acrylic resin and is a good catalyst that provides biodiesel yield higher than 90% with vegetable oil or waste cooking oil as feedstock [31]. The other commercialized lipase is known as *Candida sp.* 99–125 lipase immobilized on textile membrane, which can catalyze lard, waste oil and vegetable oils with higher yields that is more than 87% [31]. Besides many advantages of immobilization by adsorption method, the main disadvantage is that desorption of the lipase from the carrier occurs because of the weak interactions between the enzyme and support.

2.1.2. Covalent binding technique

Another approach is covalent binding technique, which is the formation of covalent bonds between the aldehyde groups of support surface and active amino acid residues on the surface of the enzyme [29]. A variety of supports have been used such inorganic materials, natural polymers (agarose, chitin and chitosan), synthetic polymers (hydrophobic polypeptides,nylon fibers) and Eupergit® (made by copolymerization of N,N′-methylene-bis-(methacrylamide), glycidyl methacrylate, allyl glycidyl ether and methacrylamide) for immobilization of lipases by covalent binding [56].The main advantage of covalent binding method is obtaining thermal and operational stable enzymes because of strong interactions between the lipase and the carrier [31]. The comparison of biodiesel production performance using immobilized lipase via covalent binding method is summarized in Table 4. Chitosan is a promising carrier as a natural polymer due to its membrane forming and adhesion ability, high mechanical strength and facility of forming insoluble in water thermally and chemically inert films [57]. Xie and Wang [58], reported a technique for immobilization of *Candida rugosa* lipase on magnetic chitosan microspheres for transesterification of soybean oil.The immobilized enzyme was determined as an effective biocatalyst for the transesterification reaction due to giving a good conversion of soybean oil and retaining its activity during the four cycles [58].

Table 3. Comparison of biodiesel production performance using immobilized lipase via adsorption method

Lipase Source	Carrier	Acid/Oil Source	Alcohol	Maximum Performance (%)	Reference
Burkholderia sp. C20	Alkyl-functionalized Fe_3O_4–SiO_2	Olive	Methanol	92 (conversion)	[32]
Candida antartica	Acrylic resin	Soybean	Methanol	92(yield)	[33]
Candida antartica	Acrylic resin	Soybean and rapeseed	Methanol	98.4 (conversion)	[34]
Candida antartica	Acrylic resin	Soybean and rapeseed	Methanol	"/>95 (conversion)	[35]
Candida antarctica B	Granular activated carbon	Palm	Isobutanol	100 (conversion)	[36]
Candida sp. 99–125	Textile membrane	Lard	Methanol	87.4 (yield)	[37]
Candida sp. 99–125	Textile (cotton) membrane	Salad	Methanol	96 (conversion)	[38]
Candida sp. 99–125	Textile membrane	Crude rice bran	Methanol	87.4 (yield)	[39]
Candida rugosa and Pseudomonas fluorescens	Acurel	Palm	Ethanol	89 (yield)	[40]
Chromoacterium viscosum	Celite-545	Jatropha	Ethanol	92 (yield)	[41]
Geobacillus *thermocatenulatus*	Poly-hydroxybutyrate beads	Babassu	Ethanol	100 (yield)	[42]
Pseudomonas aeroginosa	Celite	Soybean	Methanol	80(yield)	[43]
Pseudomonas cepacia	Celite	Jatropha	Ethanol	98 (yield)	[44]
Pseudomonas cepacia	Electrospun polyacrylonitrile fibers	Rapeseed	n-butanol	94 (conversion)	[45]
Pseudomonas cepacia	Polystyrene	Sapium sebiferum	Methanol	96.22 (yield)	[46]
Pseudomonas cepacia	Ceramic beads	Waste cooking	Methanol	40 (yield)	[47]
Pseudomonas fluorescens	Porous kaolinite particle	Triglyceride triolein	1-propanol	"/>90 (conversion)	[48]
Pseudomonas fluorescens and Pseudomonas cepacia	Polypropylene powder	Soybean	Methanol	58 37 (yield)	[49]
Penicillium expansum	Resin D4020	Waste	Methanol	92.8 (yield)	[50]
Rhizomucor miehei	Hydrophilic resins	Olive husk	Ethanol	-	[51]
Rhizomucor miehei	Silica	Waste cooking	Methanol	91.08 (yield)	[52]
Rhizopus oryzae	Macroporous resin HPD-400	Pistacia chinensis bge seed	Methanol	94 (yield)	[53]
Saccharomyces cerevisiae	Mg–Al hydrotalcite	Rape	Methanol	96 (conversion)	[5454]
Thermomyces lanuginosus (Lipozyme TL IM)	Hydrotalcite	Waste cooking	Methanol	95 (yield)	[55]

Table 4. Comparison of biodiesel production performance using immobilized lipase via covalent binding method

Lipase Source	Carrier	Acid/Oil Source	Alcohol	Maximum Performance (%)	Reference
Burkholderia cepacia	Niobium Oxide (Nb_2O_5)	Babassu	Ethanol	74.13 (yield)	[60]
Burkholderia cepacia	Polysiloxane–Polyvinyl Alcohol (SiO_2–Pva)	Babassu Beef Tallow	Ethanol	100 89.7 (yield)	[60]
Candida rugosa	Chitosan Microspheres	Soybean	Methanol	87 (conversion)	[58]
Candida rugosa	Chitosan Powder	Rapeseed Soapstock	Methanol	95 (conversion)	[61]
Enterobacter aerogenes	Silica	Jatropha	Methanol	94 (yield)	[62]
Porcine pancreatic	Chitosan Beads	Salicornia	Methanol	55 (conversion)	[63]
Pseudomonas fluorescens	Toyopearl Af-Amino-650m Resin	Babassu	Ethanol	94.9 (yield)	[64]
Rhizopus oryzae	Resin Amberlite Ira-93	Pistacia Chinensis Bge Seed	Methanol	92 (yield)	[63]
Pseudomonas fluorescens	Toyopearl Af-Amino-650m Resin	Babassu	Ethanol	94.9 (yield)	[64]
Rhizopus oryzae	Resin Amberlite Ira-93	Pistacia Chinensis Bge Seed	Methanol	92 (yield)	[63]
Rhizopus oryzae	Polystyrene Polymer(Amberlite Ira-93)	Soybean	Methanol	90.05 (yield)	[65]
Rhizopus Orizae +Candida rugosa	Silica	-	Methanol	*/>98 (conversion)	[66]
Rhizopus orizae +Candida rugosa	Silica	Crude Canola	Methanol	88.9 (conversion)	[59]
Thermomyces lanuginosus	Olive Pomace	Pomace	Methanol	93 (yield)	[67]
Thermomyces lanuginosus	Polyglutaraldehyde Activated Styrene-Divinylbenzene Copolymer	Canola	Methanol	97 (yield)	[68]
Thermomyces lanuginosus	Toyopearl Af-Amino-650m Resin	Palm	Ethanol	100 (yield)	[64]
Thermomyces lanuginosus	Polyurethane Foam	Canola	Methanol	90 (yield)	[69]
Thermomyces lanuginosus	Aldehyde-Lewatit	Soybean	Ethanol	100 (conversion)	[70]
Thermomyces lanuginosus	Magnetic Fe_3O_4 Nano-Particles	Soybean	Methanol	90 (conversion)	[71]

Using two immobilized lipases with complementary position specificity instead of one lipase is a new approach to produce a cost effective biodiesel [19]. Lipase from *Rhizopus orizae* and *Candida rugosa* was covalently bound to the silica, which was used to produce biodiesel from crude canola oil. Under optimum conditions, the conversion rate of degummed crude canola oil to fatty acid methyl esters was 88.9%, which is higher than the conversion obtained by free enzyme mixture (84.25%) [59].

Table 5. Comparison of biodiesel production performance using immobilized lipase via entrapment method

Lipase Source	Carrier	Acid/Oil Source	Alcohol	Maximum Performance (%)	Reference
Burkholderia cepacia	*K*-Carrageenan	Palm	Methanol	100 (conversion)	[76]
Burkholderia cepacia	Phyllosilicate Sol–Gel	Tallow and Grease	Ethanol	94 (yield)	[77]
Burkholderia cepacia	Mtms-Based Silica Monolith Coated With Butyl-Substituted Silicates	Jatropha	Methanol	95 (yield)	[78]
Candida antarctica	Celite®	Triolein	Methanol	60 (conversion)	[79]
Candida rugosa	Calcium Alginate Matrix	Palm	Ethanol	83 (yield)	[80]
Candida rugosa	Activated Carbon	Palm	Ethanol	85 (conversion)	[81]
Pseudomonas cepacia	Hydrophobic Sol–Gel	Soybean	Methanol	67 (conversion)	[82]
Pseudomonas fluorescens Mtcc 103	Alginate	Jatropha	Methanol	72 (yield)	[83]
Via Encapsulation Method					
Burkholderia cepacia	Silica Aerogels	Sunflower Seed	-	56 (conversion)	[84]
Burkholderia cepacia	*K*-Carrageenan	Palm	Methanol	100 (conversion)	[85]
Candida antartica	Silica Aerogels	Sunflower Seed	Methanol	90 (conversion)	[86]

2.1.3. Entrapment technique

Entrapment method is based on capturing of the lipase within a polymer network that retains the enzyme but allows the substrate and products to pass through [72]. This method can be simply defined as mixing an enzyme with a polymer solution and then crosslinking the polymer to form a lattice structure that captures the enzyme [29]. Entrapment is often used for industrial applications because the method is fast, cheap and can be carried out under mild conditions [73]. Entrapment can be divided into three categories such as gel or fiber entrapping and microencapsulation [74]. A number of supports have been investigated such as alginate, celite, carrageenan, resins, acrylic polymers etc. Some carriers used for entrapment and the biodiesel production yields obtained by these enzymes are displayed in Table 5. A disadvantage of entrapment method is the mass transfer problem due to the act of support as a barrier, so the lipase became effective only for low molecular weight substrates [19,75].

2.1.4. Cross linking technique

Cross-linking is another method for immobilization that can be defined as the interaction of a three dimensional network within enzyme, coupling reagent, and carrier [19]. The advantage of cross-linking is obtaining stable lipases due to the strong interaction between the lipase and the carrier. On the other hand, the cross-linking conditions are intense and the immobilized lipase shows lower activity [31].

The high free fatty acid content of waste cooking oil form water by esterification with alcohol which cause agglomeration of lipase and lowering biocatalysis efficiency [87]. Hence, free *Geotrichum sp.* lipase was not a suitable enzyme catalyst for transesterification of waste cooking oil. Yan et al. [87], report a modification procedure for preparation of cross-linked *Geotrichum sp.* The obtained lipase exhibited improved pH and thermostable stability compared to free lipase. The relative biodiesel yield was 85% for transesterification of waste cooking oil with methanol.

Kumari et al. [88] studied the preparation of *Pseudomonas cepacia* lipase cross-linked enzyme aggregates. It was shown that cross linked lipases has a greater stability than free enzymes to the denaturing conditions. The enzyme also used to catalyze madhuca indica oil, which's transesterification is difficult by chemical routes due to its high free fatty acid content. As a result, 92% conversion was obtained after 2.5 h.

Immobilization of *Candida rugosa* lipase on fine powder of Scirpus grossus L.f. by glutaraldehyde by cross linked technique for biodiesel production from palm oil, as already investigated by Kensingh et al. [89]. It was concluded that immobilized lipase yielded higher conversion of biodiesel than that of free lipase.

Lorena et al. [90] investigated the immobilization of the *Alcaligenes spp.* lipase on polyethylenimine agarose, glutaraldehyde agarose, octyl agarose, glyoxyl agarose, Sepabeads® by the aggregation and crosslinking method. The transesterification of canola oil was achieved with a yield 80% using a six-step addition of methanol and lipase immobilized on Sepabeads® by the aggregation method.

All these methods are shown schematically in Figure 2.

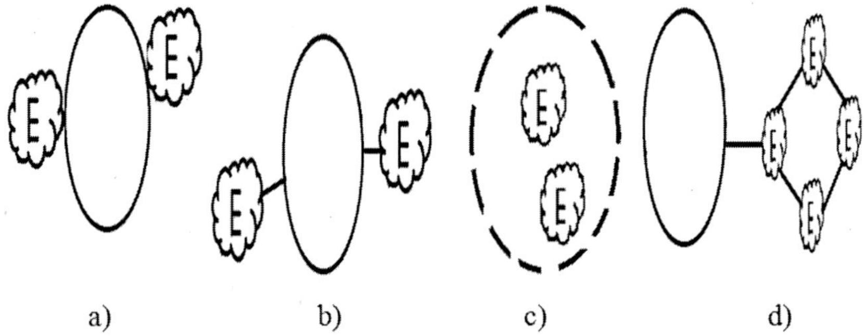

Figure 2. Schematic diagram of enzyme immobilization methods: a)Adsorption method b)Covalent binding method c)Entrapment method d) Crosslinking method

2.1.5. Whole cell immobilization

The applicability of lipases for the bulk production of fuels was limited significantly by the high cost of lipases [91]. Utilizing microbial cells such as fungi, bacteria, and yeasts cells containing intracellular lipase instead of extracellular lipases (free and immobilized lipase) is an easier and a cost effective way of enzymatic biodiesel production. Compared to conventional enzymatic processes, the use of whole cells provides excellent operational stability and avoids the complex procedures of isolation, purification and immobilization [91,92]. The general preparation steps for immobilized extracellular enzymes and whole cell enzymes showed in Figure 3. Biomass support particles have been used for immobilization of whole cells.

Aspergillus and Rhizopus have been most widely used as whole cell biocatalyst. Ban et al. [93], used first a whole cell biocatalyst, immobilized Rhizopus oryzae IFO4697 (a 1,3-positional specificity lipase) cells within biomass support particles, for the production of biodiesel and 91.1% methyl ester content was attained which was a similar result as that using the extracellular lipase. Many researchers have experimented on the use of whole cells to catalyze transesterification reaction summarized in Table 6.

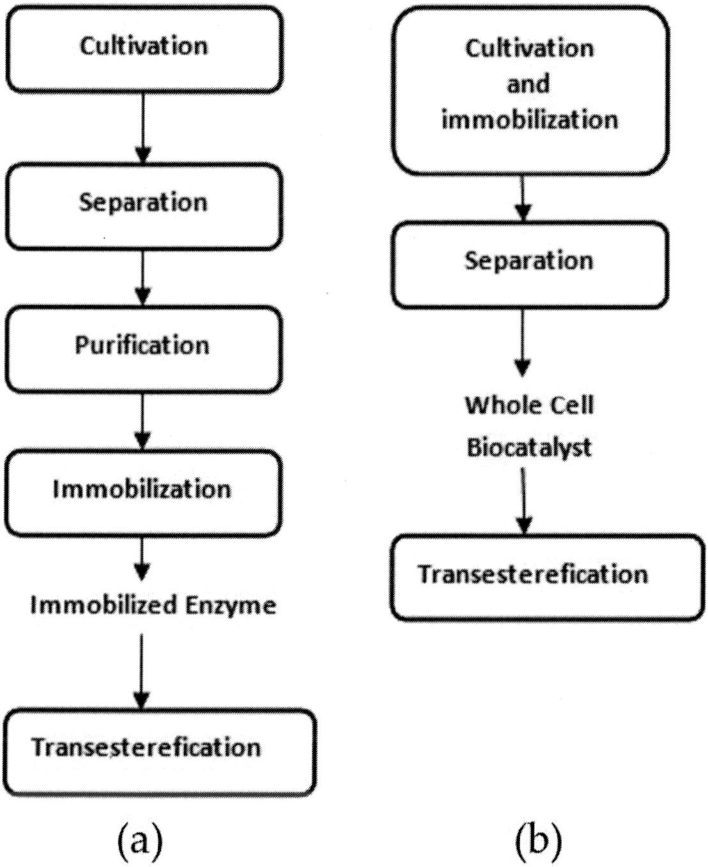

Figure 3. The preparation steps of a) immobilized extracellular lipase and b) whole cell biocatalyst

A technique using glutaraldehyde cross-linking treatment on whole cell catalyst for methanolysis of soybean oil was developed by Sun et al. [94]. The glutaraldehyde cross linking treatment resulted in higher methanol tolerance and high catalytic activity (with the ratio of methanol to oil reaching 3). Also, a novel methanol addition strategy was proposed as stepwise addition of different amounts of methanol (1.0, 1.2, 1.5, and 2.0M equivalent of oil) every 24 h. It was found that the highest methyl ester yield could reach 94.1% after 24 h reaction by 1.2 mol, 1.5 M and 1.2 mol methanol additions at 0, 8, and 14 h. In general, the whole cell catalyzed process is slower than extracellular lipase catalyzed process. Sun et al. [94], also reported that the reaction time could be shortened by this way. It is clear that significant reduction in the cost of biodiesel production can be achieved by combining the whole cell biocatalyst process with stepwise addition of methanol.

Table 6. Comparison of biodiesel production performance using whole cell biocatalysts

Lipase Source	Carrier	Acid/Oil Source	Alcohol	Maximum Performance (%)	Reference
Aspergillus niger	BSPs[a]	Waste Cooking	Methanol	86.4 (yield)	[96]
Aspergillus niger	Polyurethane BSPs[a]	Palm	Methanol	>90 (yield)	[97]
Aspergillus niger	BSPs[a]	Palm	Methanol	87 (yield)	[98]
Aspergillus oryzae NS4	BSPs[a]	Soybean	Methanol	98 (conversion)	[99]
A. oryzae carrying r-CALB[b]	BSPs[a]	Palm Soybean	Methanol	85 90 (conversion)	[100]
Aspergillus oryzae expressing r-FHL[c]	BSPs[a]	Rapeseed	Methanol Ethanol 1-Propanol 1-Butanol	96 (yield) 94 (yield) 96 (yield) 97 (yield)	[101]
Escherichia coli BL21	-	Rapeseed	Methanol	97.7 (conversion)	[102]
Rhizopus chinensis CCTCC M201021	-	Soybean	Methanol	*/>86 (yield)	[103]
Rhizomucor miehei displaying Pichia pastoris	-	Soybean	Methanol	83.14 (yield)	[104]
Rhizopus oryzae IFO 4697	BSPs[a]	Soybean	Methanol	~90(yield)	[106]
Rhizopus oryzae IFO 4697	BSPs[a]	Soybean	Methanol	~85(yield)	[107]
Rhizopus oryzae IFO4697	-	Soybean	Methanol	71 (conversion)	[108]
Rhizopus oryzae IFO 4697 and Aspergillus oryzae niaD300 (combined use)	BSPs[a]	Soybean	Methanol	~100 (conversion)	[109]

Lipase Source	Carrier	Acid/Oil Source	Alcohol	Maximum Performance (%)	Reference
Rhizopus oryzae ATCC 24563	-	Soybean (Free Fatty Acid Content 5.5%)	Methanol	97 (conversion)	[110]
Rhizopus oryzae IFO 4697	BSPs[a]	Soybean	Methanol	72 (yield)	[111]
Rhizopus oryzae	Polyurethane foam BSPs[a]	Soybean	Methanol	90 (conversion)	[112]
Rhizopus oryzae	BSPs[a]	Jatropha Curcas	Methanol	80 (conversion)	[113]
Rhizopus oryzae IFO 4697	-	Soybean	Methanol	86 (yield)	[114]
Rhizopus oryzae	BSPs[a]	Rapeseed	Methanol, Ethanol, 1-Propanol, 1-Butanol	83, 79, 93, 69 (yield)	[101]
Serratia marcescens YXJ-1002	-	Grease	Methanol	97 (yield)	[115]

[a]BSPs:Biomass support particles

[b]r-CALB: Candida antarctica lipase B

[c]r-FHL: Fusarium heterosporum lipase

Whole cell biocatalysts will be a way to industrialization of biodiesel production but the limited mass transfer efficiency of product and substrate is a hurdle to further investigations [95].

3. FEEDSTOCKS

The main aim of researches is to obtain a biodiesel, which will have a competitive price compared to other conventional sources of energy [116]. At this point, selecting the feedstock, represents more than 75-80% of the overall biodiesel production cost, is a vital step to ensure a cost effective biodiesel production. Different kinds of feedstock with varied range of edible and inedible vegetable oil, animal fats, waste oil, microbial oil and microalgae oil can be used for enzyme catalyzed transesterification [117].

3.1. Vegetable Oils

Vegetable oils are candidates as alternative fuels for diesel engines with their high heat content [118]. But, direct use of vegetable oils is not possible because of the high kinematics viscosity of them which are varies in the range of 30–40 cSt at 38 °C and are about 10 times higher than of diesel fuel (Grade No. 2D) leads to many problems [118,119]. Therefore, modification of vegetable oil is necessary and the valuable product of this modification is named "biodiesel". The edible vegetable oils such as soybean [120,121], sunflower [122-124], palm [81,125], corn [126], cottonseed [127], canola [68,69,128] and olive [129,130] oils have been widely used in enzymatic transesterification. In developed countries, edible oils constitute more than 95% of biodiesel production feedstock because the produced biodiesel from these oils have properties very similar to petroleum-based diesel [131]. Also, the country and its climate, the oil percentage and the yield per hectare are effective parameters in selecting the potential renewable feedstock of fuel [118,132]. For example, while rapeseed oil prevailing the EU production, soybean oil prevailing the US and Latin American production, and palm oil mainly being used in Asia [133].

Inedible oils do not find a place in human consumption due to including toxic components. Therefore, inedible oils do not compete with food crops. Thus, inedible vegetable oils are an alternative feedstock for biodiesel production. Babassu (Orbinya martiana), Jatropha curcas (Linnaeus), neem (Azadiracta indica), polanga (Calophyllum inophyllum),karanja (Pongamia pinnata), rubber seed tree (Hevea brasiliensis), mahua (Madhuca indica and Madhuca longifolia), tobacco (Nicotina tabacum), silk cotton tree, etc. are promising inedible vegetable oil sources. Jatropha curcas is an attractive feedstock between various oil bearing seeds as it has been developed scientifically and found to give better biodiesel yield and productivity [134]. Crude Jatropha oil contains about 14% of free fatty acid that is too high for alkaline catalyzed biodiesel production [118]. However, high free acid content is not a problem in the production process of biodiesel via using enzyme catalysts. Besides Jatropha curcas, 26 species of fatty acid methyl ester of oils of including Azadirachta indica, Calophyllum inophyllum, and Pongamia pinnata were found most suitable for use as biodiesel, which adjust to the major specification of biodiesel standards of European Standard Organization, Germany, and USA [135]. Modi et al. reported conversion of crude oils of Pongamia pinnata (karanj), Jatropha curcas (jatropha) via immobilized Novozym 435 to biodiesel fuel with yield 90, and 92.7%, respectively [136].

3.2. Animal Oils/Fats

Animal fats are another group of feedstock for biodiesel production. Animal fats used to produce biodiesel via enzymatic route include lard [137], lamb meet [138] and beef tallow [139]. Animal fats are economically feasible feedstocks compared to vegetable oils. Animal fat methyl ester also has many favorably properties such as non-corrosive, high cetane number, and renewable [140,141]. However, animal fats saturated compounds lead to a tendency to oxidation and crystallization unacceptably at high temperatures [142].

3.3. Waste Oils/Fats

In general, around the world only half of the discharged edible oils recycled as animal feed or as raw material for lubricant and paint and the remainder isdischarged into the environment [143]. Hence, the use of waste oils/fats for biodiesel production is very important to reduce and recycle the waste oil [143], to eliminate the environment and human health risk caused by waste oils [144] and to lower the biodiesel production cost. Waste cooking oil, animal fats, yellow grease, brown grease obtained from highly oxidized yellow grease or recovered waste grease from plumbing trap and waste sludge or soap-stock from the vegetable oil refining process were the major sources of waste oil have been used for biodiesel production [145]. The selection of a catalyst to be used for the production of biodiesel fuel is mostly influenced by the amount of free fatty acid content in various feedstocks [146]. The lipase-catalyzed reaction is a promising method for converting waste oils which contains high percentage of free fatty acids and high water content, into biodiesel with high yield [145]. It has been reported that Novozym 435 is capable of converting the used olive oils [129].

3.4. Algae Oils

There is a considerable interest in the use of algae (micro and macro) oils for synthesis of biodiesel. Because these oils are cheap raw materials besides animal fats and have rapid growth rate and productivity when compared to conventional forestry, agricultural crops, high lipid content, tolerance for poor quality water, smaller land usage up to 49 or 132 times less when compared to rapeseed or soybean crops [142,147]. The smaller land usage brings the advantage of reducing the competition for arable soil with other crops, in particular for human consumption [147]. However, there are still some drawbacks for utilization of algae for biodiesel production. A considerable investment in technological development and technical expertise is needed to optimize the microalgae harvesting and oil extraction processes, to use cheap sources of CO_2 for culture enrichment [147]. Algae oils contain about 20-40% oil [148]. Several researchers have been experimented on microalgal oils as raw material for biodiesel production. Tran et al. [130], investigated the conversion of microalgal oil from Chlorella vulgaris ESP-31 to biodiesel by using immobilized Burkholderia lipase and a high fatty acid methyl esters conversion efficiency of 97.25 wt% oil (or 58.35 wt% biomass) was obtained for 48 h reaction. It is proposed that microalgal oil has good potential for application in the commercial production of biodiesel. The enzymatic conversion of microalgal oils to biodiesel in ionic liquids was firstly studied by Lai et al. [149]. Four microalgae two strains of Botryococcus braunii (BB763 and BB764), Chlorella vulgaris, and Chlorella pyrenoidosa have been catalyzed by two immobilized lipases, Penicillium expansum lipase and Candida antarctica lipase B (Novozym 435), in two solvent systems: an ionic liquid (1-butyl-3-methylimidazolium hexafluorophosphate, [BMIm][PF6]) and an organic solvent (tert-butanol). Penicillium expansum lipase was found more efficient for this application and the ionic liquid [BMIm] [PF6] showed a greater conversion yield (90.7% and 86.2%) obtained relative to the one obtained in the commonly used organic solvent tert-butanol (48.6% and 44.4%).

4. THE EFFECT OF REACTION PARAMETERSON ENZYMATIC TRANSESTERIFICATION

4.1. The effect of temperature on enzymatic transesterification

Enzymatic transesterification takes place at low temperatures varying from 25 to 60°C. In general, initially the rate of reaction increases with rise in reaction temperature, because of an increase in rate constants with temperature and less mass transfer limitations [150,151]. Nevertheless, increased temperature after the optimum temperature promotes to denaturation and higher thermal deactivation of the enzyme, since it decreased the catalytic activity [152].

Various researches have been carried out to find out the effect of temperature on biodiesel production with immobilized enzymes. It is clear that immobilization provide more temperature resistance compared to free enzymes due to supplying a more rigid external backbone for lipase molecule [150,151]. However, optimum temperature is specific for each production. The studies about the effect of temperature for enzymatic transesterification are shown in Table 7.

4.2. The Effect of Water Content On Enzymatic Transesterification

Water content is one of the key factors for enzymatic transesterification reaction that have a strong effect on lipase's active three-dimensional conformational state [21,180]. Biocatalysts, needs a small amount of water to retain their activities [181]. Lipase has an unique feature on the water-oil interface, and the lipase activity depends on this interface. The presence of an oil–water interface required because it provides a suitable environment

Table 7. Data on optimum temperature for enzymatic biodiesel production

Lipase	Oil Source	Alcohol	Performed Temperatures In The Range (°C)	Optimum Temperature (°C)	Reference
Immobilized Aspergillus niger	Palm	Methanol	25-50	40	[153]
Immobilized Aspergillus niger	Waste Cooking	Methanol	25-50	30	[154]
Immobilized Burkholderia cepacia	Babassu	Ethanol	39-56	39	[155]
Candida antarctica	Cotton Seed	T-Butanol	30-50	50	[156]
Candida antarctica	Acid	Methanol	30-50	30	[157]
Immobilized *Candida* Sp. 99–125	Salad	Methanol	27-50	40	[158]
Candida Sp. 99–125	Waste Cooking	Methanol	35-50	40-50	[159]
Immobilized Enterobacter aerogenes	Jatropha	T-Butanol	30-55	55	[160]
Immobilized Enterobacter aerogenes	Crude Rapeseed	Ethanol	25-50	35	[161]
Lipozyme RM IM	Soybean	Butanol	20-50	30	[162]
Lipozyme RM IM	Soybean	Methanol and Ethanol	40–60	50	[163]
Lipozyme RM IM	Soybean Oil Deodorizer Distillate	Ethanol	45-78	50	[164]
Lipozyme TL IM	Rapeseed	N-Butanol	30-60	40	[165]
Lipozyme TL IM	Soybean	Ethanol	20-50	35	[162]

Lipase	Oil Source	Alcohol	Performed Temperatures In The Range (°C)	Optimum Temperature (°C)	Reference
Lipozyme TL IM	Palm	Ethanol	30-78	50	[166]
Novozyme 435	Rapeseed	Methanol	25-55	40	[167]
Novozyme 435	Tung and Palm	Methanol and Ethanol	45-55	55	[168]
Novozym 435	Cottonseed	-(Dimethyl Carbonate As Organic Solvent)	30-55	50	[169]
Novozym 435	Canalo	Methanol	25-65	38	[170]
Novozym 435	Olive	Methanol	30-70	40	[129]
Novozym 435	Soybean	T-Amyl	30-60	40	[171]
Novozym 435	Sunflower	Methanol	25-65	45	[172]
Novozym 435	Stillingia	Methanol	30-60	40	[173]
Novozym 435	Cotton Seed	Methanol	30-70	50	[174]
Novozym 435, Lipozyme TL IM and Lipozyme RM IM	Soybean	Ethanol	25-60	25	[175]
Immobilized Penicillium expansum	Waste	T-Amyl	25-55	35	[176]
Immobilized Pseudomonas cepacia	Soybean	Methanol and Ethanol	25–60	35	[177]
Pseudomonas cepacia	Soybean	Methanol	20–60	30	[178]
Immobilized Pseudomonas fluorescens	Triolein	1-Propanol	40-70	60	[48]
Pseudomonas fluorescens	Soybean	Methanol	30-60	40	[49]
Rhizopus chinensis CCTCC M201021	Soybean	Methanol	30-40	30	[179]
Thermomyces lanuginosus	Canola	Methanol	30-70	40	[69]

for enzyme activation which occurs due to the unmasking and restructuring the active site through conformational changes of the lipase molecule [182,183]. When the addition of water increased, the amount of water available for oil to form oil–water droplets also increases, hence increasing the available interfacial area [182]. Thus, enzymatic activity can not be possible in a water free media. However, excess water cause reverse reaction of hydrolysis. The amount of required water, to provide an optimum enzyme activity, differs according to the type of enzyme and reaction medium composition. Enzymes, substrates, organic solvent and also immobilized support have a crucial role on optimal water activity for lipase [184]. Optimum water content not only provides keeping the hydrolysis of ester linkages at the minimum level, but also ensures the highest degree of transesterification [24]. Thus, a better control of water content is very important for enzymatic process.

Water activity (a_w) is defined as free (boundness) water in the system, which is a ratio of vapor pressure over the given system versus that over pure water [24]. Thermodynamic water activity is the best predictor of reaction rate that can be determined in any phase by different kinds of sensors such as holographic sensor, Weiss LiCl humidity sensor [180,185]. Also, several methods have been developed for control of water activity, for example, equilibration with saturated salt solutions [186], addition of salt hydrate pairs [187,188] and introduction of air or nitrogen into the reactor [189]. Recently, Peterson et. al. developed a practical way for control of water activity in large-scale enzymatic reactions by using a programmable logic controller. On the other hand, percentage water content is another expression which is used widely in transesterification, generally assayed by Karl-Fischer coulometer.

In general, lipases show higher activity with higher water activities in solvent free systems instead of Candida antarctica lipase (Novozym 435) [184]. For Candida sp. 99–125 lipase, the optimum water content is 10–20% based on the oil weight to maintain the highest transesterification activity [31].

Salis et al., investigated production of oleic acid alkyl esters by using Pseudomonas cepacia and determined that a_w in the range 0.4–0.6, 1-butanol:triolein 3:1 – were the best conditions to reach maximum enzymatic activity. It was also found that at the higher values of water activity, no hydrolysis reaction was occurred [190].

Noureddini and Philkana [82] tested immobilized Pseudomonas cepacia for the transesterification of soybean oil with methanol and ethanol and observed that increased addition of water provide a considerable increase in the ester yield. The optimal conditions were determined for processing 10 g of soybean oil by 475 mg lipase in 1 h as 1:7.5 oil/methanol molar ratio, 0.5 g water in the presence of methanol that resulted in 67 % yield and 1:15.2 oil/ethanol molar ratio, 0.3 g water in the presence of ethanol that resulted in 65% yield.

Al-Zuhair et al. studied the esterification of n-butyric acid with methanol in the presence of Mucor miehei lipase, and found similar results with literature [191] that higher water content, makes lipase more efficient [182].

Shah and Gupta used immobilized Pseudomonas cepacia lipase for ethanolysis of Jatropha oil and noted that the best yield 98% gained by in the presence of 4–5% (w/w) water in 8 h. The yield was only 70% in absence of water [44].

Kawakami et al. determined the effect of water content for transesterification of Jatropha oil and methanol to characterize Burkholderia cepacia lipase immobilized in an n-butyl-substituted hydrophobic silica monolith. The authors reported that biodiesel yield reached 90% with water content of 0.6% (w/w) after 12 h using a stoichiometric mixture of methanol and oil (3:1) [78].

Chen et al. investigated the effect of water content for production of biodiesel with oleic acid with methanol catalyzed by soluble lipase NS81020, produced by modification of Aspergillus oryzae microorganism, in the biphasic aqueous-oil systems and found that the esterification yield is low if the water was scant. The higher reaction rate and fatty acid methyl ester yield was obtained with 10 wt % water by oleic acid weight [192].

It is clear that during the past decade numerous investigations have been made to determine the optimal water content for transesterification. As a result, the necessary amount of water content is an important factor to create an interfacial surface between oil and water and to ensure optimal enzymatic activity. Also, water has a strong influence on structural integrity, active site polarity, and protein stability of lipase [21,193]. However, it differs from enzyme to reaction conditions.

4.3. The Effect of Acyl Acceptors on Enzymatic Transesterification

Methanol, short chain alcohol, usually used as an acyl acceptor due to its low price and availability. Insoluble and a relatively high amount of methanol with respect to oil, have a negative influence on the stability of lipases and could be solved by a stepwise addition of the alcohols [15, 194]. To eliminate inhibitory effects of methanol some co-solvents are added to the reaction mixture. Tert-butanol is one of the important co-solvents which is added to enzymatic reaction. Usage of tert-butanol, a polar solvent, is also a possible solution for eliminating the inhibitory effects of methanol and glycerol (both of them soluble in tert-butanol) and suggested instead of using butanol [195]. Liu et al. [196], transesterified waste baked duck oil by three different commercial immobilized lipases (Novozym 435, Lipozyme TLIM and Lipozyme RMIM) with different monohydric alcohols (methanol, ethanol, propanol, isopropanol, isobutanol, isoamyl alcohol) and fusel oil-like alcohol mixture (containing 15% isobutanol, 80% isoamyl alcohol, 5%

methanol) in solvent-free and tert-butanol systems. It was reported that each lipase presented a different kinetic pattern depending on the monohydric alcohols. The results showed that Lipozyme TL IM and Novozym 435 gave high conversion rate with isobutanol and isoamyl alcohol either in solvent-free or in tert-butanol system. Thus, the combined use of lipases, Novozym 435 and Lipozyme TLIM, as catalyst and fusel oil-like mixture as raw material for biodiesel synthesis was found effective in view of cost saving of biodiesel production [195].

Recently, novel acyl acceptors were investigated such as ethyl acetate, methyl acetate, butyl acetate, vinyl acetate [197], dimethyl carbonate [198]. Du and coworkers demonstrated the positive effect of methyl acetate, on enzymatic activity of Novozym 435 and found that lipase could be reused directly without any additional treatment [199]. The advantage of using methyl acetate is that the cost of the catalyst can be reduced dramatically due to the longer operational life and reusability of lipase. The byproduct of the system is triacetylglycerol, which does not have any negative effect on the fuel property, and also no glycerol produced [200]. Hence, these advantages will provide industrial implementation of enzymatic biodiesel production. Dimethyl carbonate is another promising alternative acyl acceptor, which is eco-friendly, odorless, cheap, non-corrosive, and non-toxic [200]. The transesterification reaction is irreversible, because carbonic acid monoacyl ester, the intermediate compound, immediately decomposes to carbon dioxide and alcohol [200]. The fatty acid methyl ester yield is higher for lipase-catalyzed transesterification of vegetable oils with dimethyl carbonate besides conventional acyl acceptors (methanol and methyl acetate) [200]. Only, the higher price of acyl acceptor besides alcohols is a disadvantage [194].

4.4. Effects of the Solvent on Enzymatic Transesterification Reaction

In enzymatic transesterification reaction, excess of alcohol increases reaction efficiency, but if alcohol doesn't dissolve in reaction medium it can disrupt the enzyme activity. Methanol and vegetable oil in the values close to 1:1 molar ratio forms a solution in 40°C. Solvent is added into the reaction medium to increase the solubility of alcohol and thus it allows first step enzymatic transesterification by blocking degradation lipase catalytic activity [24]. To overcome deactivation of lipase activity and improve the lipase activity, various organic solvents have been used for enzymatic biodiesel synthesis. These solvents have been listed in Table 8. Cyclohexane, n-hexane, tert-butanol, petroleum ether, isooctane and 1,4-dioxane are mainly studied hydrophilic and hydrophobic organic solvents in enzymatic biodiesel production. In organic solvent medium, overall alcohol is added at the beginning of the reaction. In solvent free reaction medium, alcohol is added in several portions to prevent enzyme activity with high alcohol concentration [24].

Hexane is generally preferred because of its low cost and easily availability in the market. Some studies were performed in hexane solvent systems with soybean and tallow oil using monohydric alcohols [70,201, 202]. Nelson et al. performed transesterification of tallow with monohydric alcohols by Lipozyme IM 60 (M. miehei) and Novozyme SP435 (C. antarctica) in hexane and a solvent-free system. They compared the transesterification yields of two different systems. The yields with higher than 95% were obtained with methanol, ethanol and butanol with Lipozyme IM 60 lipase under hexane system (Table 8) while reaction yields under solvent-free system were 19% for methanol, 65.5% for ethanol, and 97.4% for isobutanol [201]. Similar results were found by Rodrigues et al. [70]. They compared the yields of transesterification of soybean with ethanol by Lipozyme TL IM. In the presence of n-hexane with 7.5:1 molar ratio of ethanol:soybean oil, the transesterification conversion was found to be as 100% while in solvent-free system the yield was 75%. At stoichiometric molar ratio, the yield was 70% conversion after 10 h of reaction in both systems. Transesterification conversion was obtained as 80% by three stepwise addition of ethanol, while a two step ethanolysis produced 100% conversion after 10 h of reaction in both solvent and solvent-free systems.

In enzyme catalyzed reaction, both alcohol amount and low glycerol solubility in biodiesel have negative effects on enzyme activity. Deposit of glycerol coating the immobilized catalyst is formed during the process, which reduces the enzymes activity [203]. The solubility of methanol and glycerol in hydrophobic solvents is low. For this reason, this problem may occur in hydrophobic solvent system.

The enzymatic alcoholysis of triglyceride also was studied with petroleum ether, isooctane, cyclo hexane,1,4-dioxane (Table 4-2) [16,48,204]. Iso et al. [48], reported that when methanol and ethanol were used as alcohol in enzymatic transesterification, the reactions need an appropriate organic solvent [48]. On the other hand, the reaction could be performed without solvent when 1-propanol and 1-butanol was used. They also used, benzene, chloroform and tetrahydrofuran as solvent and immobilized P. fluorescens lipase as catalyst at 50°C to compare the results that of the 1,4 dioxane. The highest enzymatic activity was observed with 1,4-dioxane. The enzymatic activity increased with the high amount of 1,4-dioxane. But high conversion of oil (app.90%) to biodiesel was obtained with high proportion of 1,4 dioxane(90%). Although usage of high amount of solvents is not preferable in industry solvents can be recovered together with methanol after transesterifiation reaction.

Hydrophilic organic solvents can interact with water molecule in enzyme and this may affect the catalytic activity of enzyme. However, as shown in Table 8 high performance was ensured with hydrophilic solvents such as 1,4-dioxane and tert-butanol [48,156, 205-208]. Some studies were performed in the presence of t-butanol solvent because of positive effects on enzymatic catalyzed reaction. T-butanol has moderate polarity so methanol and glycerol are easily soluble in tertiary butanol. Solubility of methanol prevents enzyme inhibition and solubility of glycerol prevents accumulation on the enzyme carrier material. Another advantage of this solvent is sinteric hindrance. Due to this property, tert-butanol is not accepted by the lipase. High yield and conversions were obtained in the presence of t-butanol with various vegetable oils and immobilized lipases shown in Table 4-2. For example, Liu et al., [196] studied biodiesel synthesis by immobilized lipases in solvent-free and tert-butanol media. Each lipase showed a different conversion depending on the monohydric alcohols and immobilized lipase in solvent-free medium and tert-butanol system. For methanolysis, regardless of the lipase type, the conversion rate is higher in tert-butanol than that in solvent-free medium. Novozym 435 showed higher conversion rate with straight monoalcohols in tert-butanol medium. Lipozym RM IM and Lipozyme TL IM showed lower conversion with straight and branched monoalcohols (except methanol) in solvent free system. Similar results were obtained by Halim and Kamaruddin [208], in transesterification of waste cooking palm oil using various commercial lipases (Lipozyme RM IM, Lipozyme TL IM and Novozyme 435) in tert-butanol as reaction medium. Novozyme 435 was found to be more effective in catalyzing the transesterification with methanol in in-tert-butanol medium. It was also been demonstrated that even 3:1 methanol to oil molar ratio didn't inhibit the Novozyme 435 in tert-butanol system. Du et al. [209], showed that Lipozyme TL IM could be used without loss of lipase activity for 200 batches in tert-butanol system. Li et al. [210], used acetonitrileand tert-butanol mixture as co-solvent in transesterfication of stillingia oil with methanol. The highest biodiesel yield (90.57%) was obtained in co-solvent with 40% tert-butanol and 60% acetonitrile (v/v) with co-solvent. They also reported that co-solvent (as a mixture)enhance the tolerance of lipase to the methanol than the pure tert-butanol.

Table 8. Effect of the solvent on the performance of enzymatic transesterification reaction

Solvent	Oil	Alcohol	Lipase	Temp/ Time	Reaction mixture	Performance (%)	Ref.
Tert-butanol	Cotton seed	Methanol	Novozyme 435 *(Candida antarctica)*	50 °C / 24h	13.5% meth., 54% oil 32.5% tert-butanol, Lipase:1.7% (wt of oil)	97 (yield)	[156]
Tert-butanol	Cotton seed	Methanol	Pancreatic lipase	37 °C / 4 h	Methanol :oil mol ratio:1:15 Lipase:0.5% enzyme (wt of oil) water conc.5% (wt of oil)	75–80 (conversion)	[205]
Tert-butanol	Rapeseed	Methanol	Novozyme435 & Lipozyme TL IM	35 °C / 12 h	Methanol: oil mol. ratio 4:1 tert-butanol/oil vol. 1:1 Lipase: 3% Lipozyme TL IM 1% Novozym 435 (wt of oil)	95 (conversion)	[206]
Tert-butanol	Soybean and deodorizer distillate	Methanol	Lipozyme TL IM Novozym 435	40°C/ 12 h	Methanol:oil molar ratio 3.6:1 Lipase :3% Lipozyme TL IM 2% Novozym 435 tert-butanol: 80% (wt of oil)	84 (yield)	[207]
Tert-butanol	Waste cooking palm	Methanol	Novozyme 435	40°C / 12 h	Methanol:oil mol. ratio 4:1, Lipase:4% (wt of oil)	88(yield)	[208]
Tert-butanol	Waste baked duck	Methanol	Novozym 435 Lipozyme TL IM	45 °C / 20 h	Methanol:oil mol. ratio 4:1, Lipase: 5 wt%(wt of oil)	85.4, 78.5, (conversion)	[196]
Hexane	Tallow	Methanol Ethanol Propanol	Lipozyme IM 60	45 C/ 5 h	0.34 M tallow in hexane (8 mL), Lipase: 10 (wt of oil) 200rpm	94.8, 98.0, 98.5 (conversion)	[201]

Solvent	Oil	Alcohol	Lipase	Temp/ Time	Reaction mixture	Performance (%)	Ref.
Hexane	Soybean	Methanol	Lipozyme IM 77	36.5°C/ 3h	Methanol :oil mol ratio:3.4:1 Lipase:0.9BAUN*of lipase; water 5.8% (wt% of oil)	92.2 (yield)	[202]
Hexane	Soybean	Ethanol	Lipozyme TL IM	30 °C/ 10 h	Ethanol:oil mol.ratio:7.5:1 Lipase: 15 %(wt of oil). 4% water	100 (conversion)	[70]
Cyclo hexane	Sunflower	Methanol	Lipase AK Lipozyme TL IM Lipozyme RM IM	40°C/ 24 h	Volume of organic solvent/ oil: 2 ml/0.2 mmol Lipase: 10% (wt of oil)	65, 75, 35 (conversion)	[204]
Acetonitrile 60%and 40% t-butanol (v/v)	Stillingia	Methanol	Novozym 435 and Lipozyme TL IM	40°C/ 24h	Methanol:oil mol ratio: 6.4:1 Lipase: 4% (w/w) of multiple-lipase (1.96% Novozym 435+2.04% Lipozyme TL IM)	90.57 (yield)	[210]
Petroleum ether	Sunflower	Ethanol	Lipozyme IM Lipase AK	45°C / 5h	Ethanol:oil mol. ratio:11:1 Lipase:20% (wt of oil)	82, 99, (yield)	[16]
I-octane	Sunflower	Methanol	Lipase AK Lipozyme TL, IM Lipozyme RM,IM	40 °C	Methanol: oil mol ratio::3:1 Vol. of organic solvent/oil: 2 ml/0.2 mmol	80, 65, 60, (yield)	[204]
1,4-dioxane	Triolein	Methanol	Lipase AK	50°C / 80h	Methanol:oil mol. ratio: 3:1 90% solvent	~70 (conversion)	[48]

*BAUN:Batch Acidolysis Units Novo

Although positive effects of the usage of the solvents on the transesterification reaction, some drawbacks has also been known) such as; extra reactor volume, solvent toxicity and emissions, solvent recovery and loss cost [133].

4.5. The Effect of Molar Ratio of Alcohol to Oil on Enzymatic Transesterification

Biodiesel yield always increased due to the molar excess of alcohol over fatty acids in triglycerides in traditional transesterification system [15]. The transesterification reaction is reversible and so, an increase in the amount of one of the reactants will result in higher ester yield and minimally 3 molar equivalents of methanol are required for the complete conversion of methyl ester [174]. Conversely, for enzyme catalyzed transesterification, insoluble excess methanol which exists as fine droplets demonstrates negative effects on enzyme activity and also decrease the production yield [211]. The reaction medium is an important factor during the determination of the optimum molar of alcohol to oil. The inactivation of lipases occurs by contact with insoluble alcohol because the highly hydrophilic alcohol eliminates the layer of essential water from the enzymes [212]. Thus, stepwise addition of alcohol is a potential approach for ratio optimizing the molar ratio in solvent free systems [15]. Whilst, higher reaction rates could be obtained with a slight excess of alcohol in organic solvent systems [15].

The two-step reaction system was reported to avoid the inactivation of the lipase by addition of excess amounts of methanol in the first-step reaction, and by addition of vegetable oil and glycerol in the second-step reaction [213]. Watanabe et al. [213], used a two-step reaction system for methyl esterification of free fatty acids and methanolysis of triacylglycerols using immobilized Candida antarctica lipase. The first step reaction was methyl esterification of free fatty acids that was performed by treating a mixture of 66 wt % acid oil and 34 wt % methanol with 1 wt % immobilized lipase. The second step reaction was conducted to convert triacylglycerols to fatty acid methyl esters. In this step, a mixture of 52.3 wt % dehydrated first-step product, 42.2 wt% rapeseed oil, and 5.5 wt% methanol using 6 wt% immobilized lipase in the presence of additional 10 wt % glycerol was treated. The contents of fatty acid methyl esters was 91.1wt.% after the second step reaction was repeated by the use of immobilized lipase for 50 cycles using recovered glycerol.

Moreno-Pirajan and Giraldo [81], added different amounts of alcohol varied from 2.7 to 13.7 molar equivalents for methanol and from 5.7 to 26.7 molar equivalents for ethanol, based on the moles of triglycerides toward the transesterification of palm oil catalyzed by Candida rugosa lipase and 10.4 molar ratio for all alcohols to palm oil was determined as optimal alcohol requirement resulted in 85 mol% of methyl esters yield with n-butanol.

Lipase catalyzed esterification of palmitic acid with ethanol in the presence of Lipozyme IM 20 in a solvent free medium was investigated by Vieira et al. [212]. Different acid/alcohol molar ratios were tried as 0.16, 0.5, 1.0, 1.5, and 1.84. The best result was obtained with 0.5 acid/alcohol molar ratio.

Zaidi et al. [214], explained the correlation existing between the kinetic parameters and the chain-length of the substrates in esterification of oleic acid using nylon-immobilized lipase in n-hexane. It is observed that the inhibition coefficient of the alcohol increased from 0.034 to 0.42 mol l^{-1}, when the number of carbon atoms increased from 1(methanol) to 18 (oleyl alcohol), respectively.

Dizge and Keskinler [69], used immobilized Thermomyces lanuginosus lipase to produce biodiesel with canola oil with methanol and investigated the role of substrate molar ratio. The biodiesel production was conducted at 1:1, 1:2,1:3,1:4,1:5;1:6 and 1:10 oil/alcohol molar ratios at 40°C. The highest methyl ester yield (85.8%) was obtained at the oil/methanol molar ratio of 1:6. Two important result from this study can be concluded as (i) an increase in the number of moles of methanol resulted in an increase in the ester production, (ii) when the formation of esters reached a maximum level the further increases in the methanol concentrations cause a decrease in the formation of esters due to enzyme inactivation.

Thus, the actual amount of alcohol needed varies significantly depending on the origin of the lipase and fat.

5. REACTORS FOR ENZYMATIC TRANSESTERIFICATION

Through the industrialization of enzymatic biodiesel production, it is necessary to show the applicability of enzymes in reactor systems. Various reactors, including batch reactors, packed bed reactors and supercritical reactors have been investigated by researchers. Most of the investigations on enzymatic synthesis of biodiesel have been performed in batch reactors and packed bed reactors.

Batch reactors are simple designs used in the laboratory. In batch reactors, methanol shows a good dispersion in the oil phase. But the physical agitation caused by shear stress from the stirring would disrupt the enzyme carrier which shortens the enzymes life [31]. On the other hand, batch operation is labor intensive, and not suitable for automation [215]. Packed bed reactors are alternative of batch reactors which are substantially faster and more economical continuous reactors [216]. A packed-bed reactor system is most widely used in biotechnology, as it is easy to operate and scale up these systems. In addition, these systems have high bed volume. The most important advantage of these systems is that it is lowering shear stress on immobilized enzymes which leads to long-term enzyme stability [217]. Furthermore, stepwise addition of alcohol can be performed to reduce the inactivation of the enzyme caused by excess alcohol. One of the encountered problems with an immobilized lipase is the inhibition of the enzyme due to the cloggage of the catalyst by accumulation of the glycerol by-product inside the reactor [218]. Also, the separation of glycerol which remains in the bottom of the reactor can be achieved in a simple way by using more than one column. Recently, a packed-bed reactor system, in which a reactant solution is pumped through a column containing biomass support particles immobilized recombinant Aspergillus oryzae and the effluent from the column is recycled into the same column with a stepwise addition of methanol was developed by Yoshida et al. [219]. In this system, lipase retains its activity for five batch cycles and 96.1% methyl ester content was obtained with a residence time of 140 min per pass and stepwise addition of 4.25 molar equivalents of methanol to oil for 6 passes. The methanolysis of soybean oil in packed bed reactor system using Rhizopus oryzae whole cell was studied by Hama et al. [112]. The final methyl ester content was over 90% at a flow rate of 25 l/h in the first cycle and also, after 10 cycles approximately 80% conversion was achieved. Wang et al. [216], developed Pseudomonas cepacia lipase $-Fe_3O_4$ nanoparticle biocomposite based packed bed reactors. A single-packed-bed reactor and the four-packed-bed reactor were used to produce biodiesel by using refined soybean oil. A high conversion rate (over 88%, 192 h) and great stability was achieved with the four-packed-bed reactor compared to single-packed-bed reactor. It is considered that the four-packed-bed reactor supplied a longer residence time of the reaction mixture in the reactor and lowered the inhibition of the lipase by products [216]. By this way, the reaction efficiency was improved. Additionally, the cost of biodiesel production can be reduced by the effective recycling of the enzyme catalysts [184].

Supercritical reactors also have been investigated by researchers for enzymatic biodiesel production. D. Oliveira and J. V. Oliveira [220], produced biodiesel from palm kernel oil in the presence of Novozym 435 and Lipozyme IM in supercritical carbon dioxide in the temperature range of 40–70 °C and from 60 to 200 bar using a water concentration of 0–10 wt % and oil/ethanol molar ratios from 1:3 to 1:10. Lipozyme IM showed better results and the highest reaction conversion was obtained as 77.5 %. It was observed that lipase structure changed at pressures beyond 200 bar. Madras et al. [221], synthesized biodiesel from sunflower oil in supercritical carbon dioxide catalyzed by Novozym. However, the obtained conversions, when the reaction was conducted in supercritical methanol and ethanol at the optimum conditions, were 23 and 27%, respectively [221]. Enzymatic transesterification of lamb meat fat in supercritical carbon dioxide was investigated by Taher et al. [222].The maximum conversion (49.2%) was obtained at 50°C, with 50% Novozym 435 loading, 4:1 molar ratio, within 25 h reaction. Supercritical reactors could not commercialized according to the low conversion rate and cost of the system.

Consequently, packed bed reactor systems seem to be a practical transesterification reactor system with high transesterification efficiency. These systems will bring industrial scale up enzymatic biodiesel production in an economic way.

6. CONCLUSION

Today, the growing energy necessity and environmental pollution problem requires the use of renewable alternative energy sources to become less dependent on fossil resources. As known, biodiesel is an important alternative energy resource and seems to be the fuel of future because it is an environmentally friendly, nontoxic, renewable, and biodegradable fuel.

Conventionally, biodiesel production is achieved by mainly alkaline or acid catalysts. The interest in the use of biocatalyst for biodiesel production has been an increasing trend due to its many advantages.

Biodiesel have been shown to be effectively produced by enzymatic catalyst and also, numerous researches have been performed to obtain highly active lipases and to optimize process conditions for biodiesel production. Besides many advantages, to produce biodiesel by enzyme catalysts on an industrial scale, it is necessary to reduce the high cost of enzymes and obtain lipases with better features. The immobilization of lipases and genetic engineering methods seems to be an attractive way to obtain more active, stable, and reusable lipases in organic solvents and alcohols. Also, selection of alternative acyl-acceptors is an option for eliminating the negative effects of methanol on lipase activity.

It can be concluded that in enzyme catalyzed biodiesel production significant progresses have been made but further improvements such as novel reactor design should be addressed and emphasized in the future research in order to ensure industrial enzymatic biodiesel production. By making novel improvements, much attention will be focused on enzyme usage in biodiesel production, and especially lipase reactions will be applied much more in this area.

REFERENCES

1. Demirbas A.Biodiesel: a realistic fuel alternative for diesel engine. London:Springer Publishing Co.;2008.
2. Demirbas A. Progress and recent trends in biodiesel fuels. Energy Conversion and Management 2009; 50: 14–34.
3. Demirbas M F, Balat M.Recent advances on the production and utilization trends of bio-fuels: a global perspective. Energy Convers Manage. 2006; 47: 2371–81.
4. Chetri A B, Watts K C, Islam M R. Waste cooking oil as an alternate feedstock for biodiesel production. Energies 2008; 1: 3–18.
5. Bamgboye A I, Hansen A C. Prediction of cetane number of biodiesel fuel from the fatty acid methyl ester (FAME) composition. Int Agrophys 2008; 22: 21–9.
6. Balat M, Balat H. A critical review of biodiesel as a vehicular fuel. Energy Convers Manage, 2008; 49: 2727–41.
7. Kulkarni BM, Bujar BG.Shanmukhappa S. Investigation of acid oil as a source of biodiesel. Indian J Chem Tech 2008; 15: 467–71.
8. Gerpen JV, Shanks B, Pruszko R, Clements D, Knothe G.Biodiesel Production Technology. National Renewable Energy Laboratory. CO, USA; 2004.
9. Capehart B L. Encyclopedia of energy engineering and technology. CRC Press, Taylor & Francis LLC; 2007.
10. Pandey A. Handbook of plant-based biofuels.CRC Press, Taylor & FrancisLLC; 2008.
11. Ghaly A E, Dave D, Brooks M S, Budge S.Production of Biodiesel by Enzymatic Transesterification: Review. American Journal of Biochemistry and Biotechnology 2010; 6 (2): 54-76.
12. Mittelbach M, Remschmidt C. Biodiesel-the comprehensive handbook. Boersedruck Ges.m.b.H.69–80; 2004.
13. Jegannathan KR, Abang S, Poncelet D, Chan E S, Ravindra P.Production of Biodiesel Using Immobilized Lipase—A Critical Review. Critical Reviews in Biotechnology 2008; 28: 253–264.
14. Akoh C C, Chang S W, Lee G C, Shaw J F. Enzymatic approach to biodiesel production. J. Agric. Food. Chem. 2007; 55: 8995–9005.
15. Fan X. Enzymatic biodiesel production – the way of the Future, Lipid Technology 2012; 24(2).

16. Mittelbach M. Lipase catalyzed alcoholysis of sunflower oil. J. Am. Oil Chem. Soc.1990;67:168-170.
17. Hasan F, Shah A A, Javed S, Hameed A. Enzymes used in detergents: lipases. African Journal of Biotechnology 2010; 9(31) : 4836-4844.
18. Villeneuve P, Muderhwa J M, Graille J, Haas MJ. Customizing lipases for biocatalysis: a survey of chemical, physical and molecular biological approaches. Journal of Molecular Catalysis B: Enzymatic 2000; 9 :113–148.
19. Zhang B, Weng Y, Xu H, Mao Z. Enzyme immobilization for biodiesel production. Applied Microbiol Biotechnol 2012; 93:61–70. DOI 10.1007/s00253-011-3672-x
20. Ribeiro B D, De Castro A M, Coelho M A Z, Freire D M G. Production and Use of Lipases in Bioenergy: A Review from the Feedstocks to Biodiesel Production. Enzyme Research 2011;1-16.
21. Yahya A R M, Anderson W A, Moo-Young M. Ester synthesis in lipase catalyzed reactions. Enzyme and Microbial Technology 1998; 23:438–450.
22. Freire D M G, Castilho L R. Lipases em Biocat´alise. Enzimas em Biotecnologia. Producao, Aplicacoes e Mercado 2008;1: 369–385.
23. Cho S S, Park D J, Simkhada J R, Hong J H, Sohng J K, Lee O H, Yoo J C. A neutral lipase applicable in biodiesel production from a newly isolated Streptomyces sp. CS326. Bioprocess Biosyst Eng 2012; 35:227–234.
24. Antczak M S, Kubiak A, Antczak T, Bielecki S. Enzymatic biodiesel synthesis – Key factors affecting efficiency of the process. Renewable Energy 2009; 34:1185–1194.
25. Du W, Xu Y, Liu D, Li Z. Study on acyl migration in immobilized lipozyme TL catalyzed transesterification of soybean oil for biodiesel production. J Mol Catal B: Enzym 2005;37:68–71.
26. Hartmeier W. Immobilized Biocatalysts: An Introduction. Springer-Verlag, Berlin;1988.
27. Gomes FM, De Paula A V, Silva SG, De Castro HF. Determinação das propriedades catalíticas em meio aquoso e orgânico da lipase de Candida rugosa imobilizada em celulignina quimicamente modificada por carbonildiimidazol. Quim. Nova 2006; 29, 710–718.
28. Jegannathan KR, Abang S, Poncelet D, Chan ES, Ravindra P. Production of biodiesel using immobilized lipase-a critical review. Crit Rev Biotechnol 2008; 28(4):253–264.

29. Costa S A, Azevedo H S, Reis R L. Biodegradable Systems in Tissue Engineering and Regenerative Medicine .Enzyme Immobilization in Biodegradable Polymers for Biomedical Applications, CRC Press LLC; 2005.
30. Fukuda H, Kondo A, Noda H. Biodiesel fuel production by transesterification of oils. J Biosci Bioeng 2001; 92(5):405–416. doi:10.1016/S1389-1723(01)80288-7
31. Tan T, Lu J, Nie K, Deng L, Wang F, Biodiesel production with immobilized lipase: A review. Biotechnology Advances 2010; 28: 628–634.
32. Tran D T, Chen C L, Chang J S. Immobilization of Burkholderia sp. lipase on a ferric silica nanocomposite for biodiesel production. Journal of Biotechnology 2012; 158: 112– 119.
33. Du W, Xu Y, Liu D, Zeng J. Comparative study on lipase-catalyzed transformation of soybean oil for biodiesel production with different acyl acceptors. Journal of Molecular Catalysis B: Enzymatic 2004; 30: 125–129.
34. Shimada Y, Watanabe Y, Samukawa T, Sugihara A, Noda H, Fukuda H, Tominaga Y. Conversion of Vegetable Oil to Biodiesel Using Immobilized Candida antarctica Lipase. J9045, JAOCS 1999; 76: 789–793.
35. Watanabe Y, Shimada Y, Sugihara A, Noda H, Fukuda H, Tominaga Y. Continuous Production of Biodiesel Fuel from Vegetable Oil Using Immobilized Candida antarctica Lipase. J9352, JAOCS 2000;77:355–360.
36. Naranjo J, Córdoba A, Giraldo L, García V, Moreno-Parajàn J C. Lipase supported on granular activated carbon and activated carbon cloth as a catalyst in the synthesis of biodiesel fuel. J. Mol. Catal. B: Enzymatic2010; 66: 166–171.
37. Lu J, Nie K, Xie F, Wang F, Tan T. Enzymatic synthesis of fatty acid methyl esters from lard with immobilized Candida sp. 99–125. Process Biochem 2007;42:1367–70.
38. Nie K, Xie F, Wang F, Tan T. Lipase catalyzed methanolysis to produce biodiesel: optimization of the biodiesel production. J Mol Catal B Enzym 2006;43:142–7.
39. Li Z, Deng L, Lu J, Guo X, Yang Z, Tan T. Enzymatic synthesis of fatty acid methyl esters from crude rice bran oil with immobilized

Candida sp. 99-125. ChineseJournal of Chemical Engineering 2010; 18:870-875.

40. Tongboriboon K, Cheirsilp P, H-Kittikun A. Mixed lipases for efficient enzymatic synthesis of biodiesel from used palm oil and ethanol in a solvent-free system. J. Mol. Catal. B: Enzymatic 2010; 67: 52–59.
41. Shah S, Sharma S, Gupta M N.Biodiesel preparation by lipase-catalyzed transesterification ofjatropha oil. Energy & Fuels2004;18: 154-159.
42. MendesA A, Oliveirac P C, Velezb A M, Giordanob R C, Giordanob R L C, De Castro H F.International Journal of Biological Macromolecules 2012; 50 :503– 511.
43. Ji Q, Xiao S, He B, Liu X. Purification and characterization of an organic solvent-tolerant lipase from Pseudomonas aeruginosa LX1 and its application for biodiesel production. J. Mol. Catal. B: Enzymatic 2010; 66: 264–269.
44. Shah S, Gupta MN. Lipase catalyzed preparation of biodiesel from Jatropha oil in a solvent free system. Process Biochem 2007;42:409–14.
45. Sakai S, Liu Y, Yamaguchi T, Watanabe R, Kawabe M, Kawakami K. Production of butyl-biodiesel using lipase physically-adsorbed onto electrospun polyacrylonitrile fibers. Bioresource Technology 2010; 101 : 7344–7349.
46. Li Q, Yan Y. Production of biodiesel catalyzed by immobilized Pseudomonas cepacia lipase from Sapium sebiferum oil in micro-aqueous phase. Applied Energy2010;87: 3148–3154.
47. Al-Zuhair S. Enzymatic Production of Bio-Diesel from Waste Cooking Oil Using Lipase. The Open Chemical Engineering Journal 2008; 2: 84-88.
48. Iso J M, Chen B, Eguchi M, Kudo T, Shrestha S. Production of biodiesel fuel from triglycerides and alcohol using immobilized lipase. Journal of Molecular Catalysis B: Enzymatic 2001; 16: 53–58.
49. Salis A, Pinna M, Monduzzi M, Solinas V, Comparison among immobilized lipases on macroporous polypropylene toward biodiesel synthesis. Journal of Molecular Catalysis B: Enzymatic 2008; 54: 19–26.

50. Li N, Zong M, Wu H. Highly efficient transformation of waste oil to biodiesel by immobilized lipase from Penicillium expansum. Process Biochemistry 2009; 44: 685–688.
51. De Paola M G, Ricca E, Calabrò V, Curcio S, Iorio G. Factor analysis of transesterification reaction of waste oil for biodiesel production. BioresourceTechnology 2009; 100:5126–513.
52. Chen Y, Xiao B,Chang J, Fu Y, Lv P, Wang X. Synthesis of biodiesel from waste cooking oil using immobilized lipase in fixed bed reactor. Energy Conversionand Management 2009; 50:668–673.
53. Li X, He X Y, Li Z L, Wang Y D, Wang C Y, Shi H, Wang F. Enzymatic production of biodiesel from Pistacia chinensis bge seed oil using immobilized lipase. Fuel 2012; 92: 89–93.
54. Zeng H, Liao K, Deng X, Jiang H, Zhang F. Characterization of the lipase immobilized on Mg–Al hydrotalcite for biodiesel. Process Biochemistry 2009; 44:791–798.
55. Yagiz F, Kazan D, Akin N A. Biodiesel production from waste oils by using lipase immobilized on hydrotalcite and zeolites. Chem. Eng. J. 2007; 134:262–267
56. Knezevic Z, Milosavic N, Bezbradica D, Jakovljevic Z, Prodanovic R. Immobilization of lipase from Candida rugosa on Eupergit® C supports by covalent attachment. Biochemical Engineering Journal 2006; 30: 269–278.
57. Stoytcheva M, Montero G, Toscano L, Gochev V , Valdez B. The immobilized lipases inbiodiesel production. Biodiesel – Feedstocks and Processing Technologies,InTech.
58. Xie W, Wang J. Immobilized lipase on magnetic chitosan microspheres for transesterification of soybean oil. Biomass and Bioenergy 2012; 3 6: 373-380.
59. Jang M G, Kim D K, Park S C, Lee J S, Kim S W. Biodiesel production from crude canola oil by two-step enzymatic processes. Renewable Energy 2012; 42: 99-104.
60. Da Ros P, Silva G, Mendes A, Santos J, Castro H. Evaluation of the catalytic properties of Burkholderia cepacia lipase immobilized on non-commercial matrices to be used in biodiesel synthesis from different feedstocks. Bioresource Technology 2010; 101:5508–5516.

61. Shao P, Meng X, He J, Sun P. Analysis of immobilized Candida rugosa lipase catalyzed preparation of biodiesel from rapeseed soapstock. Food and BioproductsProcessing 2008; 86: 283–289.
62. Kumari A,Mahapatra P, Garlapati V, Banerjee R. Enzymatic transesterification of Jatropha oil. Biotechnology for Biofuels2009; 2:1, doi:10.1186/1754-6834-2-1
63. Desai P D, Dave A M, Devi S. Alcoholysis of salicornia oil using free and covalently bound lipase onto chitosan beads. Food Chemistry 2006; 95: 193–199.
64. Mendes A, Giordano R C, Giordano R L C, Castro H. Immobilization and stabilization of microbial lipases by multipoint covalent attachment on aldehyderesin affinity: Application of the biocatalysts in biodiesel synthesis. J. Mol. Catal. B:Enzymatic 2011; 68: 109–115.
65. Wang Y, Shen X, Li Z, Li X, Wang F, Nie X, Jiang J. Immobilized recombinant Rhizopus oryzae lipase for the production of biodiesel in solvent free system. Journal of Molecular Catalysis B: Enzymatic 2010; 67: 45–51.
66. Ho L J, Lee D H, Lim J S, Um B H, Park C, Kang S W, Kim S W. Optimization of the process for biodiesel production using a mixture of immobilized Rhizopus oryzae and Candida rugosa lipases. J. Microbiol. Biotechnol. 2008; 18:1927–1931.
67. Yücel Y. Biodiesel production from pomace oil by using lipase immobilized onto olive pomace. Bioresource Technology 2011; 102: 3977–3980.
68. Dizge N, Keskinler B, Tanriseven A. Biodiesel production from canola oil by using lipase immobilized onto hydrophobic microporous styrene-divinylbenzene copolymer. Biochem Eng J 2009;44:220–5.
69. Dizge N, Keskinler B. Enzymatic production of biodiesel from canola oil using immobilized lipase. Biomass and Bioenergy 2008; 32:1274–1278.
70. Rodrigues R, Pessela B,Volpato G, Fernandez-Lafuente R, Guisan J, Ayub M. Two step ethanolysis: A simple and efficient way to improve the enzymatic biodiesel synthesis catalyzed by an immobilized–stabilized lipase from Thermomyces lanuginosus. Process Biochemistry 2010; 45: 1268–1273.

71. Xie W, Ma N. Enzymatic transesterification of soybean oil by using immobilized lipase on magnetic nano-particles. Biomass and Bioenergy2010; 34: 890-896.
72. O'Driscoll, K. F. (1976) Techniques of enzyme entrapment in gels. In: Methods in Enzymology, volume XLIV, (Mosbach K., ed.), Academic Press, New York, NY, 169–183.
73. Ikeda Y, Kurokawa Y, Nakane K, Ogata N.Entrap-immobilization of biocatalysts on cellulose acetate-inorganic composite gel fiber using a gel formation of cellulose acetate–metal (Ti, Zr) alkoxide.Cellulose2002; 9: 369–379.
74. Brena B M, Batista-Viera F. Immobilization of Enzymes,Methods in Biotechnology: Immobilization of Enzymes and Cells, Second Edition, Totowa, NJ: Humana Press Inc.
75. Murty V R, Bhat J, Muniswaran P K A. Hydrolysis of oils by using immobilized lipase enzyme: a review.Biotechnol. Bioprocess Eng. 2002; 7: 57-66.
76. Jegannathan K, Jun-Yee L, Chan E, Ravindra P. Design an immobilized lipase enzyme for biodiesel production. J. Renewable and Sustainable Energy 2009; 1: 063101-1 -063101-8.
77. Hsu A, Jones K, Marmer WN, Foglia TA. Production of alkyl esters from tallow and grease using lipase immobilized in a phyllosilicate sol–gel. J Am Oil Chem 2001;78:585–8.
78. Kawakami K, Oda Y, Takahashi R. Application of a Burkholderia cepacia lipase immobilized silica monolith to batch and continuous biodiesel production with a stoichiometric mixture of methanol and crude Jatropha oil. Biotechnology for Biofuels 2011; 4:42.
79. Meunier S, Legge R. Evaluation of diatomaceous earth as a support for sol–gel immobilized lipase for transesterification. J. Mol. Catal. B: Enzymatic 2010; 62, 54–58.
80. Sawangpanya N, Muangchim C, Phisalaphong M. Immobilization of lipase on CaCO3 and entrapment in calcium alginate bead for biodiesel production,Sci. J. UBU 2010; 1(2): 46-51.
81. Moreno-Pirajan J C, Giraldo L. Study of immobilized candida rugosa lipase for biodiesel fuel production from palm oil by flow microcalorimetry. Arabian Journal of Chemistry 2011; 4: 55–62.
82. Noureddini H, Gao X, Philkana R S. Immobilized Pseudomonas cepacia lipase for biodiesel fuel production from soybean oil. Bioresource Technology 2005; 96: 769–777.

83. Devanesan M G, Viruthagiri T, Sugumar N, Transesterification of jatropha oil using immobilized Pseudomonas fluorescens. African Journal of Biotechnology 2007; 6 (21): 2497-2501.
84. Orçaire O, Buisson P, Pierre A. Application of silica aerogel encapsulated lipases in the synthesis of biodiesel by transesterification reactions. J. Mol. Catal. B:Enzymatic 2006; 42: 106–113.
85. Jegannathan K R, Jun-Yee L, Chan E S, Ravindra P. Production of biodiesel from palm oil using liquid core lipase encapsulated in K-carrageenan.Fuel 2010; 89: 2272–2277.
86. Nassreddine S, Karout A, Christ M, Pierre A. Transesterification of a vegetal oil with methanol catalyzed by a silica fibre reinforced aerogel encapsulated lipase. Applied Catalysis A: General 2008; 344: 70–77.
87. Yan J, Yan Y, Liu S, Hu J, Wang G, Preparation of cross-linked lipase-coated micro-crystals for biodiesel production from waste cooking oil. Bioresource Technology 2011; 102 : 4755–4758.
88. Kumari V, Shah S, Gupta M N. Preparation of biodiesel by lipase-catalyzed transesterification of high free fatty acid containing oil from Madhuca indica. Energy & Fuels2007; 21:368-372.
89. Kensingh P, Chulalaksananukul W, Charuchinda S. Lipase immobilization on Scirpus grossus L.f. fiber support byglutaraldehyde-crosslinked technique for biodiesel synthesis. Special Abstracts / Journal of Biotechnology 2010; 150S: 1–576.
90. LorenS, Wilson L, Andrés I. Immobilization of the Alcaligenes spp. lipase to catalyze the transesterification of fatty acids to produce biodiesel. New Biotechnology 2009; 25S.
91. Fukuda H, Hama S, Tamalampudi S, Noda H. Whole-cell biocatalysts for biodiesel fuel production. Trends in Biotechnology 2008;26:12.
92. Li W, Du W, Liu D, Yao Y. Study on factors influencing stability of whole cell during biodiesel production in solvent-free and tert-butanol system. Biochemical Engineering Journal 2008; 41:111–115.
93. Ban K, Kaieda M, Matsumoto T,Kondo A, Fukuda H. Whole-cell biocatalyst for biodiesel fuel production utilizing Rhizopus oryzae cells immobilized within biomass support particles. Biochem. Eng. J. 2001; 8:39–43.

94. Sun T, Du W, Liu D, Dai L. Improved catalytic performance of GA cross-linking treated Rhizopus oryzae IFO 4697 whole cell for biodiesel production. Process Biochemistry 2010; 45: 1192–1195.
95. Gog A, Roman M, Tos M, Paizs C, Irimie F D. Biodiesel production using enzymatic transesterification- Current state and perspectives. Renewable Energy 2012; 39: 10-16.
96. Xıao M, Obbard J P. Whole cell-catalyzed transesterification of waste vegetable oil. GCB Bioenergy 2010; 2: 346–352.
97. Xıao M, Qi C, Obbard J P. Biodiesel production using Aspergillus niger as a wholecell biocatalyst in a packed-bed reactor.GCB Bioenergy 2011; 3: 293–298.
98. Xıao M, Mathew S, Obbard J P. A newly isolated fungal strain used as whole-cell biocatalyst for biodiesel production from palm oil.GCB Bioenergy 2010; 2:45–51.
99. Adachi D, Hama S, Numata T, Nakashima K, Ogino C, Fukuda H, Kondo A. Development of an Aspergillus oryzae whole-cell biocatalyst coexpressing triglyceride and partial glyceride lipases for biodiesel production. Bioresource Technology 2011; 102:6723–6729.
100. Adachi D, Hama S, Nakashima K, Bogaki T, Ogino C, Kondo A. Production of biodiesel from plant oil hydrolysates using an Aspergillus oryzae whole-cell biocatalyst highly expressing Candida antarctica lipase B. Bioresource Technology 2012;in press: http://dx.doi.org/10.1016/j.biortech.2012.06.092
101. Koda R, Numata T, Hama S, Tamalampudi S, Nakashima K, Tanaka T, Ogino C, Fukuda H, Kondo A. Ethanolysis of rapeseed oil t produce biodiesel fuel catalyzed by Fusarium heterosporum lip expressing fungus immobilized whole-cell biocatalysts. Jour Molecular Catalysis B: Enzymatic 2010; 66: 101–104.
102. Gao B, Su E, Lin J, Jiang Z, Ma Y, Wei D. Development of re Escherichia coli whole-cell biocatalyst expressing a n lipase-coding gene from Proteus sp. for biodie Journal of Biotechnology 2009; 139:169–175.
103. He Q, Xu Y, Teng Y, Wang D. Biodiesel prod whole-cell lipase from Rhizopus chinensis. Ch 41–46.
104. Huang D, Han S, Han Z, Lin Y. Biodiesel prod Rhizomucor miehei lipase-displaying Pichia pasto

an isooctane system. Biochemical Engineering Journal 2012; 63: 10– 14.

105. Li W, Du W, Liu D. Rhizopus oryzae IFO 4697 whole cell catalyzed methanolysis of crude and acidified rapeseed oils for biodiesel production in tert-butanol system. Process Biochemistry 2007; 42: 1481–1485.
106. Sun T, Du W, Zeng J, Dai L, Liu D. Exploring the effects of oil inducer on whole cell-mediated methanolysis for biodiesel production. Process Biochemistry 2010; 45: 514–518.
107. Sun T, Du W, Zeng J, Liu D. Comparative study on stability of whole cells during biodiesel production in solvent-free system. Process Biochemistry 2011; 46: 661–664.
108. Matsumoto T, Takahashi S, Kaieda M, Tanaka M U A, Fukuda H, Kondo A. Yeast whole-cell biocatalyst constructed by intracellular overproduction of Rhizopus oryzae lipase is applicable to biodiesel fuel production. Appl Microbiol Biotechnol 2001;57:515–520.
109. Arai S, Nakashima K, Tanino T, Ogino C, Kondo A, Fukuda H. Production of biodiesel fuel from soybean oil catalyzed by fungus whole-cell biocatalysts in ionic liquids. Enzyme and Microbial Technology 2010; 46: 51–55.
110. Lin Y H, Luo J J, Hwang S C J, Liau P R, Lu W J, Lee H T. The influence of free fatty acid intermediate on biodiesel production from soybean oil by whole cell biocatalyst. Biomass and Bioenergy 2011;35 : 2217-2223.
111. Li W, Du W, Liu D. Optimization of whole cell-catalyzed methanolysis of soybean oil for biodiesel production using response surface methodology, Journal of Molecular Catalysis B: Enzymatic 2007;45 :122–127.
112. Hama S, Yamaji H, Fukumizu T, Numata T, Tamalampudi S, Kondo A, Noda H, Fukuda H. Biodiesel-fuel production in a packed-bed reactor using lipase-producing Rhizopus oryzae cells immobilized within biomass support particles. Biochemical Engineering Journal 2007; 34: 273–278.
113. Tamalampudi S, Talukder M R, Hama S, Numata T, Kondo A, Fukuda H. Enzymatic production of biodiesel from jatropha oil: a comparative study of immobilized-whole cell and commercial lipases as a biocatalyst. Biochemical Engineering Journal 2008; 39: 185–189.

114. Zeng J, Du W, Liu X, Liu D, Dai L, Study on the effect of cultivation parameters and pretreatment on Rhizopus oryzae cell-catalyzed transesterification of vegetable oils for biodiesel production. Journal of Molecular Catalysis B: Enzymatic 2006; 43:15–18.
115. Li A, Ngo T P N, Yan J, Tian K, Li Z. Whole-cell based solvent-free system for one-pot production of biodiesel from waste grease. Bioresource Technology 2012; 114: 725–729.
116. Atabani A E, Silitonga A S, Badruddin I A, Mahlia TMI, Masjuki HH, Mekhilef S. A comprehensive review on biodiesel as an alternative energy resource and its characteristics. Renewable and Sustainable Energy Reviews 2012; 16: 2070– 2093.
117. Demirbas A. Progress and recent trends in biodiesel fuels. Energy Conversion and Management 2009; 50:14–34.
118. Koh M Y, Ghazi T I M. A review of biodiesel production from Jatropha curcas L. Oil. Renewable and Sustainable Energy Reviews 2011;15: 2240–2251.
119. Singh S P, Singh D. Biodiesel production through the use of different sources and characterization of oils and their esters as the substitute of biodiesel: a review. Renewable and Sustainable Energy Reviews 2010;12:200–16.
120. Mander P, Cho S S, Simkhada J R, Choi Y H, Park D J, Yoo J C. An organic solvent–tolerant lipase from Streptomyces sp. CS133 for enzymatic transesterification of vegetable oils in organic media. Process Biochemistry 2012; 47:635–642.
121. Huang D, Han S, Han Z, Lin Y. Biodiesel production catalyzed by Rhizomucor miehei lipase-displaying Pichia pastoris whole cells in an isooctane system. Biochemical Engineering Journal 2012; 63: 10– 14.
122. Karout A, Pierre A C. Partial transesterification of sunflower oil with ethanol by a silica fiber reinforced aerogel encapsulated lipase. J Sol-Gel Sci Technol 2009; 52:276–286.
123. De los Ríos A P, Hernández Fernández FJ, Gómez D, Rubio M, Víllora G. Biocatalytic transesterification of sunflower and waste cooking oils in ionic liquid Media. Process Biochemistry 2011; 46: 1475–1480.
124. Ognjanovic N, Bezbradica D, Knezevic-Jugovic Z. Enzymatic conversion of sunflower oil to biodiesel in a solvent-free system:

Process optimization and the immobilized system stability. Bioresource Technology 2009; 100: 5146–5154.

125. Talukder M M R, Das P, Fang T S, Wu J C. Enhanced enzymatic transesterification of palm oil to biodiesel. Biochemical Engineering Journal 2011; 55: 119–122.

126. Mata T M, Sousa I R B G, Vieira S S, Caetano N S. Biodiesel Production from Corn Oil via Enzymatic Catalysis with Ethanol. Energy Fuels 2012;26: 3034–3041.

127. Chattopadhyay S, Karemore A, Das S, Deysarkar A, Sen R. Biocatalytic production of biodiesel from cottonseed oil: Standardization of process parameters and comparison of fuel characteristics. Applied Energy 2011; 88: 1251–1256.

128. Jang M G, Kim D K, Park S C, Lee J S, Kim S W. Biodiesel production from crude canola oil by two-step enzymatic processes. Renewable Energy 2012; 42: 99-104.

129. Sanchez F, Vasudevan PT. Enzyme catalyzed production of biodiesel from olive oil. Appl Biochem Biotechnol2006; 135:1–14.

130. Tran D T, Yeh K L, Chen C L, Chang J S. Enzymatic transesterification of microalgal oil from Chlorella vulgaris ESP-31 for biodiesel synthesis using immobilized Burkholderia lipase. Bioresource Technology 2012; 108:119–127.

131. Leung D Y C, Wu X, Leung M K H. A review on biodiesel production using catalyzed transesterification. Applied Energy 2010;87:1083–95.

132. Knothe G, Dunn R O. Biodiesel: an alternative diesel fuel from vegetable oils or animal fats, In: Industrial uses of vegetable oils. USA, Champaign. AOCS Press, ISBN 1-893997-84-7,2005.

133. Nielsen P M, Brask J, Fjerbaek L. Enzymatic biodiesel production: Technical and economical considerations. Eur. J. Lipid Sci. Technol. 2008;110: 692–700.

134. Jain S, Sharma MP. Prospects of biodiesel from Jatropha in India: a review. Renewable and Sustainable Energy Reviews 2010;14:763–71.

135. Mohibbe A M, Amtul W, Nahar N M. Prospects and potential of fatty acid methyl esters of some non-traditional seed oils for use as biodiesel in India. Biomass Bioenergy 2005;29:293–302.

136. Modi M K, Reddy J R C, Rao B V S K, Prasad R B N. Lipase-catalyzed mediated conversion of vegetable oils into biodiesel using ethyl acetate as acyl acceptor. Bioresour Technol 2007;98:1260–4.

137. Lu J, Nie K, Xie F, Wang F, Tan T. Enzymatic synthesis of fatty acid methyl esters from lard with immobilized Candida sp. 99–125. Process Biochem 2007;42:1367–70.

138. Al-Zuhair S, Hussein A, Al-Marzouqi A H, Hashim I. Continuous production of biodiesel from fat extracted from lamb meat in supercritical CO2 media. Biochemical Engineering Journal 2012; 60: 106– 110.

139. Da Ros P C M, De Castro H F, Carvalho A K F,. Soares C M F, De Moraes F F, Zanin G M. Microwave-assisted enzymatic synthesis of beef tallow biodiesel. Ind Microbiol Biotechnol 2012; 39:529–536.

140. Guru M, Koca A, Can O, Cinar C, Sahin F. Biodiesel production from waste chicken fat based sources and evaluation with Mg based additive in a diesel engine. Renew Energy 2010;35:637–43.

141. Balat M. Potential alternatives to edible oils for biodiesel production – A review of current work. Energy Conversion and Management 2011; 52: 1479–1492.

142. Karatay S E, Donmez G. Microbial oil production from thermophile cyanobacteria for biodiesel production. Applied Energy 2011; 88: 3632–3635.

143. Watanabe Y, Shimada Y, Sugihara A, Tominaga Y. Enzymatic conversion of waste edible oil to biodiesel fuel in a fixed-bed bioreactor. J9830, JAOCS 2001; 78: 703–707.

144. Chen G, Ying M, Li W. Enzymatic conversion of waste cooking oils Intoalternative fuel-biodiesel. Applied Biochemistry and Biotechnology. 129–132: Humana press; 2006.

145. Huynh L H, Kasim N S, Ju Y H. Biodiesel production from waste oils. Biofuels: Alternative Feedstocks and Conversion Processes. 375 - 392, Elsevier Inc; 2011.

146. Atadashi I M, Aroua M K, Aziz A R A, Sulaiman N M N, Production of biodiesel using high free fatty acid feedstocks. Renewable and Sustainable Energy Reviews 2012; 16: 3275– 3285.

147. Mata T M, Martins A A, Caetano N S. Microalgae for biodiesel production and other applications: A review. Renewable and Sustainable Energy Reviews 2010;14: 217–232.

148. Borugadda V B, Goud V V. Biodiesel production from renewable feedstocks: Status and opportunities. Renewable and Sustainable Energy Reviews 2012; 16: 4763–4784.
149. Lai J Q, Hu Z L, Wang P W, Yang Z, Enzymatic production of microalgal biodiesel in ionic liquid [BMIm][PF6].Fuel 2012; 95: 329–333.
150. Leca M, Tcacenco L, Micutz M, Staicu T.Optimization of biodiesel production by transesterification of vegetable oils using lipases.Romanian Biotechnological Letters 2010; 15(5): 5618-5630.
151. Al-Zuhair S. Production of biodiesel: possibilities and challenges.Biofuels, Bioprod. Bioref.2007;1: 57–66.
152. Luković N, Knežević-Jugović Z, Bezbradica D. Biodiesel fuel production by enzymatic transesterification of oils: recent trends, Challenges and Future Perspectives, Alternative Fuel,Intech
153. Xıao M, Mathew S, Obbard JP. A newly isolated fungal strain used as whole-cell biocatalyst for biodiesel production from palm oil. GCB Bioenergy 2010; 2: 45–51.
154. Xıao M, Obbard JP. Whole cell-catalyzed transesterification of waste vegetable oil. GCB Bioenergy 2010; 2 : 346–352.
155. Freitas L, Da Ros PCM, Santos JC, Castro HF. An integrated approach to produce biodiesel and monoglycerides by enzymatic interestification of babassu oil (Orbinya sp). Process Biochemistry 2009; 44: 1068–1074.
156. Royon D, Daz M, Ellenrieder G, Locatelli S. Enzymatic production of biodiesel from cotton seed oil using t-butanol as a solvent. Bioresour Technol2007; 98:648-653.
157. Watanabe Y, Nagao T, Nishida Y, Takagi Y, Shimada Y. Enzymatic production of fatty acid methyl esters by hydrolysis of acid oil followed by esterification. J Am Oil Chem Soc 2007; 84:1015–1021. DOI 10.1007/s11746-007-1143-4
158. Nie K, Xie F, Wang F, Tan T. Lipase catalyzed methanolysis to produce biodiesel: Optimization of the biodiesel production. Journal of Molecular Catalysis B: Enzymatic 2006; 43: 142–147.
159. Chen Y, Xiao B, Chang J, Fu Y, Lv P, Wang X. Synthesis of biodiesel from waste cooking oil using immobilized lipase in fixed bed reactor. Energy Conversion and Management 2009; 50: 668–673.

160. Kumari A, Mahapatra P, Garlapati V K, Banerjee R. Enzymatic transesterification of Jatropha oil.Biotechnology for Biofuels2009; 2:1.
161. Zhang Z, Guan C. A promising alternate lipase for biodiesel fuel production. Front. Agric. China 2010; 4(2): 129–136, DOI 10.1007/s11703-010-0005-2.
162. Rodrigues R C, Volpato G, Wada K, Ayub M A Z. Enzymatic synthesis of biodiesel from transesterification reactions of vegetable oils and short chain alcohols. J Am Oil Chem Soc 2008; 85:925–930 DOI 10.1007/s11746-008-1284-0
163. Bernardes O L, Bevilaqua J V, Leal M C M R, Freire D M G, Langone M A P. Biodiesel fuel production by the transesterification reaction of soybean oil using immobilized lipase.Applied Biochemistry and Biotechnology 2007;136–140.
164. Souza M S, Aguieiras E C G, Silva M A P, Langone M A P. Biodiesel synthesis via esterification of feedstock with high content of free fatty acids. Appl Biochem Biotechnol 2009;154:253–267.
165. Kazanceva I, Makarevičienė V, Kazancev K. Application of biotechnological method to biodiesel fuel production using n-butanol.Environmental Research, Engineering and Management 2011;2(56): 35 – 42.
166. Matassoli A L F, Corrêa I N S, Portilho M F, Veloso C O, Langone M A P. Enzymatic synthesis of biodiesel via alcoholysis of palm oil. Appl Biochem Biotechnol 2009; 155:347–355.
167. Jeong G T, Park D H. Lipase-catalyzed transesterification of rapeseed oil for biodiesel production with tert-butanol. Appl Biochem Biotechnol 2008; 148:131–139, DOI 10.1007/s12010-007-8050-x
168. Wang Y N, Chen M H, Ko C H, Lu P J, Chern J M, Wu C H, Chang F C. Lipase catalyzed transesterification of tung and palm oil for biodiesel. World Renewable Energy Congress 2011;Sweden,8-13 May.
169. Su E Z, Zhang M J, Zhang J G, Gao J F, Wei D Z. Lipase-catalyzed irreversible transesterification of vegetable oils for fatty acid methyl esters production with dimethyl carbonate as the acyl acceptor. Biochemical Engineering Journal 2007; 36:167–173.

170. Chang H M, Liao H F, Lee C C, Shieh C J. Optimised synthesis of lipase-catalyzed biodiesel by Novozym 435. J Chem Technol Biotechnol 2005; 80:307–312.
171. Zheng Y, Quan J, Ning X, Zhu L M, Jiang B, He Z Y. Lipase-catalyzed transesterification of soybean oil for biodiesel production in tert-amyl alcohol. World J Microbiol Biotechnol 2009; 25:41–46.
172. Ognjanovic N, Bezbradica D, Knezevic-Jugovic Z. Enzymatic conversion of sunflower oil to biodiesel in a solvent-free system: Process optimization and the immobilized system stability. Bioresource Technology 2009; 100: 5146–5154.
173. Liu Y, Xin H l, Yan Y J. Physicochemical properties of stillingia oil: Feasibility for biodiesel production by enzyme transesterification. Industrial Crops and Products 2009; 30: 431–436.
174. Köse O, Tüter M, Aksoy H A. Immobilized Candida antarctica lipase-catalyzed alcoholysis of cotton seed oil in a solvent-free medium. Bioresource Technology 2002; 83: 125–129.
175. Hernandez-Martın E, Otero C. Different enzyme requirements for the synthesis of biodiesel: Novozym 435 and Lipozyme TL IM. Bioresource Technology 2008; 99: 277–286.
176. Li N W, Zong M H, Wu H. Highly efficient transformation of waste oil to biodiesel by immobilized lipase from Penicillium expansum.Process Biochemistry 2009; 44: 685–688.
177. Noureddini H, Gao X, Philkana RS.Immobilized Pseudomonas cepacia lipase for biodiesel fuel production from soybean oil. Bioresource Technology 2005; 96: 769–777.
178. Gamba M, Lapis A A M, Dupont J. Supported ionic liquid enzymatic catalysis for the production of biodiesel. Adv. Synth. Catal. 2008; 350: 160 – 164.
179. Qin H, Xu Y, Teng Y, Wang D. Biodiesel Production Catalyzed by whole-cell lipase from Rhizopus chinensis.Chinese Journal of Catalysis 2008; 29: 1.
180. Petersson A E V, Adlercreutz P, Mattiasson B. A water activity control system for enzymatic reactions in organic media. Biotechnology and Bioengineering. 2007;97 (2):1, ,DOI 10.1002/bit.21229.
181. Bommarius A S, Riebel-Bommarius B R. Biocatalysts:Fundamentals and Applications. John Wiley & Sons, 266; 2000.

182. Al-Zuhair S, Jayaraman K V, Krishnan S, Chan W H. The effect of fatty acid concentration and water content on the production of biodiesel by lipase. Biochemical Engineering Journal 2006; 30: 212–217.
183. Al-Zuhair S, Hasan M, Ramachandran K B. Kinetic hydrolysis of palm oil using lipase. Proc Biochem2003; 38:1155–1163.
184. Fjerbaek L, Christensen K V, Norddahl B. A review of the current state of biodiesel production using enzymatic transesterification. Biotechnology and Bioengineering 2009; 102 (5):1. DOI 10.1002/bit.22256
185. Vermue M H, Tramper J. Biocatalysis in non-conventional media: medium engineering aspects. Pure &Appl. Chem. 1995; 67(2): 345-373.
186. Ujang Z, Vaidya A M. Stepped water activity control for efficient enzymatic interesterification. Appl Microbiol Biotechnol 1998; 50(3):318–322.
187. Bell G, Halling P J, May L, Moore BD, Robb DA, Ulijn R, Valivety RH. 2001. Methods for measurement and control of water in non-aqueous biocatalysis. In: Vulfson EN, Halling PJ, Holland HL, editors. Methods in Biotechnology. Totowa, NJ: Humana Press. p 105–126.
188. Rosell C M, Vaidya A M. Twin-core packed-bed reactors for organic phase enzymatic esterification with water activity control. Appl Microbiol Biotechnol 1995; 44(3–4):283–286.
189. Won K, Lee S B. On-line conversion estimation for solvent-free enzymatic esterification systems with water activity control. Biotechnol Bioprocess Eng 2002; 7(2):76–84.
190. Salis A, Pinna M, Monduzzi M, Solinas V. Biodiesel production from triolein and short chain alcohols through biocatalysis. Journal of Biotechnology 2005; 119: 291–299.
191. Tweddell R J, Kermasha S, Combes D, Marty A. Esterification, interesterification activities of lipase from Rhizopus niveus and Mucor miehei in three different types of organic media: a comparitive study. Ezyme Microb. Technol. 1998; 22: 439–445.
192. Chen X, Du W, Liu D. Effect of several factors on soluble lipase-mediated biodiesel preparation in the biphasic aqueous-oil systems. World J Microbiol Biotechnol 2008; 24:2097–2102,DOI 10.1007/s11274-008-9714-6

193. Atadashi I M, Aroua M K, Abdul Aziz A R, Sulaiman N M N. The effects of water on biodiesel production and refining technologies: a review. Renewable and Sustainable Energy Reviews 2012; 16: 3456– 3470.
194. Robles-Medina A, González-Moreno P A, Esteban-Cerdán L, Molina-Grima E. Biocatalysis: Towards ever greener biodiesel production. Biotechnology Advances 2009; 27: 398–408.
195. Ranganathan S V, Narasimhan S L, Muthukumar K. An overview of enzymatic production of biodiesel.Bioresource Technology 2008; 99:3975–3981.
196. Liu Y, Tan H, Zhang X, Yan Y, Hameed B H. Effect of monohydric alcohols on enzymatic transesterification for biodiesel production. Chemical Engineering Journal 2010;157:223–229.
197. Jeong G T,Park D H. Synthesis of rapeseed biodiesel using short-chained alkyl acetates as acyl acceptor. Appl Biochem Biotechnol 2010; 161:195–208. DOI 10.1007/s12010-009-8777-7.
198. Su E Z, Zhang M J, Zhang J G, Gaoa J F, Wei D Z. Lipase-catalyzed irreversible transesterification of vegetable oils for fatty acid methyl esters production with dimethyl carbonate as the acyl acceptor. Biochemical Engineering Journal 2007; 36: 167–173.
199. Du W, Xu Y, Liu D, Zeng J. Comparative study on lipase-catalyzed transformation of soybean oil for biodiesel production with different acyl acceptors. Journal of Molecular Catalysis B: Enzymatic 2004; 30: 125–129.
200. Xu Y, Du W, Liu D, Zeng J.A novel enzymatic route for biodiesel production from renewable oils in a solvent-free medium.Biotechnology Letters2003; 25:1239–1241.
201. Nelson LA, Foglia A, Marmer WN. Lipase-catalyzed production of biodiesel. J Am Oil Chem Soc 1996;73:1191–5.
202. Shieh C J, Liao H F, Lee C C. Optimization of lipase-catalyzed biodiesel by response surface methodology. Bioresource Technology 2003; 88: 103-106.
203. Dossat V, Combes D, Marty A. Continuous enzymatic transesterication of high oleic sunfower oil in a packed bed reactor:Infuence of the glycerol production. Enzyme Microb. Technol. 1999 ; 25: 194–200.

204. Soumanou MM, Bornscheuer UT. Improvement in lipase-catalyzed synthesis of fatty acid methyl esters from sunflower oil. Enz Microb Technol 2003;33(1):97–103.

205. Chattopadhyay S, Karemore A, Das S, Deysarkar A, Sen R.Biocatalytic production of biodiesel from cottonseed oil: Standardization of process parameters and comparison of fuel characteristics.Applied Energy 2011; 88: 1251–1256.

206. Li L, Du W, Liu D, Wang L, Li Z. Lipase-catalyzed transesterification of rapeseed oils for biodiesel production with a novel organic solvent as the reaction medium. J. Mol. Catal. B:Enzym. 2006 ;43 : 58–62.

207. Wang L, Du W, Liu D, Li L, Dai N. Lipase-catalyzed biodiesel production from soybean oil deodorizer distillate with absorbent present in tert-butanol system. Journal of Molecular Catalysis B: Enzymatic 2006;43: 29–32.

208. Halim S F Al, Kamaruddin A H. Catalytic studies of lipase on FAME production from waste cooking palm oil in a tert-butanol system. Process Biochemistry 2008; 43: 1436–1439.

209. Du W, Liu D, Li L, Dai L. Mechanism exploration during lipase-mediated methanolysis of renewable oils for biodieselproduction in a tert-butanol system. Biotechnol Prog. 2007 ;23: 1087–1090.

210. Li Q, Zheng J, Yan Y. Biodiesel preparation catalyzed by compound-lipase in co-solvent. Fuel Processing Technology 2010; 91:1229–1234.

211. Ganesan D, Rajendran A, Thangavelu V. An overview on the recent advances in the transesterification of vegetable oils for biodiesel production using chemical and biocatalysts. Rev Environ Sci Biotechnol 2009;8:367–394.

212. De Vieira A P A, Da Sılva MAP, Langone MAP, Biodiesel production via esterification reactions catalyzed by lipase. Latin American Applied Research 2006;36:283-288.

213. Watanabe Y, Pinsirodom P, Nagao T, Asao Y, Kobayashi T , Nishida Y, Takagi Y, Shimada Y, Conversion of acid oil by-produced in vegetable oil refining to biodiesel fuel by immobilized Candida antarctica lipase, Journal of Molecular Catalysis B: Enzymatic 2007; 44 : 99–105.

214. Zaidi A, Gainer J L, Carta G, Mrani A, Kadiri T, Belarbi Y, Mir A, Esterification of fatty acids using nylon-immobilized lipase in n-

hexane: kinetic parameters and chain-length effects. Journal of Biotechnology 2002; 93: 209–216.
215. Chen Y H, Huang Y H, Lin R H, Shang N C. A continous-flow biodiesel production process using a rotating packed bed. Bioresource Technology 2010; 101: 668–673.
216. Wang X, Liu X, Zhao C, Ding Y, Xu P. Biodiesel production in packed-bed reactors using lipase–nanoparticle biocomposite.Bioresource Technology 2011; 102: 6352–6355.
217. Hama S, Tamalampudi S, Yoshida A, Tamadani N, Kuratani N, Nodaa H, Fukuda H, Kondo A. Enzymatic packed-bed reactor integrated with glycerol-separating system for solvent-free production of biodiesel fuel.Biochemical Engineering Journal 2011;55: 66–71.
218. Xu Y, Nordblad M,Woodley JM. A two-stage enzymatic ethanol-based biodiesel production in a packed bed reactor. Journal of Biotechnology2010; doi:10.1016/j.jbiotec.2012.05.017
219. Yoshida A, Hama S, Tamadani N, Fukuda H, Kondo A. Improved performance of a packed-bed reactor for biodiesel production through whole-cell biocatalysis employing a high-lipase-expression system.Biochemical Engineering Journal 2012; 63: 76– 80.
220. Oliveira D, Oliveira J V. Enzymatic alcoholysis of palm kernel oil in n-hexane and SC CO2. Journal of Supercritical Fluids 2001; 19 (2): 141–148.
221. Madras G, Kolluru C, Kumar R. Synthesis of biodiesel in supercritical fluids. Fuel 2004;83 :2029–2033.
222. Taher H, Al-Zuhair S, AlMarzouqui A, Hashim I. Extracted fat from lamb meat by supercritical CO2 as feedstock for biodiesel production. Biochemical Engineering Journal 2011; 55: 23–31.

INDEX

U

V

W